Applied Linear Regression

Applied Linear Regression

Third Edition

SANFORD WEISBERG

University of Minnesota
School of Statistics
Minneapolis, Minnesota

WILEY-INTERSCIENCE

A JOHN WILEY & SONS, INC., PUBLICATION

Library of Congress Cataloging-in-Publication Data:

Weisberg, Sanford, 1947–
    Applied linear regression / Sanford Weisberg.—3rd ed.
      p. cm.—(Wiley series in probability and statistics)
    Includes bibliographical references and index.
    ISBN 0-471-66379-4 (acid-free paper)
      1. Regression analysis. I. Title. II. Series.

    QA278.2.W44 2005
    519.5'36—dc22

                                        2004050920

Printed in the United States of America.

10 9 8 7 6 5

To Carol, Stephanie
and
to the memory of my parents

Contents

Preface   **xiii**

1 Scatterplots and Regression   **1**

   1.1   Scatterplots, 1
   1.2   Mean Functions, 9
   1.3   Variance Functions, 11
   1.4   Summary Graph, 11
   1.5   Tools for Looking at Scatterplots, 12
        1.5.1   Size, 13
        1.5.2   Transformations, 14
        1.5.3   Smoothers for the Mean Function, 14
   1.6   Scatterplot Matrices, 15
   Problems, 17

2 Simple Linear Regression   **19**

   2.1   Ordinary Least Squares Estimation, 21
   2.2   Least Squares Criterion, 23
   2.3   Estimating σ^2, 25
   2.4   Properties of Least Squares Estimates, 26
   2.5   Estimated Variances, 27
   2.6   Comparing Models: The Analysis of Variance, 28
        2.6.1   The F-Test for Regression, 30
        2.6.2   Interpreting p-values, 31
        2.6.3   Power of Tests, 31
   2.7   The Coefficient of Determination, R^2, 31
   2.8   Confidence Intervals and Tests, 32
        2.8.1   The Intercept, 32
        2.8.2   Slope, 33

2.8.3 Prediction, 34

2.8.4 Fitted Values, 35

2.9 The Residuals, 36

Problems, 38

3 Multiple Regression 47

3.1 Adding a Term to a Simple Linear Regression Model, 47

3.1.1 Explaining Variability, 49

3.1.2 Added-Variable Plots, 49

3.2 The Multiple Linear Regression Model, 50

3.3 Terms and Predictors, 51

3.4 Ordinary Least Squares, 54

3.4.1 Data and Matrix Notation, 54

3.4.2 Variance-Covariance Matrix of e, 56

3.4.3 Ordinary Least Squares Estimators, 56

3.4.4 Properties of the Estimates, 57

3.4.5 Simple Regression in Matrix Terms, 58

3.5 The Analysis of Variance, 61

3.5.1 The Coefficient of Determination, 62

3.5.2 Hypotheses Concerning One of the Terms, 62

3.5.3 Relationship to the t-Statistic, 63

3.5.4 t-Tests and Added-Variable Plots, 63

3.5.5 Other Tests of Hypotheses, 64

3.5.6 Sequential Analysis of Variance Tables, 64

3.6 Predictions and Fitted Values, 65

Problems, 65

4 Drawing Conclusions 69

4.1 Understanding Parameter Estimates, 69

4.1.1 Rate of Change, 69

4.1.2 Signs of Estimates, 70

4.1.3 Interpretation Depends on Other Terms in the Mean Function, 70

4.1.4 Rank Deficient and Over-Parameterized Mean Functions, 73

4.1.5 Tests, 74

4.1.6 Dropping Terms, 74

4.1.7 Logarithms, 76

4.2 Experimentation Versus Observation, 77

4.3 Sampling from a Normal Population, 80

4.4 More on R^2, 81

    4.4.1 Simple Linear Regression and R^2, 83

    4.4.2 Multiple Linear Regression, 84

    4.4.3 Regression through the Origin, 84

4.5 Missing Data, 84

    4.5.1 Missing at Random, 84

    4.5.2 Alternatives, 85

4.6 Computationally Intensive Methods, 87

    4.6.1 Regression Inference without Normality, 87

    4.6.2 Nonlinear Functions of Parameters, 89

    4.6.3 Predictors Measured with Error, 90

Problems, 92

5 Weights, Lack of Fit, and More   **96**

5.1 Weighted Least Squares, 96

    5.1.1 Applications of Weighted Least Squares, 98

    5.1.2 Additional Comments, 99

5.2 Testing for Lack of Fit, Variance Known, 100

5.3 Testing for Lack of Fit, Variance Unknown, 102

5.4 General F Testing, 105

    5.4.1 Non-null Distributions, 107

    5.4.2 Additional Comments, 108

5.5 Joint Confidence Regions, 108

Problems, 110

6 Polynomials and Factors   **115**

6.1 Polynomial Regression, 115

    6.1.1 Polynomials with Several Predictors, 117

    6.1.2 Using the Delta Method to Estimate a Minimum or a Maximum, 120

    6.1.3 Fractional Polynomials, 122

6.2 Factors, 122

    6.2.1 No Other Predictors, 123

    6.2.2 Adding a Predictor: Comparing Regression Lines, 126

    6.2.3 Additional Comments, 129

6.3 Many Factors, 130

6.4 Partial One-Dimensional Mean Functions, 131

6.5 Random Coefficient Models, 134

Problems, 137

7   Transformations   147

　　7.1   Transformations and Scatterplots, 147
　　　　7.1.1   Power Transformations, 148
　　　　7.1.2   Transforming Only the Predictor Variable, 150
　　　　7.1.3   Transforming the Response Only, 152
　　　　7.1.4   The Box and Cox Method, 153
　　7.2   Transformations and Scatterplot Matrices, 153
　　　　7.2.1   The 1D Estimation Result and Linearly Related
　　　　　　　Predictors, 156
　　　　7.2.2   Automatic Choice of Transformation of Predictors, 157
　　7.3   Transforming the Response, 159
　　7.4   Transformations of Nonpositive Variables, 160
　　Problems, 161

8   Regression Diagnostics: Residuals   167

　　8.1   The Residuals, 167
　　　　8.1.1   Difference Between $\hat{e}$ and e, 168
　　　　8.1.2   The Hat Matrix, 169
　　　　8.1.3   Residuals and the Hat Matrix with Weights, 170
　　　　8.1.4   The Residuals When the Model Is Correct, 171
　　　　8.1.5   The Residuals When the Model Is Not Correct, 171
　　　　8.1.6   Fuel Consumption Data, 173
　　8.2   Testing for Curvature, 176
　　8.3   Nonconstant Variance, 177
　　　　8.3.1   Variance Stabilizing Transformations, 179
　　　　8.3.2   A Diagnostic for Nonconstant Variance, 180
　　　　8.3.3   Additional Comments, 185
　　8.4   Graphs for Model Assessment, 185
　　　　8.4.1   Checking Mean Functions, 186
　　　　8.4.2   Checking Variance Functions, 189
　　Problems, 191

9   Outliers and Influence   194

　　9.1   Outliers, 194
　　　　9.1.1   An Outlier Test, 194
　　　　9.1.2   Weighted Least Squares, 196
　　　　9.1.3   Significance Levels for the Outlier Test, 196
　　　　9.1.4   Additional Comments, 197
　　9.2   Influence of Cases, 198
　　　　9.2.1   Cook's Distance, 198

9.2.2 Magnitude of D_i, 199

9.2.3 Computing D_i, 200

9.2.4 Other Measures of Influence, 203

9.3 Normality Assumption, 204

Problems, 206

10 Variable Selection   **211**

10.1 The Active Terms, 211

    10.1.1 Collinearity, 214

    10.1.2 Collinearity and Variances, 216

10.2 Variable Selection, 217

    10.2.1 Information Criteria, 217

    10.2.2 Computationally Intensive Criteria, 220

    10.2.3 Using Subject-Matter Knowledge, 220

10.3 Computational Methods, 221

    10.3.1 Subset Selection Overstates Significance, 225

10.4 Windmills, 226

    10.4.1 Six Mean Functions, 226

    10.4.2 A Computationally Intensive Approach, 228

Problems, 230

11 Nonlinear Regression   **233**

11.1 Estimation for Nonlinear Mean Functions, 234

11.2 Inference Assuming Large Samples, 237

11.3 Bootstrap Inference, 244

11.4 References, 248

Problems, 248

12 Logistic Regression   **251**

12.1 Binomial Regression, 253

    12.1.1 Mean Functions for Binomial Regression, 254

12.2 Fitting Logistic Regression, 255

    12.2.1 One-Predictor Example, 255

    12.2.2 Many Terms, 256

    12.2.3 Deviance, 260

    12.2.4 Goodness-of-Fit Tests, 261

12.3 Binomial Random Variables, 263

    12.3.1 Maximum Likelihood Estimation, 263

    12.3.2 The Log-Likelihood for Logistic Regression, 264

12.4 Generalized Linear Models, 265
Problems, 266

Appendix   **270**

A.1 Web Site, 270
A.2 Means and Variances of Random Variables, 270
    A.2.1 E Notation, 270
    A.2.2 Var Notation, 271
    A.2.3 Cov Notation, 271
    A.2.4 Conditional Moments, 272
A.3 Least Squares for Simple Regression, 273
A.4 Means and Variances of Least Squares Estimates, 273
A.5 Estimating $E(Y|X)$ Using a Smoother, 275
A.6 A Brief Introduction to Matrices and Vectors, 278
    A.6.1 Addition and Subtraction, 279
    A.6.2 Multiplication by a Scalar, 280
    A.6.3 Matrix Multiplication, 280
    A.6.4 Transpose of a Matrix, 281
    A.6.5 Inverse of a Matrix, 281
    A.6.6 Orthogonality, 282
    A.6.7 Linear Dependence and Rank of a Matrix, 283
A.7 Random Vectors, 283
A.8 Least Squares Using Matrices, 284
    A.8.1 Properties of Estimates, 285
    A.8.2 The Residual Sum of Squares, 285
    A.8.3 Estimate of Variance, 286
A.9 The **QR** Factorization, 286
A.10 Maximum Likelihood Estimates, 287
A.11 The Box-Cox Method for Transformations, 289
    A.11.1 Univariate Case, 289
    A.11.2 Multivariate Case, 290
A.12 Case Deletion in Linear Regression, 291

References   **293**

Author Index   **301**

Subject Index   **305**

Preface

Regression analysis answers questions about the dependence of a response variable on one or more predictors, including prediction of future values of a response, discovering which predictors are important, and estimating the impact of changing a predictor or a treatment on the value of the response. At the publication of the second edition of this book about 20 years ago, regression analysis using least squares was essentially the only methodology available to analysts interested in questions like these. Cheap, widely available high-speed computing has changed the rules for examining these questions. Modern competitors include *nonparametric regression*, *neural networks*, *support vector machines*, and *tree-based methods*, among others. A new field of computer science, called *machine learning*, adds diversity, and confusion, to the mix. With the availability of software, using a neural network or any of these other methods seems to be just as easy as using linear regression.

So, a reasonable question to ask is: Who needs a *revised* book on linear regression using ordinary least squares when all these other newer and, presumably, better methods exist? This question has several answers. First, most other modern regression modeling methods are really just elaborations or modifications of linear regression modeling. To *understand*, as opposed to *use*, neural networks or the support vector machine is nearly impossible without a good understanding of linear regression methodology. Second, linear regression methodology is relatively transparent, as will be seen throughout this book. We can draw graphs that will generally allow us to see relationships between variables and decide whether the models we are using make any sense. Many of the more modern methods are much like a black box in which data are stuffed in at one end and answers pop out at the other, without much hope for the nonexpert to understand what is going on inside the box. Third, if you know how to do something in linear regression, the same methodology with only minor adjustments will usually carry over to other regression-type problems for which least squares is not appropriate. For example, the methodology for comparing response curves for different values of a treatment variable when the response is continuous is studied in Chapter 6 of this book. Analogous methodology can be used when the response is a possibly censored survival time, even though the method of fitting needs to be appropriate for the censored response and not least squares. The methodology of Chapter 6 is useful both in its own right when applied to linear regression problems and as a set of core ideas that can be applied in other settings.

Probably the most important reason to learn about linear regression and least squares estimation is that even with all the new alternatives most analyses of data continue to be based on this older paradigm. And why is this? The primary reason is that it works: least squares regression provides good, and useful, answers to many problems. Pick up the journals in any area where data are commonly used for prediction or estimation and the dominant method used will be linear regression with least squares estimation.

What's New in this Edition

Many of the examples and homework data sets from the second edition have been kept, although some have been updated. The fuel consumption data, for example, now uses 2001 values rather than 1974 values. Most of the derivations are the same as in the second edition, although the order of presentation is somewhat different. To keep the length of the book nearly unchanged, methods that failed to gain general usage have been deleted, as have the separate chapters on prediction and missing data. These latter two topics have been integrated into the remaining text.

The continuing theme of the second edition was the need for *diagnostic methods*, in which fitted models are analyzed for deficiencies, through analysis of residuals and influence. This emphasis was unusual when the second edition was published and important quantities like Studentized residuals and Cook's distance were not readily available in the commercial software of the time.

Times have changed, and so has the emphasis of this book. This edition stresses *graphical methods* including looking at data both *before* and *after* fitting models. This is reflected immediately in the new Chapter 1, which introduces the key idea of looking at data with scatterplots and the somewhat less universal tool of scatterplot matrices. Most analyses and homework problems start with drawing graphs. We tailor analyses to correspond to what we see in the graphs, and this additional step can make modeling easier and fitted models reflect the data more closely. Remarkably, this also lessens the need for diagnostic methods.

The emphasis on graphs leads to several additional methods and procedures that were not included in the second edition. The use of smoothers to help summarize a scatterplot is introduced early, although only a little of the theory of smoothing is presented (in Appendix A.5). Transformations of predictors and the response are stressed, and relatively unfamiliar methods based both on smoothing and on generalization of the Box–Cox method are presented in Chapter 7.

Another new topic included in the book is computationally intensive methods and simulation. The key example of this is the bootstrap, in Section 4.6, which can be used to make inferences about fitted models in small samples. A somewhat different computationally intensive method is used in an example in Chapter 10, which is a completely rewritten chapter on variable selection.

The book concludes with two expanded chapters on nonlinear and logistic regression, both of which are generalizations of the linear regression model. I have included these chapters to provide instructors and students with enough information for basic usage of these models and to take advantage of the intuition gained about them from an in-depth study of the linear regression model. Each of these can be treated at book-length, and appropriate references are given.

Mathematical Level

The mathematical level of this book is roughly the same as the level of the second edition. Matrix representation of data is used, particularly in the derivation of the methodology in Chapters 2–4. Derivations are less frequent in later chapters, and so the necessary mathematics is less. Calculus is generally not required, except for an occasional use of a derivative, for the discussion of the delta method, Section 6.1.2, and for a few topics in the Appendix. The discussions requiring calculus can be skipped without much loss.

Computing and Computer Packages

Like the second edition, only passing mention is made in the book to computer packages. To help the reader make a connection between the text and a computer package for doing the computations, we provide several web companions for *Applied Linear Regression* that discuss how to use standard statistical packages for linear regression analysis. The packages covered include JMP, SAS, SPSS, R, and S-plus; others may be included after publication of the book. In addition, all the data files discussed in the book are also on the website. The web address for this material is

$$\texttt{http://www.stat.umn.edu/alr}$$

Some readers may prefer to have a book that integrates the text more closely with a computer package, and for this purpose, I can recommend R. D. Cook and S. Weisberg (1999), *Applied Regression Including Computing and Graphics*, also published by John Wiley. This book includes a very user-friendly, free computer package called *Arc* that does everything that is described in that book and also nearly everything in *Applied Linear Regression*.

Teaching with this Book

The first ten chapters of the book should provide adequate material for a one-quarter course on linear regression. For a semester-length course, the last two chapters can be added. A teacher's manual, primarily giving solutions to all the homework problems, can be obtained from the publisher by instructors.

Acknowledgments

I am grateful to several people who generously shared their data for inclusion in this book; they are cited where their data appears. Charles Anderson and Don Pereira suggested several of the examples. Keija Shan, Katherine St. Clair, and Gary Oehlert helped with the website and its content. Brian Sell helped with the examples and with many administrative chores. Several others helped with earlier editions: Christopher Bingham, Morton Brown, Cathy Campbell, Dennis Cook, Stephen Fienberg, James Frane, Seymour Geisser, John Hartigan, David Hinkley, Alan Izenman, Soren Johansen, Kenneth Koehler, David Lane, Michael Lavine, Kinley Larntz, John Rice, Donald Rubin, Joe Shih, Pete Stewart, Stephen Stigler, Douglas Tiffany, Carol Weisberg, and Howard Weisberg.

SANFORD WEISBERG

St. Paul, Minnesota
April 13, 2004

CHAPTER 1

Scatterplots and Regression

Regression is the study of *dependence*. It is used to answer questions such as Does changing class size affect success of students? Can we predict the time of the next eruption of Old Faithful Geyser from the length of the most recent eruption? Do changes in diet result in changes in cholesterol level, and if so, do the results depend on other characteristics such as age, sex, and amount of exercise? Do countries with higher per person income have lower birth rates than countries with lower income? Regression analysis is a central part of many research projects. In most of this book, we study the important instance of regression methodology called *linear regression*. These methods are the most commonly used in regression, and virtually all other regression methods build upon an understanding of how linear regression works.

As with most statistical analyses, the goal of regression is to summarize observed data as simply, usefully, and elegantly as possible. In some problems, a theory may be available that specifies how the response varies as the values of the predictors change. In other problems, a theory may be lacking, and we need to use the data to help us decide on how to proceed. In either case, an essential first step in regression analysis is to draw appropriate graphs of the data.

In this chapter, we discuss the fundamental graphical tool for looking at regression data, a two-dimensional *scatterplot*. In regression problems with one predictor and one response, the scatterplot of the response versus the predictor is the starting point for regression analysis. In problems with many predictors, several simple graphs will be required at the beginning of an analysis. A *scatterplot matrix* is a convenient way to organize looking at many scatterplots at once. We will look at several examples to introduce the main tools for looking at scatterplots and scatterplot matrices and extracting information from them. We will also introduce the notation that will be used throughout the rest of the book.

1.1 SCATTERPLOTS

We begin with a regression problem with one predictor, which we will generically call X and one response variable, which we will call Y. Data consists of

Applied Linear Regression, Third Edition, by Sanford Weisberg
ISBN 0-471-66379-4 Copyright © 2005 John Wiley & Sons, Inc.

values $(x_i, y_i), i = 1, \ldots, n$, of (X, Y) observed on each of n units or *cases*. In any particular problem, both X and Y will have other names such as *Temperature* or *Concentration* that are more descriptive of the data that is to be analyzed. The goal of regression is to understand how the values of Y change as X is varied over its range of possible values. A first look at how Y changes as X is varied is available from a scatterplot.

Inheritance of Height

One of the first uses of regression was to study inheritance of traits from generation to generation. During the period 1893–1898, E. S. Pearson organized the collection of $n = 1375$ heights of mothers in the United Kingdom under the age of 65 and one of their adult daughters over the age of 18. Pearson and Lee (1903) published the data, and we shall use these data to examine inheritance. The data are given in the data file `heights.txt`[1].

Our interest is in inheritance *from* the mother *to* the daughter, so we view the mother's height, called *Mheight*, as the predictor variable and the daughter's height, *Dheight*, as the response variable. Do taller mothers tend to have taller daughters? Do shorter mothers tend to have shorter daughters?

A scatterplot of *Dheight* versus *Mheight* helps us answer these questions. The scatterplot is a graph of each of the n points with the response *Dheight* on the vertical axis and predictor *Mheight* on the horizontal axis. This plot is shown in Figure 1.1. For regression problems with one predictor X and a response Y, we call the scatterplot of Y versus X a *summary graph*.

Here are some important characteristics of Figure 1.1:

1. The range of heights appears to be about the same for mothers and for daughters. Because of this, we draw the plot so that the lengths of the horizontal and vertical axes are the same, and the scales are the same. If all mothers and daughters had *exactly* the same height, then all the points would fall exactly on a 45° line. Some computer programs for drawing a scatterplot are not smart enough to figure out that the lengths of the axes should be the same, so you might need to resize the plot or to draw it several times.

2. The original data that went into this scatterplot was rounded so each of the heights was given to the nearest inch. If we were to plot the original data, we would have substantial *overplotting* with many points at exactly the same location. This is undesirable because we will not know if one point represents one case or many cases, and this can be very misleading. The easiest solution is to use *jittering*, in which a small uniform random number is added to each value. In Figure 1.1, we used a uniform random number on the range from -0.5 to $+0.5$, so the jittered values would round to the numbers given in the original source.

3. One important function of the scatterplot is to decide if we might reasonably assume that the response on the vertical axis is *independent* of the predictor

---

[1] See Appendix A.1 for instructions for getting data files from the Internet.

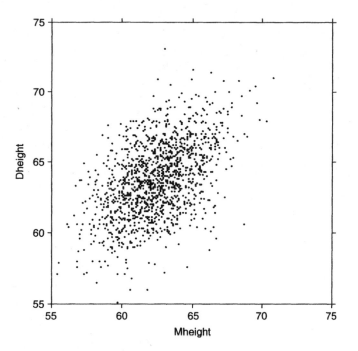

FIG. 1.1   Scatterplot of mothers' and daughters' heights in the Pearson and Lee data. The original data have been jittered to avoid overplotting, but if rounded to the nearest inch would return the original data provided by Pearson and Lee.

on the horizontal axis. This is clearly not the case here since as we move across Figure 1.1 from left to right, the scatter of points is different for each value of the predictor. What we mean by this is shown in Figure 1.2, in which we show only points corresponding to mother–daughter pairs with *Mheight* rounding to either 58, 64 or 68 inches. We see that within each of these three strips or *slices*, even though the number of points is different within each slice, (a) the mean of *Dheight* is increasing from left to right, and (b) the vertical variability in *Dheight* seems to be more or less the same for each of the fixed values of *Mheight*.

4. The scatter of points in the graph appears to be more or less elliptically shaped, with the axis of the ellipse tilted upward. We will see in Section 4.3 that summary graphs that look like this one suggest use of the simple linear regression model that will be discussed in Chapter 2.

5. Scatterplots are also important for finding *separated points*, which are either points with values on the horizontal axis that are well separated from the other points or points with values on the vertical axis that, given the value on the horizontal axis, are either much too large or too small. In terms of this example, this would mean looking for very tall or short mothers or, alternatively, for daughters who are very tall or short, given the height of their mother.

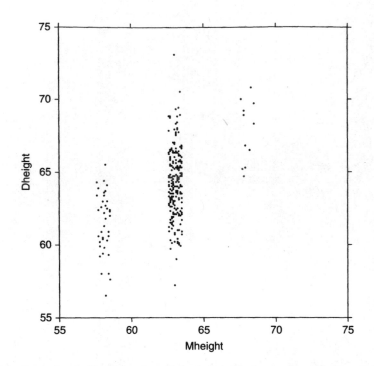

FIG. 1.2   Scatterplot showing only pairs with mother's height that rounds to 58, 64 or 68 inches.

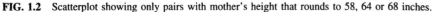

These two types of separated points have different names and roles in a regression problem. Extreme values on the left and right of the horizontal axis are points that are likely to be important in fitting regression models and are called *leverage* points. The separated points on the vertical axis, here unusually tall or short daughters give their mother's height, are potentially *outliers*, cases that are somehow different from the others in the data.

While the data in Figure 1.1 do include a few tall and a few short mothers and a few tall and short daughters, given the height of the mothers, none appears worthy of special treatment, mostly because in a sample size this large we expect to see some fairly unusual mother–daughter pairs.

We will continue with this example later.

Forbes' Data

In an 1857 article, a Scottish physicist named James D. Forbes discussed a series of experiments that he had done concerning the relationship between atmospheric pressure and the boiling point of water. He knew that altitude could be determined from atmospheric pressure, measured with a barometer, with lower pressures corresponding to higher altitudes. In the middle of the nineteenth century, barometers were fragile instruments, and Forbes wondered if a simpler measurement of the boiling point of water could substitute for a direct reading of barometric pressure. Forbes

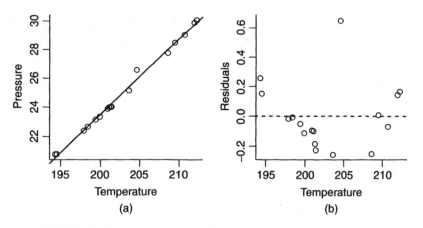

FIG. 1.3 Forbes data. (a) *Pressure* versus *Temp*; (b) Residuals versus *Temp*.

collected data in the Alps and in Scotland. He measured at each location pressure in inches of mercury with a barometer and boiling point in degrees Fahrenheit using a thermometer. Boiling point measurements were adjusted for the difference between the ambient air temperature when he took the measurements and a standard temperature. The data for $n = 17$ locales are reproduced in the file forbes.txt.

The scatterplot of *Pressure* versus *Temp* is shown in Figure 1.3a. The general appearance of this plot is very different from the summary graph for the heights data. First, the sample size is only 17, as compared to over 1300 for the heights data. Second, apart from one point, all the points fall almost exactly on a smooth curve. This means that the variability in pressure for a given temperature is extremely small.

The points in Figure 1.3a appear to fall very close to the straight line shown on the plot, and so we might be encouraged to think that the mean of pressure given temperature could be modelled by a straight line. Look closely at the graph, and you will see that there is a small systematic error with the straight line: apart from the one point that does not fit at all, the points in the middle of the graph fall below the line, and those at the highest and lowest temperatures fall above the line. This is much easier to see in Figure 1.3b, which is obtained by removing the linear trend from Figure 1.3a, so the plotted points on the vertical axis are given for each value of *Temp* by

$$Residual = Pressure - \text{point on the line}$$

This allows us to gain resolution in the plot since the range on the vertical axis in Figure 1.3a is about 10 inches of mercury while the range in Figure 1.3b is about 0.8 inches of mercury. To get the same resolution in Figure 1.3a, we would need a graph that is $10/0.8 = 12.5$ as big as Figure 1.3b. Again ignoring the one point that clearly does not match the others, the curvature in the plot is clearly visible in Figure 1.3b.

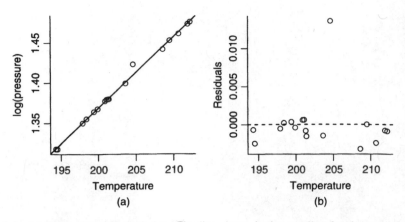

FIG. 1.4 (a) Scatterplot of Forbes' data. The line shown is the OLS line for the regression of log(*Pressure*) on *Temp*. (b) Residuals versus *Temp*.

While there is nothing at all wrong with curvature, the methods we will be studying in this book work best when the plot can be summarized by a straight line. Sometimes we can get a straight line by transforming one or both of the plotted quantities. Forbes had a physical theory that suggested that log(*Pressure*) is linearly related to *Temp*. Forbes (1857) contains what may be the first published summary graph corresponding to his physical model. His figure is redrawn in Figure 1.4. Following Forbes, we use base ten common logs in this example, although in most of the examples in this book we will use base-two logarithms. The choice of base has no material effect on the appearance of the graph or on fitted regression models, but interpretation of parameters can depend on the choice of base, and using base-two often leads to a simpler interpretation for parameters.

The key feature of Figure 1.4a is that apart from one point the data appear to fall very close to the straight line shown on the figure, and the residual plot in Figure 1.4b confirms that the deviations from the straight line are not systematic the way they were in Figure 1.3b. All this is evidence that the straight line is a reasonable summary of these data.

Length at Age for Smallmouth Bass

The smallmouth bass is a favorite game fish in inland lakes. Many smallmouth bass populations are managed through stocking, fishing regulations, and other means, with a goal to maintain a healthy population.

One tool in the study of fish populations is to understand the growth pattern of fish such as the dependence of a measure of size like fish length on age of the fish. Managers could compare these relationships between different populations with dissimilar management plans to learn how management impacts fish growth.

Figure 1.5 displays the *Length* at capture in mm versus *Age* at capture for $n = 439$ small mouth bass measured in West Bearskin Lake in Northeastern Minnesota in 1991. Only fish of age seven or less are included in this graph. The data were provided by the Minnesota Department of Natural Resources and are given in the

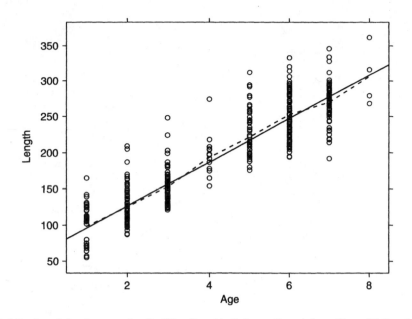

FIG. 1.5 *Length* (mm) versus *Age* for West Bearskin Lake smallmouth bass. The solid line shown was estimated using ordinary least squares or OLS. The dashed line joins the average observed length at each age.

file wblake.txt. Fish scales have annular rings like trees, and these can be counted to determine the age of a fish. These data are *cross-sectional*, meaning that all the observations were taken at the same time. In a *longitudinal* study, the same fish would be measured each year, possibly requiring many years of taking measurements. The data file gives the *Length* in mm, *Age* in years, and the *Scale* radius, also in mm.

The appearance of this graph is different from the summary plots shown for last two examples. The predictor *Age* can only take on integer values corresponding to the number of annular rings on the scale, so we are really plotting seven distinct populations of fish. As might be expected, length generally increases with age, but the longest fish at age-one fish exceeds the length of the shortest age-four fish, so knowing the age of a fish will not allow us to predict its length exactly; see Problem 2.5.

Predicting the Weather

Can early season snowfall from September 1 until December 31 predict snowfall in the remainder of the year, from January 1 to June 30? Figure 1.6, using data from the data file ftcollinssnow.txt, gives a plot of *Late* season snowfall from January 1 to June 30 versus *Early* season snowfall for the period September 1 to December 31 of the previous year, both measured in inches at Ft. Collins, Colorado[2]. If *Late* is related to *Early*, the relationship is considerably weaker than

---

[2]The data are from the public domain source http://www.ulysses.atmos.colostate.edu.

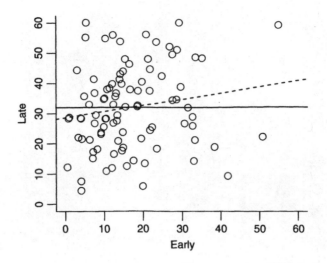

FIG. 1.6 Plot of snowfall for 93 years from 1900 to 1992 in inches. The solid horizontal line is drawn at the average late season snowfall. The dashed line is the best fitting (ordinary least squares) line of arbitrary slope.

in the previous examples, and the graph suggests that early winter snowfall and late winter snowfall may be completely unrelated, or *uncorrelated*. Interest in this regression problem will therefore be in testing the hypothesis that the two variables are uncorrelated versus the alternative that they are not uncorrelated, essentially comparing the fit of the two lines shown in Figure 1.6. Fitting models will be helpful here.

Turkey Growth

This example is from an experiment on the growth of turkeys (Noll, Weibel, Cook, and Witmer, 1984). Pens of turkeys were grown with an identical diet, except that each pen was supplemented with a *Dose* of the amino acid methionine as a percentage of the total diet of the birds. The methionine was provided using either a standard source or one of two experimental sources. The response is average weight gain in grams of all the turkeys in the pen.

Figure 1.7 provides a summary graph based on the data in the file turkey.txt. Except at *Dose* = 0, each point in the graph is the average response of five pens of turkeys; at *Dose* = 0, there were ten pens of turkeys. Because averages are plotted, the graph does not display the variation between pens treated alike. At each value of *Dose* > 0, there are three points shown, with different symbols corresponding to the three sources of methionine, so the variation between points at a given *Dose* is really the variation between sources. At *Dose* = 0, the point has been arbitrarily labelled with the symbol for the first group, since *Dose* = 0 is the same treatment for all sources.

For now, ignore the three sources and examine Figure 1.7 in the way we have been examining the other summary graphs in this chapter. Weight gain seems

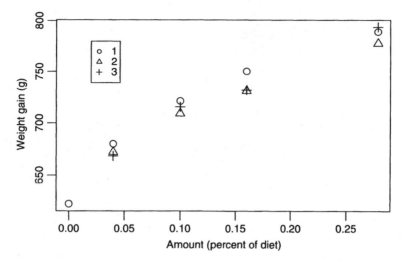

FIG. 1.7 Weight gain versus *Dose* of methionine for turkeys. The three symbols for the points refer to three different sources of methionine.

to increase with increasing *Dose*, but the increase does not appear to be linear, meaning that a straight line does not seem to be a reasonable representation of the average dependence of the response on the predictor. This leads to study of mean functions.

1.2 MEAN FUNCTIONS

Imagine a generic summary plot of Y versus X. Our interest centers on how the distribution of Y changes as X is varied. One important aspect of this distribution is the *mean function*, which we define by

$$E(Y|X = x) = \text{a function that depends on the value of } x \qquad (1.1)$$

We read the left side of this equation as "the expected value of the response when the predictor is fixed at the value $X = x$;" if the notation "E( )" for expectations and "Var( )" for variances is unfamiliar, please read Appendix A.2. The right side of (1.1) depends on the problem. For example, in the heights data in Example 1.1, we might believe that

$$E(Dheight|Mheight = x) = \beta_0 + \beta_1 x \qquad (1.2)$$

that is, the mean function is a straight line. This particular mean function has two *parameters*, an intercept β_0 and a slope β_1. If we knew the values of the βs, then the mean function would be completely specified, but usually the βs need to be estimated from data.

Figure 1.8 shows two possibilities for βs in the straight-line mean function (1.2) for the heights data. For the dashed line, $\beta_0 = 0$ and $\beta_1 = 1$. This mean function

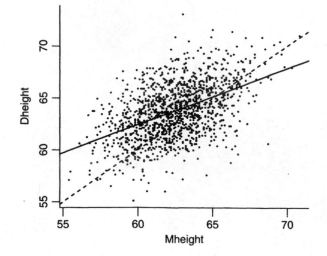

FIG. 1.8 The heights data. The dashed line is for E(*Dheight*|*Mheight*) = *Mheight*, and the solid line is estimated by OLS.

would suggest that daughters have the same height as their mothers on average. The second line is estimated using ordinary least squares, or OLS, the estimation method that will be described in the next chapter. The OLS line has slope less than one, meaning that tall mothers tend to have daughters who are taller than average because the slope is positive but shorter than themselves because the slope is less than one. Similarly, short mothers tend to have short daughters but taller than themselves. This is perhaps a surprising result and is the origin of the term *regression*, since extreme values in one generation tend to revert or regress toward the population mean in the next generation.

Two lines are shown in Figure 1.5 for the smallmouth bass data. The dashed line joins the average length at each age. It provides an estimate of the mean function E(*Length*|*Age*) without actually specifying any functional form for the mean function. We will call this a *nonparametric* estimated mean function; sometimes we will call it a *smoother*. The solid line is the OLS estimated straight line (1.1) for the mean function. Perhaps surprisingly, the straight line and the dashed lines that join the within-age means appear to agree very closely, and we might be encouraged to use the straight-line mean function to describe these data. This would mean that the increase in length per year is the same for all ages. We cannot expect this to be true if we were to include older-aged fish because eventually the growth rate must slow down. For the range of ages here, the approximation seems to be adequate.

For the Ft. Collins weather data, we might expect the straight-line mean function (1.1) to be appropriate but with $\beta_1 = 0$. If the slope is zero, then the mean function is parallel to the horizontal axis, as shown in Figure 1.6. We will eventually test for independence of *Early* and *Late* by testing the hypothesis that $\beta_1 = 0$ against the alternative hypothesis that $\beta_1 \neq 0$.

Not all summary graphs will have a straight-line mean function. In Forbes' data, to achieve linearity we have replaced the measured value of *Pressure* by log(*Pressure*). Transformation of variables will be a key tool in extending the usefulness of linear regression models. In the turkey data and other growth models, a *nonlinear* mean function might be more appropriate, such as

$$\mathrm{E}(Y|Dose = x) = \beta_0 + \beta_1[1 - \exp(-\beta_2 x)] \tag{1.3}$$

The βs in (1.3) have a useful interpretation, and they can be used to summarize the experiment. When $Dose = 0$, $\mathrm{E}(Y|Dose = 0) = \beta_0$, so β_0 is the baseline growth without supplementation. Assuming $\beta_2 > 0$, when the *Dose* is large, $\exp(-\beta_2 Dose)$ is small, and so $\mathrm{E}(Y|Dose)$ approaches $\beta_0 + \beta_1$ for large *Dose*. We think of $\beta_0 + \beta_1$ as the limit to growth with this additive. The rate parameter β_2 determines how quickly maximum growth is achieved. This three-parameter mean function will be considered in Chapter 11.

1.3 VARIANCE FUNCTIONS

Another characteristic of the distribution of the response given the predictor is the *variance function*, defined by the symbol $\mathrm{Var}(Y|X = x)$ and in words as the variance of the response distribution given that the predictor is fixed at $X = x$. For example, in Figure 1.2 we can see that the variance function for *Dheight|Mheight* is approximately the same for each of the three values of *Mheight* shown in the graph. In the smallmouth bass data in Figure 1.5, an assumption that the variance is constant across the plot is plausible, even if it is not certain (see Problem 1.1). In the turkey data, we cannot say much about the variance function from the summary plot because we have plotted treatment means rather than the actual pen values, so the graph does not display the information about the variability between pens that have a fixed value of *Dose*.

A frequent assumption in fitting linear regression models is that the variance function is the same for every value of x. This is usually written as

$$\mathrm{Var}(Y|X = x) = \sigma^2 \tag{1.4}$$

where σ^2 (read "sigma squared") is a generally unknown positive constant. We will encounter later in this book other problems with complicated variance functions.

1.4 SUMMARY GRAPH

In all the examples except the snowfall data, there is a clear dependence of the response on the predictor. In the snowfall example, there might be no dependence at all. The turkey growth example is different from the others because the average value of the response seems to change nonlinearly with the value of the predictor on the horizontal axis.

TABLE 1.1 Four Hypothetical Data Sets. The Data Are Given in the File
`anscombe.txt`

X_1	Y_1	Y_2	Y_3	X_2	Y_4
10	8.04	9.14	7.46	8	6.580
8	6.95	8.14	6.77	8	5.760
13	7.58	8.74	12.74	8	7.710
9	8.81	8.77	7.11	8	8.840
11	8.33	9.26	7.81	8	8.470
14	9.96	8.1	8.84	8	7.040
6	7.24	6.13	6.08	8	5.250
4	4.26	3.1	5.39	19	12.500
12	10.84	9.13	8.15	8	5.560
7	4.82	7.26	6.42	8	7.910
5	5.68	4.74	5.73	8	6.890

The scatterplots for these examples are all typical of graphs one might see in problems with one response and one predictor. Examination of the summary graph is a first step in exploring the relationships these graphs portray.

Anscombe (1973) provided the artificial data given in Table 1.1 that consists of 11 pairs of points (x_i, y_i), to which the simple linear regression mean function $E(y|x) = \beta_0 + \beta_1 x$ is fit. Each data set leads to an identical summary analysis with the same estimated slope, intercept, and other summary statistics, but the visual impression of each of the graphs is very different. The first example in Figure 1.9a is as one might expect to observe if the simple linear regression model were appropriate. The graph of the second data set given in Figure 1.9b suggests that the analysis based on simple linear regression is incorrect and that a smooth curve, perhaps a quadratic polynomial, could be fit to the data with little remaining variability. Figure 1.9c suggests that the prescription of simple regression may be correct for most of the data, but one of the cases is too far away from the fitted regression line. This is called the *outlier problem*. Possibly the case that does not match the others should be deleted from the data set, and the regression should be refit from the remaining ten cases. This will lead to a different fitted line. Without a context for the data, we cannot judge one line "correct" and the other "incorrect". The final set graphed in Figure 1.9d is different from the other three in that there is not enough information to make a judgment concerning the mean function. If the eighth case were deleted, we could not even estimate a slope. We must distrust an analysis that is so heavily dependent upon a single case.

1.5 TOOLS FOR LOOKING AT SCATTERPLOTS

Because looking at scatterplots is so important to fitting regression models, we establish some common vocabulary for describing the information in them and some tools to help us extract the information they contain.

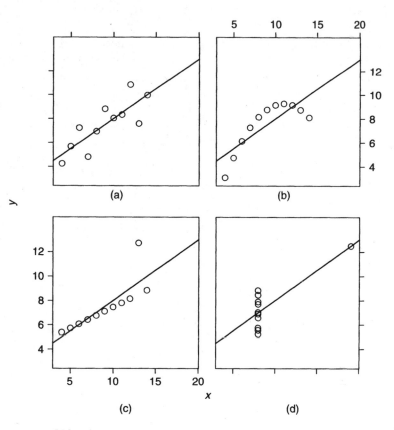

FIG. 1.9 Four hypothetical data sets (from Anscombe, 1973).

The summary graph is of the response Y versus the predictor X. The mean function for the graph is defined by (1.1), and it characterizes how Y changes on the average as the value of X is varied. We may have a parametric model for the mean function and will use data to estimate the parameters. The variance function also characterizes the graph, and in many problems we will assume at least at first that the variance function is constant. The scatterplot also will highlight separated points that may be of special interest because they do not fit the trend determined by the majority of the points.

A *null plot* has constant mean function, constant variance function and no separated points. The scatterplot for the snowfall data appears to be a null plot.

1.5.1 Size

To extract all the available information from a scatterplot, we may need to interact with it by changing scales, by resizing, or by removing linear trends. An example of this is given in Problem 1.2.

1.5.2 Transformations

In some problems, either or both of Y and X can be replaced by transformations so the summary graph has desirable properties. Most of the time, we will use *power transformations*, replacing, for example, X by X^λ for some number λ. Because logarithmic transformations are so frequently used, we will interpret $\lambda = 0$ as corresponding to a log transform. In this book, we will generally use logs to the base two, but if your computer program does not permit the use of base-two logarithms, any other base, such as base-ten or natural logarithms, is equivalent.

1.5.3 Smoothers for the Mean Function

In the smallmouth bass data in Figure 1.5, we computed an estimate of $E(Length|Age)$ using a simple nonparametric smoother obtained by averaging the repeated observations at each value of *Age*. Smoothers can also be defined when we do not have repeated observations at values of the predictor by averaging the observed data for all values of X *close to*, but not necessarily equal to, x. The literature on using smoothers to estimate mean functions has exploded in recent years, with good fairly elementary treatments given by Härdle (1990), Simonoff (1996), Bowman and Azzalini (1997), and Green and Silverman (1994). Although these authors discuss nonparametric regression as an end in itself, we will generally use smoothers as *plot enhancements* to help us understand the information available in a scatterplot and to help calibrate the fit of a parametric mean function to a scatterplot.

For example, Figure 1.10 repeats Figure 1.1, this time adding the estimated straight-line mean function and smoother called a *loess* smooth (Cleveland, 1979). Roughly speaking, the *loess* smooth estimates $E(Y|X = x)$ at the point x by fitting

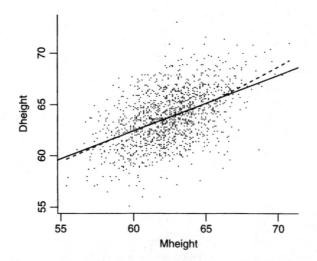

FIG. 1.10 Heights data with the OLS line and a loess smooth with span $= 0.10$.

a straight line to a fraction of the points closest to x; we used the fraction of 0.20 in this figure because the sample size is so large, but it is more usual to set the fraction to about 2/3. The smoother is obtained by joining the estimated values of $E(Y|X = x)$ for many values of x. The *loess* smoother and the straight line agree almost perfectly for *Mheight* close to average, but they agree less well for larger values of *Mheight* where there is much less data. Smoothers tend to be less reliable at the edges of the plot. We briefly discuss the *loess* smoother in Appendix A.5, but this material is dependent on the results in Chapters 2–4.

1.6   SCATTERPLOT MATRICES

With one potential predictor, a scatterplot provides a summary of the regression relationship between the response and the potential predictor. With many potential predictors, we need to look at many scatterplots. A *scatterplot matrix* is a convenient way to organize these plots.

Fuel Consumption

The goal of this example is to understand how fuel consumption varies over the 50 United States and the District of Columbia, and, in particular, to understand the effect on fuel consumption of state gasoline tax. Table 1.2 describes the variables to be used in this example; the data are given in the file `fuel2001.txt`. The data were collected by the US Federal Highway Administration.

Both *Drivers* and *FuelC* are state totals, so these will be larger in states with more people and smaller in less populous states. *Income* is computed per person. To make all these comparable and to attempt to eliminate the effect of size of the state, we compute rates *Dlic = Drivers/Pop* and *Fuel = FuelC/Pop*. Additionally, we replace *Miles* by its (base-two) logarithm before doing any further analysis. Justification for replacing *Miles* with log(*Miles*) is deferred to Problem 7.7.

TABLE 1.2   Variables in the Fuel Consumption Data[a]

Drivers	Number of licensed drivers in the state
FuelC	Gasoline sold for road use, thousands of gallons
Income	Per person personal income for the year 2000, in thousands of dollars
Miles	Miles of Federal-aid highway miles in the state
Pop	2001 population age 16 and over
Tax	Gasoline state tax rate, cents per gallon
State	State name
Fuel	$1000 \times$ *Fuelc/Pop*
Dlic	$1000 \times$ *Drivers/Pop*
log(*Miles*)	Base-two logarithm of *Miles*

Source: "Highway Statistics 2001," http://www.fhwa.dot.gov/ohim/hs01/index.htm.

[a] All data are for 2001, unless otherwise noted. The last three variables do not appear in the data file but are computed from the previous variables, as described in the text.

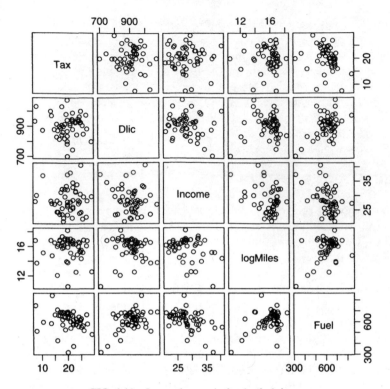

FIG. 1.11   Scatterplot matrix for the fuel data.

The scatterplot matrix for the fuel data is shown in Figure 1.11. Except for the diagonal, a scatterplot matrix is a 2D array of scatterplots. The variable names on the diagonal label the axes. In Figure 1.11, the variable log(*Miles*) appears on the horizontal axis of the all the plots in the fourth column from the left and on the vertical axis of all the plots in the fourth row from the top[3].

Each plot in a scatterplot matrix is relevant to a particular one-predictor regression of the variable on the vertical axis, given the variable on the horizontal axis. For example, the plot of *Fuel* versus *Tax* in the last plot in the first column of the scatterplot matrix is relevant for the regression of *Fuel* on *Tax*; this is the first plot in the last row of Figure 1.11. We can interpret this plot as we would a scatterplot for simple regression. We get the overall impression that *Fuel* decreases on the average as *Tax* increases, but there is lot of variation. We can make similar qualitative judgments about the each of the regressions of *Fuel* on the other variables. The overall impression is that *Fuel* is at best weakly related to each of the variables in the scatterplot matrix.

---

[3]The scatterplot matrix program used to draw Figure 1.11, which is the `pairs` function in R, has the diagonal running from the top left to the lower right. Other programs, such as the `splom` function in R, has the diagonal from lower-left to upper-right. There seems to be no strong reason to prefer one over the other.

Does this help us understand how *Fuel* is related to all four predictors simultaneously? The marginal relationships between the response and each of the variables are *not* sufficient to understand the *joint* relationship between the response and the predictors. The interrelationships among the predictors are also important. The pairwise relationships between the predictors can be viewed in the remaining cells of the scatterplot matrix. In Figure 1.11, the relationships between all pairs of predictors appear to be very weak, suggesting that for this problem the marginal plots including *Fuel* are quite informative about the multiple regression problem. General considerations for other scatterplot matrices will be developed in later chapters.

PROBLEMS

1.1. Smallmouth bass data Compute the means and the variances for each of the eight subpopulations in the smallmouth bass data. Draw a graph of average length versus *Age* and compare to Figure 1.5. Draw a graph of the standard deviations versus age. If the variance function is constant, then the plot of standard deviation versus *Age* should be a null plot. Summarize the information.

1.2. Mitchell data The data shown in Figure 1.12 give average soil temperature in degrees C at 20 cm depth in Mitchell, Nebraska, for 17 years beginning January 1976, plotted versus the month number. The data were collected by K. Hubbard and provided by O. Burnside.

1.2.1. Summarize the information in the graph about the dependence of soil temperature on month number.

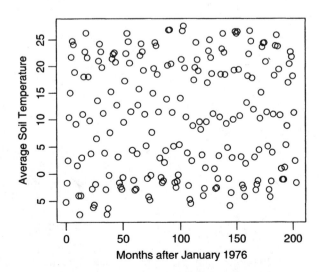

FIG. 1.12 Monthly soil temperature data.

1.2.2. The data used to draw Figure 1.12 are in the file `Mitchell.txt`. Redraw the graph, but this time make the length of the horizontal axis at least four times the length of the vertical axis. Repeat Problem 1.2.1.

1.3. United Nations The data in the file `UN1.txt` contains *PPgdp*, the 2001 gross national product per person in US dollars, and *Fertility*, the birth rate per 1000 females in the population in the year 2000. The data are for 193 localities, mostly UN member countries, but also other areas such as Hong Kong that are not independent countries; the third variable on the file called *Locality* gives the name of the locality. The data were collected from http://unstats.un.org/ unsd/demographic. In this problem, we will study the conditional distribution of *Fertility* given *PPgdp*.

1.3.1. Identify the predictor and the response.

1.3.2. Draw the scatterplot of *Fertility* on the vertical axis versus *PPgdp* on the horizontal axis and summarize the information in this graph. Does a straight-line mean function seem to be a plausible for a summary of this graph?

1.3.3. Draw the scatterplot of log(*Fertility*) versus log(*PPgdp*), using logs to the base two. Does the simple linear regression model seem plausible for a summary of this graph?

1.4. Old Faithful The data in the data file `oldfaith.txt` gives information about eruptions of Old Faithful Geyser during October 1980. Variables are the *Duration* in seconds of the current eruption, and the *Interval*, the time in minutes to the next eruption. The data were collected by volunteers and were provided by R. Hutchinson. Apart from missing data for the period from midnight to 6 AM, this is a complete record of eruptions for that month.

Old Faithful Geyser is an important tourist attraction, with up to several thousand people watching it erupt on pleasant summer days. The park service uses data like these to obtain a prediction equation for the time to the next eruption.

Draw the relevant summary graph for predicting interval from duration, and summarize your results.

1.5. Water run-off in the Sierras Can Southern California's water supply in future years be predicted from past data? One factor affecting water availability is stream run-off. If run-off could be predicted, engineers, planners and policy makers could do their jobs more efficiently. The data in the file `water.txt` contains 43 years' worth of precipitation measurements taken at six sites in the Sierra Nevada mountains (labelled *APMAM, APSAB, APSLAKE, OPBPC, OPRC*, and *OPSLAKE*), and stream run-off volume at a site near Bishop, California, labelled *BSAAM*. The data are from the UCLA Statistics WWW server.

Draw the scatterplot matrix for these data and summarize the information available from these plots.

CHAPTER 2

Simple Linear Regression

The *simple linear regression model* consists of the mean function and the variance function

$$E(Y|X = x) = \beta_0 + \beta_1 x$$
$$\text{Var}(Y|X = x) = \sigma^2$$

(2.1)

The parameters in the mean function are the intercept β_0, which is the value of $E(Y|X = x)$ when x equals zero, and the slope β_1, which is the rate of change in $E(Y|X = x)$ for a unit change in X; see Figure 2.1. By varying the parameters, we can get all possible straight lines. In most applications, parameters are unknown and must be estimated using data. The variance function in (2.1) is assumed to be constant, with a positive value σ^2 that is usually unknown.

Because the variance $\sigma^2 > 0$, the observed value of the ith response y_i will typically not equal its expected value $E(Y|X = x_i)$. To account for this difference between the observed data and the expected value, statisticians have invented a quantity called a *statistical error*, or e_i, for case i defined implicitly by the equation $y_i = E(Y|X = x_i) + e_i$ or explicitly by $e_i = y_i - E(Y|X = x_i)$. The errors e_i depend on unknown parameters in the mean function and so are not observable quantities. They are random variables and correspond to the *vertical distance between the point y_i and the mean function $E(Y|X = x_i)$*. In the heights data, page 2, the errors are the differences between the heights of particular daughters and the average height of all daughters with mothers of a given fixed height.

If the assumed mean function is incorrect, then the difference between the observed data and the incorrect mean function will have a non random component, as illustrated in Figure 2.2.

We make two important assumptions concerning the errors. First, we assume that $E(e_i|x_i) = 0$, so if we could draw a scatterplot of the e_i versus the x_i, we would have a null scatterplot, with no patterns. The second assumption is that the errors are all *independent*, meaning that the value of the error for one case gives

Applied Linear Regression, Third Edition, by Sanford Weisberg
ISBN 0-471-66379-4   Copyright © 2005 John Wiley & Sons, Inc.

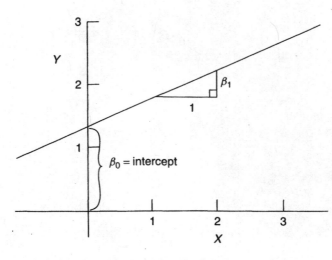

FIG. 2.1 Equation of a straight line $E(Y|X = x) = \beta_0 + \beta_1 x$.

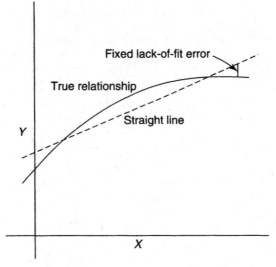

FIG. 2.2 Approximating a curved mean function by straight line cases adds a fixed component to the errors.

no information about the value of the error for another case. This is likely to be true in the examples in Chapter 1, although this assumption will not hold in all problems.

Errors are often assumed to be normally distributed, but normality is much stronger than we need. In this book, the normality assumption is used primarily to obtain tests and confidence statements with small samples. If the errors are thought to follow some different distribution, such as the Poisson or the Binomial, other methods besides OLS may be more appropriate; we return to this topic in Chapter 12.

2.1 ORDINARY LEAST SQUARES ESTIMATION

Many methods have been suggested for obtaining estimates of parameters in a model. The method discussed here is called *ordinary least squares*, or OLS, in which parameter estimates are chosen to minimize a quantity called the *residual sum of squares*. A formal development of the least squares estimates is given in Appendix A.3.

Parameters are unknown quantities that characterize a model. *Estimates of parameters* are computable functions of data and are therefore *statistics*. To keep this distinction clear, parameters are denoted by Greek letters like α, β, γ and σ, and estimates of parameters are denoted by putting a "hat" over the corresponding Greek letter. For example, $\hat{\beta}_1$, read "beta one hat," is the estimator of β_1, and $\hat{\sigma}^2$ is the estimator of σ^2. The *fitted value* for case i is given by $\widehat{E}(Y|X = x_i)$, for which we use the shorthand notation $\hat{y}_i$,

$$\hat{y}_i = \widehat{E}(Y|X = x_i) = \hat{\beta}_0 + \hat{\beta}_1 x_i \tag{2.2}$$

Although the e_i are not parameters in the usual sense, we shall use the same hat notation to specify the residuals: the residual for the ith case, denoted $\hat{e}_i$, is given by the equation

residual

$$\hat{e}_i = y_i - \widehat{E}(Y|X = x_i) = y_i - \hat{y}_i = y_i - (\hat{\beta}_0 + \hat{\beta}_1) \quad i = 1, \ldots, n \tag{2.3}$$

which should be compared with the equation for the statistical errors,

Statistic error

$$e_i = y_i - (\beta_0 + \beta_1 x_i) \quad i = 1, \ldots, n$$

All least squares computations for simple regression depend only on averages, sums of squares and sums of cross-products. Definitions of the quantities used are given in Table 2.1. Sums of squares and cross-products have been centered by subtracting the average from each of the values before squaring or taking cross-products. Appropriate alternative formulas for computing the corrected sums of squares and cross products from uncorrected sums of squares and cross-products that are often given in elementary textbooks are useful for mathematical proofs, but they can be highly inaccurate when used on a computer and should be avoided.

Table 2.1 also lists definitions for the usual univariate and bivariate summary statistics, the sample averages $(\overline{x}, \overline{y})$, sample variances (SD_x^2, SD_y^2), and estimated covariance and correlation (s_{xy}, r_{xy}). The "hat" rule described earlier would suggest that different symbols should be used for these quantities; for example, $\hat{\rho}_{xy}$ might be more appropriate for the sample correlation if the population correlation is ρ_{xy}.

TABLE 2.1   Definitions of Symbols^a

Quantity	Definition	Description
$\bar{x}$	$\sum x_i/n$	Sample average of x
$\bar{y}$	$\sum y_i/n$	Sample average of y
SXX	$\sum(x_i - \bar{x})^2 = \sum(x_i - \bar{x})x_i$	Sum of squares for the x's
SD_x^2	$SXX/(n-1)$	Sample variance of the x's
SD_x	$\sqrt{SXX/(n-1)}$	Sample standard deviation of the x's
SYY	$\sum(y_i - \bar{y})^2 = \sum(y_i - \bar{y})y_i$	Sum of squares for the y's
SD_y^2	$SYY/(n-1)$	Sample variance of the y's
SD_y	$\sqrt{SYY/(n-1)}$	Sample standard deviation of the y's
SXY	$\sum(x_i - \bar{x})(y_i - \bar{y}) = \sum(x_i - \bar{x})y_i$	Sum of cross-products
s_{xy}	$SXY/(n-1)$	Sample covariance
r_{xy}	$s_{xy}/(SD_x SD_y)$	Sample correlation

^{a}In each equation, the symbol $\sum$ means to add over all the n values or pairs of values in the data.

This inconsistency is deliberate since in many regression situations, these statistics are not estimates of population parameters.

To illustrate computations, we will use Forbes' data, page 4, for which $n = 17$. The data are given in Table 2.2. In our analysis of these data, the response will be taken to be $Lpres = 100 \times \log_{10}(Pressure)$, and the predictor is *Temp*. We have used the values for these variables shown in Table 2.2 to do the computations.

TABLE 2.2   Forbes' 1857 Data on Boiling Point and Barometric Pressure for 17 Locations in the Alps and Scotland

Case Number	Temp (°F)	Pressure (Inches Hg)	$Lpres = 100 \times \log(Pressure)$
1	194.5	20.79	131.79
2	194.3	20.79	131.79
3	197.9	22.40	135.02
4	198.4	22.67	135.55
5	199.4	23.15	136.46
6	199.9	23.35	136.83
7	200.9	23.89	137.82
8	201.1	23.99	138.00
9	201.4	24.02	138.06
10	201.3	24.01	138.04
11	203.6	25.14	140.04
12	204.6	26.57	142.44
13	209.5	28.49	145.47
14	208.6	27.76	144.34
15	210.7	29.04	146.30
16	211.9	29.88	147.54
17	212.2	30.06	147.80

Neither multiplication by 100 nor the base of the logarithms has important effects on the analysis. Multiplication by 100 avoids using scientific notation for numbers we display in the text, and changing the base of the logarithms merely multiplies the logarithms by a constant. For example, to convert from base-ten logarithms to base-two logarithms, multiply by 3.321928. To convert natural logarithms to base-two, multiply by 1.442695.

Forbes' data were collected at 17 selected locations, so the sample variance of boiling points, $SD_x^2 = 33.17$, is not an estimate of any meaningful population variance. Similarly, r_{xy} depends as much on the method of sampling as it does on the population value ρ_{xy}, should such a population value make sense. In the heights example, page 2, if the 1375 mother–daughter pairs can be viewed as a sample from a population, then the sample correlation is an estimate of a population correlation.

The usual sample statistics are often presented and used in place of the corrected sums of squares and cross-products, so alternative formulas are given using both sets of quantities.

2.2   LEAST SQUARES CRITERION

The criterion function for obtaining estimators is based on the residuals, which geometrically are the vertical distances between the fitted line and the actual y-values, as illustrated in Figure 2.3. The residuals reflect the inherent asymmetry in the roles of the response and the predictor in regression problems.

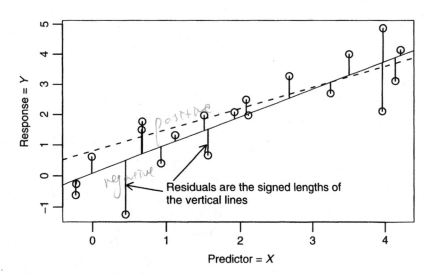

FIG. 2.3   A schematic plot for OLS fitting. Each data point is indicated by a small circle, and the solid line is a candidate OLS line given by a particular choice of slope and intercept. The solid vertical lines between the points and the solid line are the residuals. Points below the line have negative residuals, while points above the line have positive residuals.

The OLS estimators are those values β_0 and β_1 that minimize the function[1]

$$RSS(\beta_0, \beta_1) = \sum_{i=1}^{n} \left[y_i - (\beta_0 + \beta_1 x_i) \right]^2 \tag{2.4}$$

When evaluated at $(\hat{\beta}_0, \hat{\beta}_1)$, we call the quantity $RSS(\hat{\beta}_0, \hat{\beta}_1)$ the *residual sum of squares*, or just *RSS*.

The least squares estimates can be derived in many ways, one of which is outlined in Appendix A.3. They are given by the expressions

$$\hat{\beta}_1 = \frac{SXY}{SXX} = r_{xy} \frac{SD_y}{SD_x} = r_{xy} \left(\frac{SYY}{SXX} \right)^2$$

$$\hat{\beta}_0 = \bar{y} - \hat{\beta}_1 \bar{x} \tag{2.5}$$

The several forms for $\hat{\beta}_1$ are all equivalent.

We emphasize again that OLS produces *estimates* of parameters but not the actual values of the parameters. The data in Figure 2.3 were created by setting the x_i to be random sample of 20 numbers from a $N(2, 1.5)$ distribution and then computing $y_i = 0.7 + 0.8x_i + e_i$, where the errors were $N(0, 1)$ random numbers. For this graph, the true values of $\beta_0 = 0.7$ and $\beta_1 = 0.8$ are known. The graph of the true mean function is shown in Figure 2.3 as a dashed line, and it seems to match the data poorly compared to OLS, given by the solid line. Since OLS minimizes (2.4), it will always fit at least as well as, and generally better than, the true mean function.

Using Forbes' data, we will write $\bar{x}$ to be the sample mean of *Temp* and $\bar{y}$ to be the sample mean of *Lpres*. The quantities needed for computing the least squares estimators are

$$\begin{aligned} \bar{x} &= 202.95294 \quad SXX = 530.78235 \quad SXY = 475.31224 \\ \bar{y} &= 139.60529 \quad SYY = 427.79402 \end{aligned} \tag{2.6}$$

The quantity *SYY*, although not yet needed, is given for completeness. In the rare instances that regression calculations are not done using statistical software or a statistical calculator, intermediate calculations such as these should be done as accurately as possible, and rounding should be done only to final results. Using (2.6), we find

$$\hat{\beta}_1 = \frac{SXY}{SXX} = 0.895$$

$$\hat{\beta}_0 = \bar{y} - \hat{\beta}_1 \bar{x} = -42.138$$

---

[1]We abuse notation by using the symbol for a fixed though unknown quantity like β_j as if it were a variable argument. Thus, for example, $RSS(\beta_0, \beta_1)$ is a function of two variables to be evaluated as its arguments β_0 and β_1 vary. The same abuse of notation is used in the discussion of confidence intervals.

The estimated line, given by either of the equations

$$\widehat{E}(Lpres|Temp) = -42.138 + 0.895Temp$$

$$= 139.606 + 0.895(Temp - 202.953)$$

was drawn in Figure 1.4a. The fit of this line to the data is excellent.

2.3 ESTIMATING σ^2

Since the variance σ^2 is essentially the average squared size of the e_i^2, we should expect that its estimator $\hat{\sigma}^2$ is obtained by averaging the squared residuals. Under the assumption that the errors are uncorrelated random variables with zero means and common variance σ^2, an unbiased estimate of σ^2 is obtained by dividing $RSS = \sum \hat{e}_i^2$ by its *degrees of freedom* (df), where residual df = number of cases minus the number of parameters in the mean function. For simple regression, residual df $= n - 2$, so the estimate of σ^2 is given by

$$\hat{\sigma}^2 = \frac{RSS}{n-2} \tag{2.7}$$

This quantity is called the *residual mean square*. In general, any sum of squares divided by its df is called a mean square. The residual sum of squares can be computed by squaring the residuals and adding them up. It can also be computed from the formula (Problem 2.9)

$$RSS = SYY - \frac{SXY^2}{SXX} = SYY - \hat{\beta}_1^2 SXX \tag{2.8}$$

Using the summaries for Forbes' data given at (2.6), we find

$$RSS = 427.79402 - \frac{475.31224^2}{530.78235}$$

$$= 2.15493 \tag{2.9}$$

$$\sigma^2 = \frac{2.15493}{17-2} = 0.14366 \tag{2.10}$$

The square root of $\hat{\sigma}^2$, $\hat{\sigma} = \sqrt{0.14366} = 0.37903$ is often called the *standard error of regression*. It is in the same units as is the response variable.

If in addition to the assumptions made previously, the e_i are drawn from a normal distribution, then the residual mean square will be distributed as a multiple of a chi-squared random variable with df $= n - 2$, or in symbols,

$$(n-2)\frac{\hat{\sigma}^2}{\sigma^2} \sim \chi^2(n-2)$$

This is proved in more advanced books on linear models and is used to obtain the distribution of test statistics and also to make confidence statements concerning σ^2. In particular, this fact implies that $E(\hat{\sigma}^2) = \sigma^2$, although normality is not required for unbiasedness.

2.4 PROPERTIES OF LEAST SQUARES ESTIMATES

The OLS estimates depend on data only through the statistics given in Table 2.1. This is both an advantage, making computing easy, and a disadvantage, since any two data sets for which these are identical give the same fitted regression, even if a straight-line model is appropriate for one but not the other, as we have seen in Anscombe's examples in Section 1.4. The estimates $\hat{\beta}_0$ and $\hat{\beta}_1$ can both be written as linear combinations of $y_1, \ldots, y_n$, for example, writing $c_i = (x_i - \bar{x})/SXX$ (see Appendix A.3)

$$\hat{\beta}_1 = \sum \left(\frac{x_i - \bar{x}}{SXX} \right) y_i = \sum c_i y_i$$

The fitted value at $x = \bar{x}$ is

$$\widehat{E}(Y|X = \bar{x}) = \bar{y} - \hat{\beta}_1\bar{x} + \hat{\beta}_1\bar{x} = \bar{y}$$

so the fitted line must pass through the point $(\bar{x}, \bar{y})$, intuitively the center of the data. Finally, as long as the mean function includes an intercept, $\sum \hat{e}_i = 0$. Mean functions without an intercept will usually have $\sum \hat{e}_i \neq 0$.

Since the estimates $\hat{\beta}_0$ and $\hat{\beta}_1$ depend on the random e_is, the estimates are also random variables. If all the e_i have zero mean and the mean function is correct, then, as shown in Appendix A.4, the least squares estimates are unbiased,

$$E(\hat{\beta}_0) = \beta_0$$
$$E(\hat{\beta}_1) = \beta_1$$

The variance of the estimators, assuming $\text{Var}(e_i) = \sigma^2, i = 1, \ldots, n$, and $\text{Cov}(e_i, e_j) = 0, i \neq j$, are from Appendix A.4,

$$\text{Var}(\hat{\beta}_1) = \sigma^2 \frac{1}{SXX}$$

$$\text{Var}(\hat{\beta}_0) = \sigma^2 \left(\frac{1}{n} + \frac{\bar{x}^2}{SXX} \right) \tag{2.11}$$

The two estimates are correlated, with covariance

$$\text{Cov}(\hat{\beta}_0, \hat{\beta}_1) = -\sigma^2 \frac{\bar{x}}{SXX} \tag{2.12}$$

The correlation between the estimates can be computed to be

$$\rho(\hat{\beta}_0, \hat{\beta}_1) = \frac{-\bar{x}}{\sqrt{SXX/n + \bar{x}^2}} = \frac{-\bar{x}}{\sqrt{(n-1)SD_x^2/n + \bar{x}^2}}$$

This correlation can be close to plus or minus one if SD_x is small compared to $|\bar{x}|$ and can be made to equal zero if the predictor is centered to have sample mean zero.

The *Gauss–Markov theorem* provides an optimality result for OLS estimates. Among all estimates that are linear combinations of the ys and unbiased, the OLS estimates have the smallest variance. If one believes the assumptions and is interested in using linear unbiased estimates, the OLS estimates are the ones to use.

When the errors are normally distributed, the OLS estimates can be justified using a completely different argument, since they are then also maximum likelihood estimates, as discussed in any mathematical statistics text, for example, Casella and Berger (1990).

Under the assumption that errors are independent, normal with constant variance, which is written in symbols as

$$e_i \sim \text{NID}(0, \sigma^2) \qquad i = 1, \ldots, n$$

$\hat{\beta}_0$ and $\hat{\beta}_1$ are also normally distributed, since they are linear functions of the y_is and hence of the e_i, with variances and covariances given by (2.11) and (2.12). These results are used to get confidence intervals and tests. Normality of estimates also holds without normality of errors if the sample size is large enough[2].

2.5   ESTIMATED VARIANCES

Estimates of $\text{Var}(\hat{\beta}_0)$ and $\text{Var}(\hat{\beta}_1)$ are obtained by substituting $\hat{\sigma}^2$ for σ^2 in (2.11). We use the symbol $\widehat{\text{Var}}(\ )$ for an estimated variance. Thus

$$\widehat{\text{Var}}(\hat{\beta}_1) = \hat{\sigma}^2 \frac{1}{SXX}$$

$$\widehat{\text{Var}}(\hat{\beta}_0) = \hat{\sigma}^2 \left(\frac{1}{n} + \frac{\bar{x}^2}{SXX} \right)$$

The square root of an estimated variance is called a *standard error*, for which we use the symbol se( ). The use of this notation is illustrated by

$$\text{se}(\hat{\beta}_1) = \sqrt{\widehat{\text{Var}}(\hat{\beta}_1)}$$

---

[2]The main requirement for all estimates to be normally distributed in large samples is that $\max_i \left((x_i - \bar{x})^2/SXX \right)$ must get close to zero as the sample size increases (Huber, 1981).

2.6   COMPARING MODELS: THE ANALYSIS OF VARIANCE

The analysis of variance provides a convenient method of comparing the fit of two or more mean functions for the same set of data. The methodology developed here is very useful in multiple regression and, with minor modification, in most regression problems.

An elementary alternative to the simple regression model suggests fitting the mean function

$$E(Y|X = x) = \beta_0 \qquad (2.13)$$

The mean function (2.13) is the same for all values of X. Fitting with this mean function is equivalent to finding the best line parallel to the horizontal or x-axis, as shown in Figure 2.4. The OLS estimate of the mean function is $\widehat{E(Y|X)} = \hat{\beta}_0$, where $\hat{\beta}_0$ is the value of β_0 that minimizes $\sum(y_i - \beta_0)^2$. The minimizer is given by

$$\hat{\beta}_0 = \overline{y} \qquad (2.14)$$

The residual sum of squares is

$$\sum(y_i - \hat{\beta}_0)^2 = \sum(y_i - \overline{y})^2 = SYY \qquad (2.15)$$

This residual sum of squares has $n - 1$ df, n cases minus one parameter in the mean function.

Next, consider the simple regression mean function obtained from (2.13) by adding a term that depends on X

$$E(Y|X = x) = \beta_0 + \beta_1 x \qquad (2.16)$$

Fitting this mean function is equivalent to finding the best line of arbitrary slope, as shown in Figure 2.4. The OLS estimates for this mean function are given by

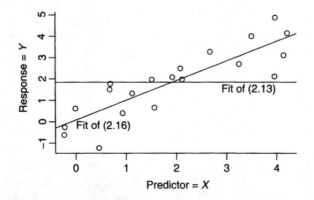

FIG. 2.4   Two mean functions compared by the analysis of variance.

(2.5). The estimates of β_0 under the two mean functions are different, just as the meaning of β_0 in the two mean functions is different. For (2.13), β_0 is the average of the y_is, but for (2.16), β_0 is the expected value of Y when $X = 0$.

For (2.16), the residual sum of squares, given in (2.8), is

$$RSS = SYY - \frac{(SXY)^2}{SXX} \tag{2.17}$$

As mentioned earlier, RSS has $n - 2$ df.

The difference between the sum of squares at (2.15) and that at (2.17) is the reduction in residual sum of squares due to enlarging the mean function from (2.13) to the simple regression mean function (2.16). This is the *sum of squares due to regression*, *SSreg*, defined by

$$SSreg = SYY - RSS$$

$$= SYY - \left(SYY - \frac{(SXY)^2}{SXX}\right)$$

$$= \frac{(SXY)^2}{SXX} \tag{2.18}$$

The df associated with *SSreg* is the difference in df for mean function (2.13), $n - 1$, and the df for mean function (2.16), $n - 2$, so the df for *SSreg* is $(n - 1) - (n - 2) = 1$ for simple regression. These results are often summarized in an analysis of variance table, abbreviated as *ANOVA*, given in Table 2.3. The column marked "Source" refers to descriptive labels given to the sums of squares; in more complicated tables, there may be many sources, and the labels given may be different in some computer programs. The df column gives the number of degrees of freedom associated with each named source. The next column gives the associated sum of squares. The mean square column is computed from the sum of squares column by dividing sums of squares by the corresponding df. The mean square on the residual line is just $\hat{\sigma}^2$, as already discussed.

The analysis of variance for Forbes' data is given in Table 2.4. Although this table will be produced by any linear regression software program, the entries in Table 2.4 can be constructed from the summary statistics given at (2.6).

The *ANOVA* is always computed relative to a specific larger mean function, here given by (2.16), and a smaller mean function obtained from the larger by setting

TABLE 2.3 The Analysis of Variance Table for Simple Regression

Source	df	SS	MS	F	p-value
Regression	1	SSreg	SSreg/1	MSreg/$\hat{\sigma}^2$	
Residual	$n - 2$	RSS	$\hat{\sigma}^2 = RSS/(n - 2)$		
Total	$n - 1$	SYY			

TABLE 2.4 Analysis of Variance Table for Forbes' Data

Source	df	SS	MS	F	p-value
Regression on *Temp*	1	425.639	425.639	2962.79	≈ 0
Residual	15	2.155	0.144		

some parameters to zero, or occasionally setting them to some other known value. For example, equation (2.13) was obtained from (2.16) by setting $\beta_1 = 0$. The line in the *ANOVA* table for the total gives the residual sum of squares corresponding to the mean function with the fewest parameters. In the next chapter, the analysis of variance is applied to a sequence of mean functions, but the reference to a fixed large mean function remains intact.

2.6.1 The *F*-Test for Regression

If the sum of squares for regression *SSreg* is large, then the simple regression mean function $E(Y|X = x) = \beta_0 + \beta_1 x$ should be a significant improvement over the mean function given by (2.13), $E(y|X = x) = \beta_0$. This is equivalent to saying that the additional parameter in the simple regression mean function β_1 is different from zero or that $E(Y|X = x)$ is not constant as X varies. To formalize this notion, we need to be able to judge how large is "large." This is done by comparing the regression mean square, *SSreg* divided by its df, to the residual mean square $\hat{\sigma}^2$. We call this ratio F:

$$F = \frac{(SYY - RSS)/1}{\hat{\sigma}^2} = \frac{SSreg/1}{\hat{\sigma}^2} \qquad (2.19)$$

F is just a rescaled version of $SSreg = SYY - RSS$, with larger values of *SSreg* resulting in larger values of F. Formally, we can consider testing the null hypothesis (NH) against the alternative hypothesis (AH)

$$\begin{array}{ll} \text{NH:} & E(Y|X = x) = \beta_0 \\ \text{AH:} & E(Y|X = x) = \beta_0 + \beta_1 x \end{array} \qquad (2.20)$$

If the errors are $NID(0, \sigma^2)$ or the sample size is large enough, then under NH (2.19) will follow an F-distribution with df associated with the numerator and denominator of (2.19), 1 and $n - 2$ for simple regression. This is written $F \sim F(1, n - 2)$. For Forbes' data, we compute

$$F = \frac{425.639}{0.144} = 2963$$

We obtain a significance level or p-value for this test by comparing F to the percentage points of the $F(1, n - 2)$-distribution. Most computer programs that fit regression models will include functions to computing percentage points of the F

and other standard distributions and will include the p-value along with the *ANOVA* table, as in Table 2.4. The p-value is shown as "approximately zero," meaning that, if the NH were true, the change of F exceeding its observed value is essentially zero. This is very strong evidence against NH and in favor of AH.

2.6.2 Interpreting p-values

Under the appropriate assumptions, the p-value is the conditional probability of observing a value of the computed statistic, here the value of F, as extreme or more extreme, here as large or larger, than the observed value, given that the NH is true. A small p-value provides evidence against the NH.

In some research areas, it has become traditional to adopt a *fixed significance level* when examining p-values. For example, if a fixed significance level of α is adopted, then we would say that an NH is rejected at level α if the p-value is less than α. The most common choice for α is 0.05, which would mean that, were the NH to be true, we would incorrectly find evidence against it about 5% of the time, or about 1 test in 20. Accept–reject rules like this are generally unnecessary for reasonable scientific inquiry. Simply reporting p-values and allowing readers to decide on significance seems a better approach.

There is an important distinction between statistical significance, the observation of a sufficiently small p-value, and scientific significance, observing an effect of sufficient magnitude to be meaningful. Judgment of the latter usually will require examination of more than just the p-value.

2.6.3 Power of Tests

When the NH is true, and all assumptions are met, the chance of incorrectly declaring an NH to be false at level α is just α. If $\alpha = 0.05$, then in 5% of tests where the NH is true we will get a p-value smaller than or equal to 0.05.

When the NH is false, we expect to see small p-values more often. The *power* of a test is defined to be the *probability of detecting a false NH*. For the hypothesis test (2.20), when the NH is false, it is shown in more advanced books on linear models (such as Seber, 1977) that the statistic F given by (2.19) has a *noncentral F* distribution, with 1 and $n - 2$ df, and with noncentrality parameter given by $SXX\beta_1^2/\sigma^2$. The larger the value of the non centrality parameter, the greater the power. The noncentrality is increased if β_1^2 is large, if SXX is large, either by spreading out the predictors or by increasing the sample size, or by decreasing σ^2.

2.7 THE COEFFICIENT OF DETERMINATION, R^2

If both sides of (2.18) are divided by SYY, we get

$$\frac{SSreg}{SYY} = 1 - \frac{RSS}{SYY} \tag{2.21}$$

The left-hand side of (2.21) is the proportion of variability of the response explained by regression on the predictor. The right-hand side consists of one minus the remaining unexplained variability. This concept of dividing up the total variability according to whether or not it is explained is of sufficient importance that a special name is given to it. We define R^2, the coefficient of determination, to be

$$R^2 = \frac{SSreg}{SYY} = 1 - \frac{RSS}{SYY} \qquad (2.22)$$

R^2 is computed from quantities that are available in the *ANOVA* table. It is a scale-free one-number summary of the strength of the relationship between the x_i and the y_i in the data. It generalizes nicely to multiple regression, depends only on the sums or squares and appears to be easy to interpret. For Forbes' data,

$$R^2 = \frac{SSreg}{SYY} = \frac{425.63910}{427.79402} = 0.995$$

and thus about 99.5% of the variability in the observed values or $100 \times \log(Pressure)$ is explained by boiling point. Since R^2 does not depend on units of measurement, we would get the same value if we had used logarithms with a different base, or if we did not multiply $\log(Pressure)$ by 100.

By appealing to (2.22) and to Table 2.1, we can write

$$R^2 = \frac{SSreg}{SYY} = \frac{(SXY)^2}{SXX \times SYY} = r_{xy}^2$$

and thus R^2 is the same as the square of the sample correlation between the predictor and the response.

2.8 CONFIDENCE INTERVALS AND TESTS

When the errors are NID$(0, \sigma^2)$, parameter estimates, fitted values, and predictions will be normally distributed because all of these are linear combinations of the y_i and hence of the e_i. Confidence intervals and tests can be based on the t-distribution, which is the appropriate distribution with normal estimates but using an estimate of variance $\hat{\sigma}^2$. Suppose we let $t(\alpha/2, d)$ be the value that cuts off $\alpha/2 \times 100\%$ in the *upper tail* of the t-distribution with d df. These values can be computed in most statistical packages or spreadsheet software[3].

2.8.1 The Intercept

The intercept is used to illustrate the general form of confidence intervals for normally distributed estimates. The standard error of the intercept is $se(\beta_0) = \hat{\sigma}(1/n + \bar{x}^2/SXX)^{1/2}$. Hence a $(1 - \alpha) \times 100\%$ confidence interval for the intercept is the set of points β_0 in the interval

$$\hat{\beta}_0 - t(\alpha/2, n - 2)se(\hat{\beta}_0) \leq \beta_0 \leq \hat{\beta}_0 + t(\alpha/2, n - 2)se(\hat{\beta}_0)$$

[3] Such as the function tinv in Microsoft Excel, or the function pt in R or S-plus.

For Forbes' data, $se(\hat{\beta}_0) = 0.37903(1/17 + (202.95294)^2/530.78235)^{1/2} = 3.340$. For a 90% confidence interval, $t(0.05, 15) = 1.753$, and the interval is

$$-42.138 - 1.753(3.340) \le \beta_0 \le -42.136 + 1.753(3.340)$$

$$-47.993 \le \beta_0 \le -36.282$$

Ninety percent of such intervals will include the true value.

A hypothesis test of

$$\text{NH:} \quad \beta_0 = \beta_0^*, \quad \beta_1 \text{ arbitrary}$$
$$\text{AH:} \quad \beta_0 \ne \beta_0^*, \quad \beta_1 \text{ arbitrary}$$

is obtained by computing the t-statistic

$$t = \frac{\hat{\beta}_0 - \beta_0^*}{se(\hat{\beta}_0)} \tag{2.23}$$

and referring this ratio to the t-distribution with $n - 2$ df. For example, in Forbes' data, consider testing the NH $\beta_0 = -35$ against the alternative that $\beta_0 \ne -35$. The statistic is

$$t = \frac{-42.138 - (-35)}{3.340} = 2.137$$

which has a p-value near 0.05, providing some evidence against NH. This hypothesis test for these data is not one that would occur to most investigators and is used only as an illustration.

2.8.2 Slope

The standard error of $\hat{\beta}_1$ is $se(\hat{\beta}_1) = \hat{\sigma}/\sqrt{SXX} = 0.0164$. A 95% confidence interval for the slope is the set of β_1 such that

$$0.8955 - 2.131(0.0164) \le \beta_1 \le 0.8955 + 2.131(0.0164)$$

$$0.867 \le \beta_1 \le 0.930$$

As an example of a test for slope equal to zero, consider the Ft. Collins snowfall data presented on page 7. One can show, Problem 2.11, that the estimated slope is $\hat{\beta}_1 = 0.2035$, $se(\hat{\beta}_1) = 0.1310$. The test of interest is of

$$\text{NH:} \quad \beta_1 = 0$$
$$\text{AH:} \quad \beta_1 \ne 0 \tag{2.24}$$

For the Ft. Collins data, $t = (0.20335 - 0)/0.1310 = 1.553$. To get a significance level for this test, compare t with the $t(91)$ distribution; the two-sided p-value is

0.124, suggesting no evidence against the NH that *Early* and *Late* season snowfalls are independent.

Compare the hypothesis (2.24) with (2.20). Both appear to be identical. In fact,

$$t^2 = \left(\frac{\hat{\beta}_1}{se(\hat{\beta}_1)}\right)^2 = \frac{\hat{\beta}_1^2}{\hat{\sigma}^2/SXX} = \frac{\hat{\beta}_1^2 SXX}{\hat{\sigma}^2} = F$$

so the square of a t statistic with d df is equivalent to an F-statistic with $(1, d)$ df. In nonlinear and logistic regression models discussed later in the book, the analog of the t test will not be identical to the analog of the F test, and they can give conflicting conclusions. For linear regression models, no conflict occurs and the two tests are equivalent.

2.8.3 Prediction

The estimated mean function can be used to obtain values of the response for given values of the predictor. The two important variants of this problem are *prediction* and *estimation of fitted values*. Since prediction is more important, we discuss it first.

In prediction we have a new case, possibly a future value, not one used to estimate parameters, with observed value of the predictor x_*. We would like to know the value y_*, the corresponding response, but it has not yet been observed. We can use the estimated mean function to predict it. We assume that the data used to estimate the mean function are relevant to the new case, so the fitted model applies to it. In the heights example, we would probably be willing to apply the fitted mean function to mother–daughter pairs alive in England at the end of the nineteenth century. Whether the prediction would be reasonable for mother–daughter pairs in other countries or in other time periods is much less clear. In Forbes' problem, we would probably be willing to apply the results for altitudes in the range he studied. Given this additional assumption, a point prediction of y_*, say $\tilde{y}_*$, is just

$$\tilde{y}_* = \hat{\beta}_0 + \hat{\beta}_1 x_*$$

$\tilde{y}_*$ predicts the as yet unobserved y_*. The variability of this predictor has two sources: the variation in the estimates $\hat{\beta}_0$ and $\hat{\beta}_1$, and the variation due to the fact that y_* will not equal its expectation, since even if we knew the parameters exactly, the future value of the response will not generally equal its expectation. Using Appendix A.4,

$$\text{Var}(\tilde{y}_*|x_*) = \sigma^2 + \sigma^2 \left(\frac{1}{n} + \frac{(x_* - \overline{x})^2}{SXX}\right) \tag{2.25}$$

Taking square roots and estimating σ^2 by $\hat{\sigma}^2$, we get the standard error of prediction (sepred) at x_*,

$$\text{sepred}(\tilde{y}_*|x_*) = \hat{\sigma} \left(1 + \frac{1}{n} + \frac{(x_* - \overline{x})^2}{SXX}\right)^{1/2} \tag{2.26}$$

A prediction interval uses multipliers from the t-distribution. For prediction of $100 \times \log(Pressure)$ for a location with $x_* = 200$, the point prediction is $\tilde{y}_* = -42.13778 + 0.89549(200) = 136.961$, with standard error of prediction

$$\text{sepred}(\tilde{y}_*|x_* = 200) = 0.37903 \left(1 + \frac{1}{17} + \frac{(200 - 202.95294)^2}{530.78235}\right)^{1/2}$$

$$= 0.393$$

Thus a 99% predictive interval is the set of all y_* such that

$$136.961 - 2.95(0.393) \leq y_* \leq 136.961 + 2.95(0.393)$$

$$135.803 \leq y_* \leq 138.119$$

More interesting would be a 99% prediction interval for *Pressure*, rather than for $100 \times \log(Pressure)$. A point prediction is just $10^{(136.961/100)} = 23.421$ inches of Mercury. The prediction interval is found by exponentiating the end points of the interval in log scale. Dividing by 100 and then exponentiating, we get

$$10^{135.803/100} \leq Pressure \leq 10^{138.119/100}$$

$$22.805 \leq Pressure \leq 24.054$$

In the original scale, the prediction interval is not symmetric about the point estimate.

For the heights data, Figure 2.5 is a plot of the estimated mean function given by the dashed line for the regression of *Dheight* on *Mheight* along with curves at

$$\hat{\beta}_0 + \hat{\beta}_1 x_* \pm t(.025, 15)\text{sepred}(\tilde{Dheight}_*|Mheight_*)$$

The vertical distance between the two solid curves for any value of *Mheight* corresponds to a 95% prediction interval for daughter's height given mother's height. Although not obvious from the graph because of the very large sample size, the interval is wider for mothers who were either relatively tall or short, as the curves bend outward from the narrowest point at $Mheight = \overline{Mheight}$.

2.8.4 Fitted Values

In rare problems, one may be interested in obtaining an estimate of $E(Y|X = x)$. In the heights data, this is like asking for the population mean height of all daughters of mothers with a particular height. This quantity is estimated by the fitted value $\hat{y} = \beta_0 + \beta_1 x$, and its standard error is

$$\text{sefit}(\tilde{y}_*|x_*) = \hat{\sigma} \left(\frac{1}{n} + \frac{(x_* - \overline{x})^2}{SXX}\right)^{1/2}$$

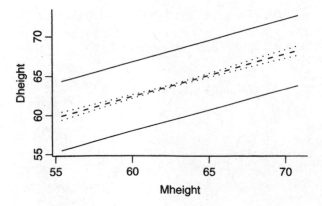

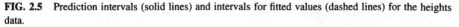

FIG. 2.5 Prediction intervals (solid lines) and intervals for fitted values (dashed lines) for the heights data.

To obtain confidence intervals, it is more usual to compute a simultaneous interval for all possible values of x. This is the same as first computing a joint confidence region for β_0 and β_1, and from these, computing the set of all possible mean functions with slope and intercept in the joint confidence set (Section 5.5). The confidence region for the mean function is the set of all y such that

$$(\hat{\beta}_0 + \hat{\beta}_1 x) - \operatorname{sefit}(\hat{y}|x)[2F(\alpha; 2, n-2)]^{1/2} \leq y$$
$$\leq (\hat{\beta}_0 + \hat{\beta}_1 x) + \operatorname{sefit}(\hat{y}|x)[2F(\alpha; 2, n-2)]^{1/2}$$

For multiple regression, replace $2F(\alpha; 2, n-2)$ by $p'F(\alpha; p', n-p')$, where p' is the number of parameters estimated in the mean function including the intercept. The simultaneous band for the fitted line for the heights data is shown in Figure 2.5 as the vertical distances between the two dotted lines. The prediction intervals are much wider than the confidence intervals. Why is this so (Problem 2.4)?

2.9 THE RESIDUALS

Plots of residuals versus other quantities are used to find failures of assumptions. The most common plot, especially useful in simple regression, is the plot of residuals versus the fitted values. A null plot would indicate no failure of assumptions. Curvature might indicate that the fitted mean function is inappropriate. Residuals that seem to increase or decrease in average magnitude with the fitted values might indicate nonconstant residual variance. A few relatively large residuals may be indicative of outliers, cases for which the model is somehow inappropriate.

The plot of residuals versus fitted values for the heights data is shown in Figure 2.6. This is a null plot, as it indicates no particular problems.

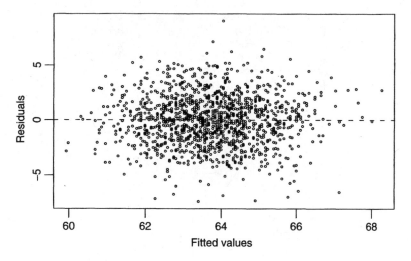

FIG. 2.6   Residuals versus fitted values for the heights data.

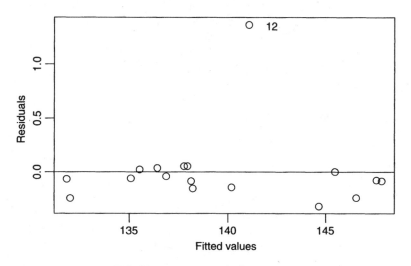

FIG. 2.7   Residual plot for Forbes' data.

    The fitted values and residuals for Forbes' data are plotted in Figure 2.7. The residuals are generally small compared to the fitted values, and they do not follow any distinct pattern in Figure 2.7. The residual for case number 12 is about four times the size of the next largest residual in absolute value. This may suggest that the assumptions concerning the errors are not correct. Either Var($100 \times$ log($Pressure$)|$Temp$) may not be constant or for case 12, the corresponding error may have a large fixed component. Forbes may have misread or miscopied the results of his calculations for this case, which would suggest that the numbers in

**TABLE 2.5 Summary Statistics for Forbes'
Data with All Data and with Case 12 Deleted**

Quantity	All Data	Delete Case 12
$\hat{\beta}_0$	−42.138	−41.308
$\hat{\beta}_1$	0.895	0.891
$se(\hat{\beta}_0)$	3.340	1.001
$se(\hat{\beta}_1)$	0.016	0.005
$\hat{\sigma}$	0.379	0.113
R^2	0.995	1.000

the data do not correspond to the actual measurements. Forbes noted this possibility himself, by marking this pair of numbers in his paper as being "evidently a mistake", presumably because of the large observed residual.

Since we are concerned with the effects of case 12, we could refit the data, this time without case 12, and then examine the changes that occur in the estimates of parameters, fitted values, residual variance, and so on. This is summarized in Table 2.5, giving estimates of parameters, their standard errors, $\hat{\sigma}^2$, and the coefficient of determination R^2 with and without case 12. The estimates of parameters are essentially identical with and without case 12. In other regression problems, deletion of a single case can change everything. The effect of case 12 on standard errors is more marked: if case 12 is deleted, standard errors are decreased by a factor of about 3.1, and variances are decreased by a factor of about $3.1^2 \approx 10$. Inclusion of this case gives the appearance of less reliable results than would be suggested on the basis of the other 16 cases. In particular, prediction intervals of *Pressure* are much wider based on all the data than on the 16-case data, although the point predictions are nearly the same. The residual plot obtained when case 12 is deleted before computing indicates no obvious failures in the remaining 16 cases.

Two competing fits using the same mean function but somewhat different data are available, and they lead to slightly different conclusions, although the results of the two analyses agree more than they disagree. On the basis of the data, there is no real way to choose between the two, and we have no way of deciding which is the correct OLS analysis of the data. A good approach to this problem is to describe both or, in general, all plausible alternatives.

PROBLEMS

2.1. Height and weight data The table below and in the data file htwt.txt gives Ht = height in centimeters and Wt = weight in kilograms for a sample of $n = 10$ 18-year-old girls. The data are taken from a larger study described in Problem 3.1. Interest is in predicting weight from height.

Ht	Wt
169.6	71.2
166.8	58.2
157.1	56.0
181.1	64.5
158.4	53.0
165.6	52.4
166.7	56.8
156.5	49.2
168.1	55.6
165.3	77.8

2.1.1. Draw a scatterplot of *Wt* on the vertical axis versus *Ht* on the horizontal axis. On the basis of this plot, does a simple linear regression model make sense for these data? Why or why not?

2.1.2. Show that $\bar{x} = 165.52$, $\bar{y} = 59.47$, $SXX = 472.076$, $SYY = 731.961$, and $SXY = 274.786$. Compute estimates of the slope and the intercept for the regression of Y on X. Draw the fitted line on your scatterplot.

2.1.3. Obtain the estimate of σ^2 and find the estimated standard errors of $\hat{\beta}_0$ and $\hat{\beta}_1$. Also find the estimated covariance between $\hat{\beta}_0$ and $\hat{\beta}_1$. Compute the t-tests for the hypotheses that $\beta_0 = 0$ and that $\beta_1 = 0$ and find the appropriate p-values using two-sided tests.

2.1.4. Obtain the analysis of variance table and F-test for regression. Show numerically that $F = t^2$, where t was computed in Problem 2.1.3 for testing $\beta_1 = 0$.

2.2. More with Forbes' data An alternative approach to the analysis of Forbes' experiments comes from the Clausius–Clapeyron formula of classical thermodynamics, which dates to Clausius (1850). According to this theory, we should find that

$$E(Lpres|Temp) = \beta_0 + \beta_1 \frac{1}{Ktemp} \tag{2.27}$$

where *Ktemp* is temperature in degrees Kelvin, which equals 255.37 plus $(5/9) \times Temp$. If we were to graph this mean function on a plot of *Lpres* versus *Ktemp*, we would get a curve, not a straight line. However, we can estimate the parameters β_0 and β_1 using simple linear regression methods by defining u_1 to be the inverse of temperature in degrees Kelvin,

$$u_1 = \frac{1}{Ktemp} = \frac{1}{(5/9)Temp + 255.37}$$

Then the mean function (2.27) can be rewritten as

$$E(Lpres|Temp) = \beta_0 + \beta_1 u_1 \tag{2.28}$$

for which simple linear regression is suitable. The notation we have used in (2.28) is a little different, as the left side of the equation says we are conditioning on *Temp*, but the variable *Temp* does not appear explicitly on the right side of the equation.

2.2.1. Draw the plot of *Lpres* versus u_1, and verify that apart from case 12 the 17 points in Forbes' data fall close to a straight line.

2.2.2. Compute the linear regression implied by (2.28), and summarize your results.

2.2.3. We now have two possible models for the same data based on the regression of *Lpres* on *Temp* used by Forbes, and (2.28) based on the Clausius–Clapeyron formula. To compare these two, draw the plot of the fitted values from Forbes' mean function fit versus the fitted values from (2.28). On the basis of these and any other computations you think might help, is it possible to prefer one approach over the other? Why?

2.2.4. In his original paper, Forbes provided additional data collected by the botanist Dr. Joseph Hooker on temperatures and boiling points measured often at higher altitudes in the Himalaya Mountains. The data for $n = 31$ locations is given in the file `hooker.txt`. Find the estimated mean function (2.28) for Hooker's data.

2.2.5. *This problem is not recommended unless you have access to a package with a programming language, like R, S-plus, Mathematica, or SAS IML.* For each of the cases in Hooker's data, compute the predicted values $\hat{y}$ and the standard error of prediction. Then compute $z = (Lpres - \hat{y})/\text{sepred}$. Each of the zs is a random variable, but if the model is correct, each has mean zero and standard deviation close to one. Compute the sample mean and standard deviation of the zs, and summarize results.

2.2.6. Repeat Problem 2.2.5, but this time predict and compute the z-scores for the 17 cases in Forbes data, again using the fitted mean function from Hooker's data. If the mean function for Hooker's data applies to Forbes' data, then each of the z-scores should have zero mean and standard deviation close to one. Compute the z-scores, compare them to those in the last problem and comment on the results.

2.3. Deviations from the mean Sometimes it is convenient to write the simple linear regression model in a different form that is a little easier to manipulate. Taking equation (2.1), and adding $\beta_1 \overline{x} - \beta_1 \overline{x}$, which equals zero, to the right-hand side, and combining terms, we can write

$$y_i = \beta_0 + \beta_1\bar{x} + \beta_1 x_i - \beta_1\bar{x} + e_i$$
$$= (\beta_0 + \beta_1\bar{x}) + \beta_1(x_i - \bar{x}) + e_i$$
$$= \alpha + \beta_1(x_i - \bar{x}) + e_i \tag{2.29}$$

where we have defined $\alpha = \beta_0 + \beta_1\bar{x}$. This is called the *deviations from the sample average form for simple regression*.

2.3.1. What is the meaning of the parameter α?

2.3.2. Show that the least squares estimates are

$$\hat{\alpha} = \bar{y}, \qquad \hat{\beta}_1 \text{ as given by (2.5)}$$

2.3.3. Find expressions for the variances of the estimates and the covariance between them.

2.4. Heights of mothers and daughters

2.4.1. For the heights data in the file `heights.txt`, compute the regression of *Dheight* on *Mheight*, and report the estimates, their standard errors, the value of the coefficient of determination, and the estimate of variance. Give the analysis of variance table that tests the hypothesis that $E(Dheight|Mheight) = \beta_0$ versus the alternative that $E(Dheight|Mheight) = \beta_0 + \beta_1 Mheight$, and write a sentence or two that summarizes the results of these computations.

2.4.2. Write the mean function in the deviations from the mean form as in Problem 2.3. For this particular problem, give an interpretation for the value of β_1. In particular, discuss the three cases of $\beta_1 = 1$, $\beta_1 < 1$ and $\beta_1 > 1$. Obtain a 99% confidence interval for β_1 from the data.

2.4.3. Obtain a prediction and 99% prediction interval for a daughter whose mother is 64 inches tall.

2.5. Smallmouth bass

2.5.1. Using the West Bearskin Lake smallmouth bass data in the file `wblake.txt`, obtain 95% intervals for the mean length at ages 2, 4 and 6 years.

2.5.2. Obtain a 95% interval for the mean length at age 9. Explain why this interval is likely to be untrustworthy.

2.5.3. The file `wblake2.txt` contains all the data for ages one to eight and, in addition, includes a few older fishes. Using the methods we have learned in this chapter, show that the simple linear regression model is not appropriate for this larger data set.

2.6. United Nations data Refer to the UN data in Problem 1.3, page 18.

2.6.1. Using base-ten logarithms, use a software package to compute the simple linear regression model corresponding to the graph in Problem 1.3.3, and get the analysis of variance table.

2.6.2. Draw the summary graph, and add the fitted line to the graph.

2.6.3. Test the hypothesis that the slope is zero versus the alternative that it is negative (a one-sided test). Give the significance level of the test and a sentence that summarizes the result.

2.6.4. Give the value of the coefficient of determination, and explain its meaning.

2.6.5. Increasing $\log(PPgdp)$ by one unit is the same as multiplying $PPgdp$ by ten. If two localities differ in $PPgdp$ by a factor of ten, give a 95% confidence interval on the difference in $\log(Fertility)$ for these two localities.

2.6.6. For a locality not in the data with $PPgdp = 1000$, obtain a point prediction and a 95% prediction interval for $\log(Fertility)$. If the interval (a, b) is a 95% prediction interval for $\log(Fertility)$, then a 95% prediction interval for $Fertility$ is given by $(10^a, 10^b)$. Use this result to get a 95% prediction interval for $Fertility$.

2.6.7. Identify (1) the locality with the highest value of $Fertility$; (2) the locality with the lowest value of $Fertility$; and (3) the two localities with the largest positive residuals from the regression when both variables are in log scale, and the two countries with the largest negative residuals in log scales.

2.7. Regression through the origin Occasionally, a mean function in which the intercept is known *a priori* to be zero may be fit. This mean function is given by

$$E(y|x) = \beta_1 x \qquad (2.30)$$

The residual sum of squares for this model, assuming the errors are independent with common variance σ^2, is $RSS = \sum(y_i - \hat{\beta}_1 x_i)^2$.

2.7.1. Show that the least squares estimate of β_1 is $\hat{\beta}_1 = \sum x_i y_i / \sum x_i^2$. Show that $\hat{\beta}_1$ is unbiased and that $Var(\hat{\beta}_1) = \sigma^2 / \sum x_i^2$. Find an expression for $\hat{\sigma}^2$. How many df does it have?

2.7.2. Derive the analysis of variance table with the larger model given by (2.16), but with the smaller model specified in (2.30). Show that the F-test derived from this table is numerically equivalent to the square of the t-test (2.23) with $\beta_0^* = 0$.

2.7.3. The data in Table 2.6 and in the file `snake.txt` give $X =$ water content of snow on April 1 and $Y =$ water yield from April to July in inches in the Snake River watershed in Wyoming for $n = 17$ years from 1919 to 1935 (from Wilm, 1950).

TABLE 2.6   Snake River Data for Problem 2.7

X	Y	X	Y
23.1	10.5	32.8	16.7
31.8	18.2	32.0	17.0
30.4	16.3	24.0	10.5
39.5	23.1	24.2	12.4
52.5	24.9	37.9	22.8
30.5	14.1	25.1	12.9
12.4	8.8	35.1	17.4
31.5	14.9	21.1	10.5
27.6	16.1		

Fit a regression through the origin and find $\hat{\beta}_1$ and σ^2. Obtain a 95% confidence interval for β_1. Test the hypothesis that the intercept is zero.

2.7.4. Plot the residuals versus the fitted values and comment on the adequacy of the mean function with zero intercept. In regression through the origin, $\sum \hat{e}_i \neq 0$.

2.8. Scale invariance

2.8.1. In the simple regression model (2.1), suppose the value of the predictor X is replaced by cX, where c is some non zero constant. How are $\hat{\beta}_0$, $\hat{\beta}_1$, $\hat{\sigma}^2$, R^2, and the t-test of NH: $\beta_1 = 0$ affected by this change?

2.8.2. Suppose each value of the response Y is replaced by dY, for some $d \neq 0$. Repeat 2.8.1.

2.9. Using Appendix A.3, verify equation (2.8).

2.10. Zipf's law Suppose we counted the number of times each word was used in the written works by Shakespeare, Alexander Hamilton, or some other author with a substantial written record (Table 2.7). Can we say anything about the frequencies of the most common words?

Suppose we let f_i be the rate per 1000 words of text for the ith most frequent word used. The linguist George Zipf (1902–1950) observed a law like relationship between rate and rank (Zipf, 1949),

$$E(f_i|i) = a/i^b$$

and further observed that the exponent is close to $b = 1$. Taking logarithms of both sides, we get approximately

$$E(\log(f_i)| \log(i)) = \log(a) - b\log(i) \qquad (2.31)$$

TABLE 2.7 The Word Count Data

Word	The word
Hamilton	Rate per 1000 words of this word in the writings of Alexander Hamilton
HamiltonRank	Rank of this word in Hamilton's writings
Madison	Rate per 1000 words of this word in the writings of James Madison
MadisonRank	Rank of this word in Madison's writings
Jay	Rate per 1000 words of this word in the writings of John Jay
JayRank	Rank of this word in Jay's writings
Ulysses	Rate per 1000 words of this word in *Ulysses* by James Joyce
UlyssesRank	Rank of this word in *Ulysses*

Zipf's law has been applied to frequencies of many other classes of objects besides words, such as the frequency of visits to web pages on the internet and the frequencies of species of insects in an ecosystem.

The data in MWwords.txt give the frequencies of words in works from four different sources: the political writings of eighteenth-century American political figures Alexander Hamilton, James Madison, and John Jay, and the book *Ulysses* by twentieth-century Irish writer James Joyce. The data are from Mosteller and Wallace (1964, Table 8.1-1), and give the frequencies of 165 very common words. Several missing values occur in the data; these are really words that were used so infrequently that their count was not reported in Mosteller and Wallace's table.

2.10.1. Using only the 50 most frequent words in Hamilton's work (that is, using only rows in the data for which *HamiltonRank* $\leq$ 50), draw the appropriate summary graph, estimate the mean function (2.31), and summarize your results.

2.10.2. Test the hypothesis that $b = 1$ against the two-sided alternative and summarize.

2.10.3. Repeat Problem 2.10.1, but for words with rank of 75 or less, and with rank less than 100. For larger number of words, Zipf's law may break down. Does that seem to happen with these data?

2.11. For the Ft. Collins snow fall data discussed in Example 1.1, test the hypothesis that the slope is zero versus the alternative that it is not zero. Show that the t-test of this hypothesis is the same as the F-test; that is, $t^2 = F$.

2.12. Old Faithful Use the data from Problem 1.4, page 18.

2.12.1. Use simple linear regression methodology to obtain a prediction equation for *interval* from *duration*. Summarize your results in a way that might be useful for the nontechnical personnel who staff the Old Faithful Visitor's Center.

2.12.2. Construct a 95% confidence interval for

$$E(interval|duration = 250)$$

2.12.3. An individual has just arrived at the end of an eruption that lasted 250 seconds. Give a 95% confidence interval for the time the individual will have to wait for the next eruption.

2.12.4. Estimate the 0.90 quantile of the conditional distribution of

$$interval|(duration = 250)$$

assuming that the population is normally distributed.

2.13. Windmills Energy can be produced from wind using windmills. Choosing a site for a *wind farm*, the location of the windmills, can be a multimillion dollar gamble. If wind is inadequate at the site, then the energy produced over the lifetime of the wind farm can be much less than the cost of building and operation. Prediction of long-term wind speed at a candidate site can be an important component in the decision to build or not to build. Since energy produced varies as the square of the wind speed, even small errors can have serious consequences.

The data in the file wm1.txt provides measurements that can be used to help in the prediction process. Data were collected every six hours for the year 2002, except that the month of May 2002 is missing. The values *Cspd* are the calculated wind speeds in meters per second at a candidate site for building a wind farm. These values were collected at tower erected on the site. The values *RSpd* are wind speeds at a *reference site*, which is a nearby location for which wind speeds have been recorded over a very long time period. Airports sometimes serve as reference sites, but in this case, the reference data comes from the National Center for Environmental Modeling; these data are described at http://dss.ucar.edu/datasets/ds090.0/. The reference is about 50 km south west of the candidate site. Both sites are in the northern part of South Dakota. The data were provided by Mark Ahlstrom and Rolf Miller of WindLogics.

2.13.1. Draw the scatterplot of the response *CSpd* versus the predictor *RSpd*. Is the simple linear regression model plausible for these data?

2.13.2. Fit the simple regression of the response on the predictor, and present the appropriate regression summaries.

2.13.3. Obtain a 95% prediction interval for *CSpd* at a time when $RSpd = 7.4285$.

2.13.4. For this problem, we revert to generic notation and let $x = CSpd$ and $y = CSpd$ and let n be the number of cases used in the regression ($n = 1116$ in the data we have used in this problem) and $\bar{x}$ and *SXX* defined from these n observations. Suppose we want to make predictions at m time points with values of wind speed $x_{*1}, .., x_{*m}$ that are different from the n cases used in constructing the prediction equation. Show that (1) the average of the m predictions is equal to the prediction taken at the average value $\bar{x}_*$ of the m values of the predictor, and (2) using the first result, the standard error of the average of m predictions is

$$\text{se of average prediction} = \sqrt{\frac{\hat{\sigma}^2}{m} + \hat{\sigma}^2 \left(\frac{1}{n} + \frac{(\bar{x}_* - \bar{x})^2}{SXX}\right)} \quad (2.32)$$

If m is very large, then the first term in the square root is negligible, and the standard error of average prediction is essentially the same as the standard error of a fitted value at $\bar{x}_*$.

2.13.5. For the period from January 1, 1948 to July 31, 2003, a total of $m = 62039$ wind speed measurements are available at the reference site, excluding the data from the year 2002. For these measurements, the average wind speed was $\bar{x}_* = 7.4285$. Give a 95% prediction interval on the long-term average wind speed at the candidate site. This long-term average of the past is then taken as an estimate of the long-term average of the future and can be used to help decide if the candidate is a suitable site for a wind farm.

CHAPTER 3

Multiple Regression

Multiple linear regression generalizes the simple linear regression model by allowing for many *terms* in a mean function rather than just one intercept and one slope.

3.1 ADDING A TERM TO A SIMPLE LINEAR REGRESSION MODEL

We start with a response Y and the simple linear regression mean function

$$E(Y|X_1 = x_1) = \beta_0 + \beta_1 x_1$$

Now suppose we have a second variable X_2 with which to predict the response. By adding X_2 to the problem, we will get a mean function that depends on both the value of X_1 and the value of X_2,

$$E(Y|X_1 = x_1, X_2 = x_2) = \beta_0 + \beta_1 x_1 + \beta_2 x_2 \tag{3.1}$$

The main idea in adding X_2 is to explain the part of Y that has not already been explained by X_1.

United Nations Data

We will reconsider the United Nations data discussed in Problem 1.3. To the regression of log(*Fertility*), the base-two log fertility rate on log(*PPgdp*), the base-two log of the per person gross domestic product, we consider adding *Purban*, the percentage of the population that lives in an urban area. The data in the file UN2.txt give values for these three variables, as well as the name of the *Locality* for 193 localities, mostly countries, for which the United Nations provides data.

Figure 3.1 presents several graphical views of these data. Figure 3.1a can be viewed as a summary graph for the simple regression of log(*Fertility*) on log(*PPgdp*). The fitted mean function using OLS is

$$\widehat{E}(\log(\textit{Fertility})|\log(\textit{PPgdp})) = 2.703 - 0.153 \log(\textit{PPgdp})$$

Applied Linear Regression, Third Edition, by Sanford Weisberg
ISBN 0-471-66379-4   Copyright © 2005 John Wiley & Sons, Inc.

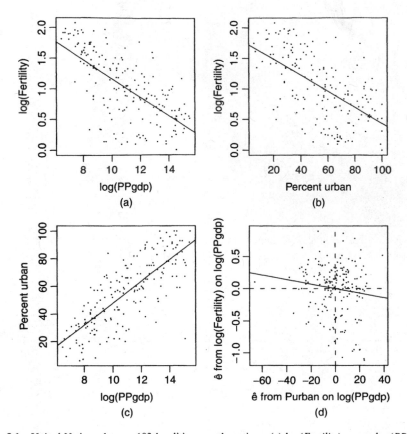

FIG. 3.1 United Nations data on 193 localities, mostly nations. (a) log(*Fertility*) versus log(*PPgdp*); (b) *log*(*Fertility*) versus *Purban*; (c) *Purban* versus log(*PPgdp*); (d) Added-variable plot for *Purban* after log(*PPgdp*).

with $R^2 = 0.459$, so about 46% of the variability in log(*Fertility*) is explained by log(*PPgdp*). An increase of one unit in log(*PPgdp*), which corresponds to a *doubling* of *PPgdp*, is estimated to decrease log(*Fertility*) by 0.153 units.

Similarly, Figure 3.1b is the summary graph for the regression of log(*Fertility*) on *Purban*. This simple regression has fitted mean function

$$\widehat{E}(\log(Fertility)|Purban) = 1.750 - 0.013\ Purban$$

with $R^2 = 0.348$, so *Purban* explains about 35% of the variability in log(*Fertility*). An increase of one percent urban implies a change on the average in log(*Fertility*) of −0.13.

To get a summary graph of the regression of log(*Fertility*) on *both* log(*PPgdp*) and *Purban* would require a three-dimensional plot of these three variables, with log(*PPgdp*) on one of the horizontal axes, *Purban* on the other horizontal axis, and log(*Fertility*) on the vertical axis. Although such plots are possible by using either perspective or motion to display the third dimension, using them is much more difficult than using two-dimensional graphics, and their successful use is not widespread. Cook and Weisberg (1999a) discuss using motion to understand three-dimensional graphics for regression.

As a partial substitute for looking at the full three-dimensional plot, we add a third plot to the first two in Figure 3.1, namely, the plot of *Purban* versus log(*PPgdp*) shown in Figure 3.1c. This graph does not include the response, so it only shows the relationship between the two potential predictors. In this problem, these two variables are positively correlated, and the mean function for Figure 3.1c seems to be well approximated by a straight line.

The inference to draw from Figure 3.1c is that to the extent that *Purban* can be predicted by log(*PPgdp*), these two potential predictors are measuring the same thing, and so the role of these two variables in predicting log(*Fertility*) will be overlapping, and they will both, to some extent, be explaining the same variability.

3.1.1 Explaining Variability

Given these graphs, what can be said about the proportion of variability in log(*Fertility*) explained by log(*PPgdp*) and *Purban*? We can say that the total explained variation must exceed 46 percent, the larger of the two values explained by each variable separately, since using both log(*PPgdp*) and *Purban* must surely be at least as informative as using just one of them. The total variation will be additive, $46\% + 35\% = 91\%$, only if the two variables are completely unrelated and measure different things. The total can be less than the sum if the terms are related and are at least in part explaining the same variation. Finally, the total can exceed the sum if the two variables act jointly so that knowing both gives more information than knowing just one of them. For example, the area of a rectangle may be only poorly determined by either the length or width alone, but if both are considered at the same time, area can be determined exactly. It is precisely this inability to predict the joint relationship from the marginal relationships that makes multiple regression rich and complicated.

3.1.2 Added-Variable Plots

The *unique* effect of adding *Purban* to a mean function that already includes log(*PPgdp*) is determined by the relationship between the part of log(*Fertility*) that is not explained by log(*PPgdp*) and the part of *Purban* that is not explained by log(*PPgdp*). The "unexplained parts" are just the residuals from these two simple regressions, and so we need to examine the scatterplot of the residuals from the regression of log(*Fertility*) on log(*PPgdp*) versus the residuals from the regression of *Purban* on log(*PPgdp*). This plot is shown in Figure 3.1d. Figure 3.1b is the summary graph for the relationship between log(*Fertility*) and *Purban* *ignoring* log(*PPgdp*), while Figure 3.1d shows this relationship, but after *adjusting* for log(*PPgdp*). If Figure 3.1d shows a stronger relationship than does Figure 3.1b, meaning that the points in the plot show less variation about the fitted straight line, then the two variables act jointly to explain extra variation, while if the relationship is weaker, or the plot exhibits more variation, then the total explained variability is less than the additive amount. The latter seems to be the case here.

If we fit the simple regression mean function to Figure3.1d, the fitted line has zero intercept, since the averages of the two plotted variables are zero, and the estimated slope via OLS is $\hat{\beta}_2 = -0.0035 \approx -0.004$. It turns out that this is exactly the estimate $\hat{\beta}_2$ that would be obtained using OLS to get the estimates using the mean function (3.1). Figure 3.1d is called an *added-variable plot*.

We now have two estimates of the coefficient β_2 for *Purban*:

$$\hat{\beta}_2 = -0.013 \; ignoring \; \log(PPgdp)$$

$$\hat{\beta}_2 = -0.004 \; adjusting \; for \; \log(PPgdp)$$

While both of these indicate that more urbanization is associated with lower fertility, adjusting for $\log(PPgdp)$ suggests that the magnitude of this effect is only about one-fourth as large as one might think if $\log(PPgdp)$ were ignored. In other problems, slope estimates for the same term but from different mean functions can be even more wildly different, changing signs, magnitude, and significance. This naturally complicates the interpretation of fitted models, and also comparing between studies fit with even slightly different mean functions.

To get the coefficient estimate for $\log(PPgdp)$ in the regression of $\log(Fertility)$ on both predictors, we would use the same procedure we used for *Purban* and consider the problem of adding $\log(PPgdp)$ to a mean function that already includes *Purban*. This would require looking at the graph of the residuals from the regression of $\log(Fertility)$ on *Purban* versus the residuals from the regression of $\log(PPgdp)$ on *Purban* (see Problem 3.2).

3.2 THE MULTIPLE LINEAR REGRESSION MODEL

The general multiple linear regression model with response Y and terms $X_1, \ldots, X_p$ will have the form

$$E(Y|X) = \beta_0 + \beta_1 X_1 + \cdots + \beta_p X_p \tag{3.2}$$

The symbol X in $E(Y|X)$ means that we are conditioning on all the terms on the right side of the equation. Similarly, when we are conditioning on specific values for the predictors $x_1, \ldots, x_p$ that we will collectively call $\mathbf{x}$, we write

$$E(Y|X = \mathbf{x}) = \beta_0 + \beta_1 x_1 + \cdots + \beta_p x_p \tag{3.3}$$

As in Chapter 2, the βs are unknown parameters we need to estimate. Equation (3.2) is a *linear function of the parameters*, which is why this is called linear regression. When $p = 1$, X has only one element, and we get the simple regression problem discussed in Chapter 2. When $p = 2$, the mean function (3.2) corresponds

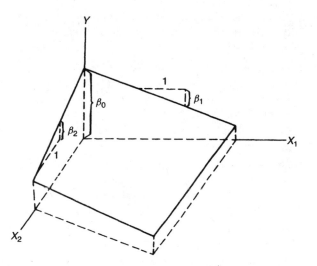

FIG. 3.2 A linear regression surface with $p = 2$ predictors.

to a plane in three dimensions, as shown in Figure 3.2. When $p > 2$, the fitted mean function is a *hyperplane*, the generalization of a p-dimensional plane in a $(p + 1)$-dimensional space. We cannot draw a general p-dimensional plane in our three-dimensional world.

3.3   TERMS AND PREDICTORS

Regression problems start with a collection of potential predictors. Some of these may be continuous measurements, like the height or weight of an object. Some may be discrete but ordered, like a doctor's rating of overall health of a patient on a nine-point scale. Other potential predictors can be categorical, like eye color or an indicator of whether a particular unit received a treatment. All these types of potential predictors can be useful in multiple linear regression.

From the pool of potential predictors, we create a set of *terms* that are the X-variables that appear in (3.2). The terms might include:

The intercept The mean function (3.2) can we rewritten as

$$E(Y|X) = \beta_0 X_0 + \beta_1 X_1 + \cdots + \beta_p X_p$$

where X_0 is a term that is always equal to one. Mean functions without an intercept would not have this term included.

Predictors The simplest type of term is equal to one of the predictors, for example, the variable *Mheight* in the heights data.

Transformations of predictors Sometimes the original predictors need to be transformed in some way to make (3.2) hold to a reasonable approximation. This was the case with the UN data just discussed, in which *PPgdp* was used in log scale. The willingness to replace predictors by transformations of them greatly expands the range of problems that can be summarized with a linear regression model.

Polynomials Problems with curved mean functions can sometimes be accommodated in the multiple linear regression model by including poly-nomial terms in the predictor variables. For example, we might include as terms both a predictor X_1 and its square X_1^2 to fit a quadratic polynomial in that predictor. Complex polynomial surfaces in several predictors can be useful in some problems[1].

Interactions and other combinations of predictors Combining several predictors is often useful. An example of this is using body mass index, given by height divided by weight squared, in place of both height and weight, or using a total test score in place of the separate scores from each of several parts. Products of predictors called *interactions* are often included in a mean function along with the original predictors to allow for joint effects of two or more variables.

Dummy variables and factors A categorical predictor with two or more levels is called a *factor*. Factors are included in multiple linear regression using *dummy variables*, which are typically terms that have only two values, often zero and one, indicating which category is present for a particular observation. We will see in Chapter 6 that a categorical predictor with two categories can be represented by one dummy variable, while a categorical predictor with many categories can require several dummy variables.

A regression with say k predictors may combine to give fewer than k terms or expand to require more than k terms. The distinction between predictors and terms can be very helpful in thinking about an appropriate mean function to use in a particular problem, and in using graphs to understand a problem. For example, a regression with one predictor can always be studied using the 2D scatterplot of the response versus the predictor, regardless of the number of terms required in the mean function.

We will use the fuel consumption data introduced in Section 1.6 as the primary example for the rest of this chapter. As discussed earlier, the goal is to understand how fuel consumption varies as a function of state characteristics. The variables are defined in Table 1.2 and are given in the file `fuel2001.txt`. From the six initial predictors, we use a set of four combinations to define terms in the regression mean function.

Basic summary statistics for the relevant variables in the fuel data are given in Table 3.1, and these begin to give us a bit of a picture of these data. First, there is quite a bit of variation in *Fuel*, with values between a minimum of about 626 gallons per year and a maximum of about 843 gallons per year. The gas *Tax* varies

---

[1]This discussion of polynomials might puzzle some readers because in Section 3.2, we said the linear regression mean function was a hyperplane, but here we have said that it might be curved, seemingly a contradiction. However, *both* of these statements are correct. If we fit a mean function like $E(Y|X = x) = \beta_0 + \beta_1 x + \beta_2 x^2$, the mean function is a quadratic curve in the plot of the response versus x but a plane in the three-dimensional plot of the response versus x and x^2.

TABLE 3.1   Summary Statistics for the Fuel Data

Variable	N	Average	Std Dev	Minimum	Median	Maximum
Tax	51	20.155	4.5447	7.5	20.	29.
Dlic	51	903.68	72.858	700.2	909.07	1075.3
Income	51	28.404	4.4516	20.993	27.871	40.64
logMiles	51	15.745	1.4867	10.583	16.268	18.198
Fuel	51	613.13	88.96	317.49	626.02	842.79

from only 7.5 cents per gallon to a high of 29 cents per gallon, so unlike much of
the world gasoline taxes account for only a small part of the cost to consumers of
gasoline. Also of interest is the range of values in *Dlic*: The number of licensed
drivers per 1000 population over the age of 16 is between about 700 and 1075.
Some states appear to have more licensed drivers than they have population over
age 16. Either these states allow drivers under the age of 16, allow nonresidents to
obtain a driver's license, or the data are in error. For this example, we will assume
one of the first two reasons.

Of course, these univariate summaries cannot tell us much about how the fuel
consumption depends on the other variables. For this, graphs are very helpful. The
scatterplot matrix for the fuel data is repeated in Figure 3.3. From our previous

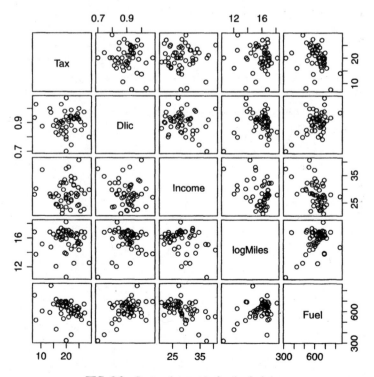

FIG. 3.3   Scatterplot matrix for the fuel data.

TABLE 3.2   **Sample Correlations for the Fuel Data**

```
Sample Correlations
                Tax      Dlic   Income  logMiles     Fuel
Tax          1.0000  -0.0858  -0.0107   -0.0437  -0.2594
Dlic        -0.0858   1.0000  -0.1760    0.0306   0.4685
Income      -0.0107  -0.1760   1.0000   -0.2959  -0.4644
logMiles    -0.0437   0.0306  -0.2959    1.0000   0.4220
Fuel        -0.2594   0.4685  -0.4644    0.4220   1.0000
```

discussion, *Fuel* decreases on the average as *Tax* increases, but there is lot of variation. We can make similar qualitative judgments about each of the regressions of *Fuel* on the other variables. The overall impression is that *Fuel* is at best weakly related to each of the variables in the scatterplot matrix, and in turn these variables are only weakly related to each other.

Does this help us understand how *Fuel* is related to all four predictors simultaneously? We know from the discussion in Section 3.1 that the marginal relationships between the response and each of the variables is *not* sufficient to understand the *joint* relationship between the response and the terms. The interrelationships among the terms are also important. The pairwise relationships between the terms can be viewed in the remaining cells of the scatterplot matrix. In Figure 3.3, the relationships between all pairs of terms appear to be very weak, suggesting that for this problem the marginal plots including *Fuel* are quite informative about the multiple regression problem.

A more traditional, and less informative, summary of the two-variable relationships is the matrix of sample correlations, shown in Table 3.2. In this instance, the correlation matrix helps to reinforce the relationships we see in the scatterplot matrix, with fairly small correlations between the predictors and *Fuel*, and essentially no correlation between the predictors themselves.

3.4   ORDINARY LEAST SQUARES

From the initial collection of potential predictors, we have computed a set of $p + 1$ terms, including an intercept, $X = (X_0, X_1, \ldots, X_p)$. The mean function and variance function for multiple linear regression are

$$
\begin{aligned}
\mathrm{E}(Y|X) &= \beta_0 + \beta_1 X_1 + \cdots + \beta_p X_p \\
\mathrm{Var}(Y|X) &= \sigma^2
\end{aligned}
\tag{3.4}
$$

Both the βs and σ^2 are unknown parameters that need to be estimated.

3.4.1   Data and Matrix Notation

Suppose we have observed data for n cases or units, meaning we have a value of Y and all of the terms for each of the n cases. We have symbols for the response and the terms using matrices and vectors; see Appendix A.6 for a brief introduction. We define

$$
Y = \begin{pmatrix} y_1 \\ y_2 \\ \vdots \\ y_n \end{pmatrix}
\qquad
X = \begin{pmatrix} 1 & x_{11} & \cdots & x_{1p} \\ 1 & x_{21} & \cdots & x_{2p} \\ \vdots & \vdots & \vdots & \vdots \\ 1 & x_{n1} & \cdots & x_{np} \end{pmatrix}
\tag{3.5}
$$

so Y is an $n \times 1$ vector and X is an $n \times (p + 1)$ matrix. We also define β to be a $(p + 1) \times 1$ vector of regression coefficients and e to be the $n \times 1$ vector of statistical errors,

$$
\beta = \begin{pmatrix} \beta_0 \\ \beta_1 \\ \vdots \\ \beta_p \end{pmatrix}
\qquad \text{and} \qquad
e = \begin{pmatrix} e_1 \\ e_2 \\ \vdots \\ e_n \end{pmatrix}
$$

The matrix X gives all of the observed values of the terms. The ith row of X will be defined by the symbol x_i', which is a $(p + 1) \times 1$ vector for mean functions that include an intercept. Even though x_i is a row of X, we use the convention that all vectors are column vectors and therefore need to write x_i' to represent a row. An equation for the mean function evaluated at x_i is

$$
E(Y|X = x_i) = x_i'\beta
$$
$$
= \beta_0 + \beta_1 x_{i1} + \cdots + \beta_p x_{ip}
\tag{3.6}
$$

In matrix notation, we will write the multiple linear regression model as

$$
Y = X\beta + e
\tag{3.7}
$$

The ith row of (3.7) is $y_i = x_i'\beta + e_i$.

For the fuel data, the first few and the last few rows of the matrix X and the vector Y are

$$
X = \begin{pmatrix} 1 & 18.00 & 1031.38 & 23.471 & 16.5271 \\ 1 & 8.00 & 1031.64 & 30.064 & 13.7343 \\ 1 & 18.00 & 908.597 & 25.578 & 15.7536 \\ \vdots & \vdots & \vdots & \vdots & \vdots \\ 1 & 25.65 & 904.894 & 21.915 & 15.1751 \\ 1 & 27.30 & 882.329 & 28.232 & 16.7817 \\ 1 & 14.00 & 970.753 & 27.230 & 14.7362 \end{pmatrix}
\qquad
Y = \begin{pmatrix} 690.264 \\ 514.279 \\ 621.475 \\ \vdots \\ 562.411 \\ 581.794 \\ 842.792 \end{pmatrix}
$$

The terms in X are in the order intercept, *Tax*, *Dlic*, *Income* and finally log(*Miles*). The matrix X is 51×5 and Y is 51×1.

3.4.2 Variance-Covariance Matrix of e

The 51×1 error vector is an unobservable random vector, as in Appendix A.6. The assumptions concerning the e_is given in Chapter 2 are summarized in matrix form as

$$E(\mathbf{e}) = \mathbf{0} \qquad \text{Var}(\mathbf{e}) = \sigma^2 \mathbf{I}_n$$

where $\text{Var}(\mathbf{e})$ means the covariance matrix of $\mathbf{e}$, $\mathbf{I}_n$ is the $n \times n$ matrix with ones on the diagonal and zeroes everywhere else, and $\mathbf{0}$ is a matrix or vector of zeroes of appropriate size. If we add the assumption of normality, we can write

$$\mathbf{e} \sim N(\mathbf{0}, \sigma^2 \mathbf{I}_n)$$

3.4.3 Ordinary Least Squares Estimators

The least squares estimate $\hat{\boldsymbol{\beta}}$ of $\boldsymbol{\beta}$ is chosen to minimize the residual sum of squares function

$$RSS(\boldsymbol{\beta}) = \sum (y_i - \mathbf{x}_i'\boldsymbol{\beta})^2 = (\mathbf{Y} - \mathbf{X}\boldsymbol{\beta})'(\mathbf{Y} - \mathbf{X}\boldsymbol{\beta}) \tag{3.8}$$

The OLS estimates can be found from (3.8) by differentiation in a matrix analog to the development of Appendix A.3. The OLS estimate is given by the formula

$$\hat{\boldsymbol{\beta}} = (\mathbf{X}'\mathbf{X})^{-1}\mathbf{X}'\mathbf{Y} \tag{3.9}$$

provided that the inverse $(\mathbf{X}'\mathbf{X})^{-1}$ exists. The estimator $\hat{\boldsymbol{\beta}}$ depends only on the sufficient statistics $\mathbf{X}'\mathbf{X}$ and $\mathbf{X}'\mathbf{Y}$, which are matrices of uncorrected sums of squares and cross-products.

Do not compute the least squares estimates using (3.9)! Uncorrected sums of squares and cross-products are prone to large rounding error, and so computations can be highly inaccurate. The preferred computational methods are based on matrix decompositions as briefly outlined in Appendix A.8. At the very least, computations should be based on *corrected* sums of squares and cross-products.

Suppose we define $\mathcal{X}$ to be the $n \times p$ matrix

$$\mathcal{X} = \begin{pmatrix} (x_{11} - \overline{x}_1) & \cdots & (x_{1p} - \overline{x}_p) \\ (x_{21} - \overline{x}_1) & \cdots & (x_{2p} - \overline{x}_p) \\ \vdots & \vdots & \vdots \\ (x_{n1} - \overline{x}_1) & \cdots & (x_{np} - \overline{x}_p) \end{pmatrix}$$

This matrix consists of the original $\mathbf{X}$ matrix, but with the first column removed and the column mean subtracted from each of the remaining columns. Similarly, $\mathcal{Y}$ is the vector with typical elements $y_i - \overline{y}$. Then

$$\mathcal{C} = \frac{1}{n-1} \begin{pmatrix} \mathcal{X}'\mathcal{X} & \mathcal{X}'\mathcal{Y} \\ \mathcal{Y}'\mathcal{X} & \mathcal{Y}'\mathcal{Y} \end{pmatrix} \tag{3.10}$$

is the matrix of sample variances and covariances. When $p = 1$, the matrix $\mathcal{C}$ is given by

$$\mathcal{C} = \frac{1}{n-1} \begin{pmatrix} SXX & SXY \\ SXY & SYY \end{pmatrix}$$

The elements of $\mathcal{C}$ are the summary statistics needed for OLS computations in simple linear regression. If we let $\boldsymbol{\beta}^*$ be the parameter vector excluding the intercept β_0, then for $p \geq 1$,

$$\hat{\boldsymbol{\beta}}^* = (\mathcal{X}'\mathcal{X})^{-1}\mathcal{X}'\mathcal{Y}$$
$$\hat{\beta}_0 = \bar{y} - \hat{\boldsymbol{\beta}}^{*\prime}\bar{\mathbf{x}}$$

where $\bar{\mathbf{x}}$ is the vector of sample means for all the terms except for the intercept.

Once $\hat{\boldsymbol{\beta}}$ is computed, we can define several related quantities. The fitted values are $\hat{\mathbf{Y}} = \mathbf{X}\hat{\boldsymbol{\beta}}$ and the residuals are $\hat{\mathbf{e}} = \mathbf{Y} - \hat{\mathbf{Y}}$. The function (3.8) evaluated at $\hat{\boldsymbol{\beta}}$ is the residual sum of squares, or RSS,

$$RSS = \hat{\mathbf{e}}'\hat{\mathbf{e}} = (\mathbf{Y} - \mathbf{X}\hat{\boldsymbol{\beta}})'(\mathbf{Y} - \mathbf{X}\hat{\boldsymbol{\beta}}) \tag{3.11}$$

3.4.4 Properties of the Estimates

Additional properties of the OLS estimates are derived in Appendix A.8 and are only summarized here. Assuming that $E(\mathbf{e}) = \mathbf{0}$ and $Var(\mathbf{e}) = \sigma^2\mathbf{I}_n$, then $\hat{\boldsymbol{\beta}}$ is unbiased, $E(\hat{\boldsymbol{\beta}}) = \boldsymbol{\beta}$, and

$$Var(\hat{\boldsymbol{\beta}}) = \sigma^2(\mathbf{X}'\mathbf{X})^{-1} \tag{3.12}$$

Excluding the intercept term,

$$Var(\hat{\boldsymbol{\beta}}^*) = \sigma^2(\mathcal{X}'\mathcal{X})^{-1} \tag{3.13}$$

and so $(\mathcal{X}'\mathcal{X})^{-1}$ is all but the first row and column of $(\mathbf{X}'\mathbf{X})^{-1}$. An estimate of σ^2 is given by

$$\hat{\sigma}^2 = \frac{RSS}{n - (p+1)} \tag{3.14}$$

which is the residual sum of squares divided by its $df = n - (p + 1)$. Several formulas for RSS can be computed by substituting the value of $\hat{\boldsymbol{\beta}}$ into (3.11) and simplifying:

$$RSS = \mathbf{Y}'\mathbf{Y} - \hat{\boldsymbol{\beta}}'(\mathbf{X}'\mathbf{X})\hat{\boldsymbol{\beta}}$$
$$= \mathbf{Y}'\mathbf{Y} - \hat{\boldsymbol{\beta}}'\mathbf{X}'\mathbf{Y}$$
$$= \mathcal{Y}'\mathcal{Y} - \hat{\boldsymbol{\beta}}^{*\prime}(\mathcal{X}'\mathcal{X})\hat{\boldsymbol{\beta}}^* \tag{3.15}$$
$$= \mathcal{Y}'\mathcal{Y} - \hat{\boldsymbol{\beta}}'(\mathbf{X}'\mathbf{X})\hat{\boldsymbol{\beta}} + n\bar{y}^2$$

Recognizing that $\mathcal{Y}'\mathcal{Y} = SYY$, (3.15) has the nicest interpretation, as it writes RSS as equal to the total sum of squares minus a quantity we will call the *regression sum of squares*, or *SSreg*. In addition, if **e** is normally distributed, then the residual sum of squares has a Chi-squared distribution,

$$(n - (p + 1))\hat{\sigma}^2/\sigma^2 \sim \chi^2(n - (p + 1))$$

By substituting $\hat{\sigma}^2$ for σ^2 in (3.12), we find the estimated variance of $\hat{\boldsymbol{\beta}}$, $\widehat{\text{Var}}(\hat{\boldsymbol{\beta}})$, to be

$$\widehat{\text{Var}}(\hat{\boldsymbol{\beta}}) = \hat{\sigma}^2(\mathbf{X}'\mathbf{X})^{-1} \tag{3.16}$$

3.4.5   Simple Regression in Matrix Terms

For simple regression, **X** and **Y** are given by

$$\mathbf{X} = \begin{pmatrix} 1 & x_1 \\ 1 & x_2 \\ \vdots & \vdots \\ 1 & x_n \end{pmatrix} \qquad \mathbf{Y} = \begin{pmatrix} y_1 \\ y_2 \\ \vdots \\ y_n \end{pmatrix}$$

and thus

$$(\mathbf{X}'\mathbf{X}) = \begin{pmatrix} n & \sum x_i \\ \sum x_i & \sum x_i^2 \end{pmatrix} \qquad \mathbf{X}'\mathbf{Y} = \begin{pmatrix} \sum y_i \\ \sum y_i^2 \end{pmatrix}$$

By direct multiplication, $(\mathbf{X}'\mathbf{X})^{-1}$ can be shown to be

$$(\mathbf{X}'\mathbf{X})^{-1} = \frac{1}{SXX} \begin{pmatrix} \sum x_i^2/n & -\bar{x} \\ -\bar{x} & 1 \end{pmatrix} \tag{3.17}$$

so that

$$\hat{\boldsymbol{\beta}} = \begin{pmatrix} \hat{\beta}_0 \\ \hat{\beta}_1 \end{pmatrix} = (\mathbf{X}'\mathbf{X})^{-1}\mathbf{X}'\mathbf{Y} = \frac{1}{SXX} \begin{pmatrix} \sum x_i^2/n & -\bar{x} \\ -\bar{x} & \sum x_i y_i \end{pmatrix} \begin{pmatrix} \sum y_i \\ \sum y_i^2 \end{pmatrix}$$

$$= \begin{pmatrix} \bar{y} - \hat{\beta}_1\bar{x} \\ SXY/SXX \end{pmatrix}$$

as found previously. Also, since $\sum x_i^2/(nSXX) = 1/n + \bar{x}^2/SXX$, the variances and covariances for $\hat{\beta}_0$ and $\hat{\beta}_1$ found in Chapter 2 are identical to those given by $\sigma^2(\mathbf{X}'\mathbf{X})^{-1}$.

The results are simpler in the deviations from the sample average form, since

$$\mathcal{X}'\mathcal{X} = SXX \qquad \mathcal{X}'\mathcal{Y} = SXY$$

and

$$\hat{\beta}_1 = (\mathcal{X}'\mathcal{X})^{-1}\mathcal{X}'\mathcal{Y} = \frac{SXY}{SXX}$$

$$\hat{\beta}_0 = \bar{y} - \hat{\beta}_1\bar{x}$$

Fuel Consumption Data

We will generally let p equal the number of terms in a mean function excluding the intercept, and $p' = p + 1$ equal if the intercept is included; $p' = p$ if the intercept is not included. We shall now fit the mean function with $p' = 5$ terms, including the intercept for the fuel consumption data. Continuing a practice we have already begun, we will write *Fuel* on *Tax Dlic Income* log(*Miles*) as shorthand for using OLS to fit the multiple linear regression model with mean function

$$\mathrm{E}(Fuel|X) = \beta_0 + \beta_1 Tax + \beta_2 Dlic + \beta_3 Income + \beta_4 \log(Miles)$$

where conditioning on X is short for conditioning on all the terms in the mean function. All the computations are based on the summary statistics, which are the sample means given in Table 3.1 and the sample covariance matrix C defined at (3.10) and given by

	Tax	Dlic	Income	logMiles	Fuel
Tax	20.6546	-28.4247	-0.2162	-0.2955	-104.8944
Dlic	-28.4247	5308.2591	-57.0705	3.3135	3036.5905
Income	-0.2162	-57.0705	19.8171	-1.9580	-183.9126
logMiles	-0.2955	3.3135	-1.9580	2.2103	55.8172
Fuel	-104.8944	3036.5905	-183.9126	55.8172	7913.8812

Most statistical software will give the sample correlations rather than the covariances. The reader can verify that the correlations in Table 3.2 can be obtained from these covariances. For example, the sample correlation between *Tax* and *Income* is $-0.2162/\sqrt{(20.6546 \times 19.8171)} = -0.0107$ as in Table 3.2. One can convert back from correlations and sample variances to covariances; the square root of the sample variances are given in Table 3.1.

The 5×5 matrix $(\mathbf{X'X})^{-1}$ is given by

	Intercept	Tax	Dlic	Income	logMiles
Intercept	9.02151	-2.852e-02	-4.080e-03	-5.981e-02	-1.932e-01
Tax	-0.02852	9.788e-04	5.599e-06	4.263e-05	1.602e-04
Dlic	-0.00408	5.599e-06	3.922e-06	1.189e-05	5.402e-06
Income	-0.05981	4.263e-05	1.189e-05	1.143e-03	1.000e-03
logMiles	-0.19315	1.602e-04	5.402e-06	1.000e-03	9.948e-03

The elements of $(\mathbf{X'X})^{-1}$ often differ by several orders of magnitude, as is the case here, where the smallest element in absolute value is $3.9 \times 10^{-6} = 0.0000039$, and the largest element is 9.02. It is the combining of these numbers of very different magnitude that can lead to numerical inaccuracies in computations.

The lower-right 4×4 sub-matrix of $(\mathbf{X}'\mathbf{X})^{-1}$ is $(\mathcal{X}'\mathcal{X})^{-1}$. Using the formulas based on corrected sums of squares in this chapter, the estimate $\hat{\boldsymbol{\beta}}^*$ is computed to be

$$\hat{\boldsymbol{\beta}}^* = (\mathcal{X}'\mathcal{X})^{-1}\mathcal{X}'\mathcal{Y} = \begin{pmatrix} \hat{\beta}_1 \\ \hat{\beta}_2 \\ \hat{\beta}_3 \\ \hat{\beta}_4 \end{pmatrix} = \begin{pmatrix} -4.2280 \\ 0.4719 \\ -6.1353 \\ 18.5453 \end{pmatrix}$$

The estimated intercept is

$$\hat{\beta}_0 = \overline{y} - \hat{\boldsymbol{\beta}}^{*'}\overline{\mathbf{x}} = 154.193$$

and the residual sum of squares is

$$RSS = \mathcal{Y}'\mathcal{Y} - \hat{\boldsymbol{\beta}}^{*'}(\mathcal{X}'\mathcal{X})\hat{\boldsymbol{\beta}}^* = 193,700$$

so the estimate of σ^2 is

$$\hat{\sigma}^2 = \frac{RSS}{n - (p+1)} = \frac{193,700}{51 - 5} = 4211$$

Standard errors and estimated covariances of the $\hat{\beta}_j$ are found by multiplying $\hat{\sigma}$ by the square roots of elements of $(\mathbf{X}'\mathbf{X})^{-1}$. For example,

$$\mathrm{se}(\hat{\beta}_2) = \hat{\sigma}\sqrt{3.922 \times 10^{-6}} = 0.1285$$

Virtually all statistical software packages include higher-level functions that will fit multiple regression models, but getting intermediate results like $(\mathbf{X}'\mathbf{X})^{-1}$ may be a challenge. Table 3.3 shows typical output from a statistical package. This output gives the estimates $\hat{\boldsymbol{\beta}}$ and their standard errors computed based on $\hat{\sigma}^2$ and the

TABLE 3.3 Edited Output from the Summary Method in R for Multiple Regression in the Fuel Data

```
Coefficients:
             Estimate Std. Error t value Pr(>|t|)
(Intercept) 154.1928    194.9062   0.791 0.432938
Tax          -4.2280      2.0301  -2.083 0.042873
Dlic          0.4719      0.1285   3.672 0.000626
Income       -6.1353      2.1936  -2.797 0.007508
logMiles     18.5453      6.4722   2.865 0.006259

Residual standard error: 64.89 on 46 degrees of freedom
Multiple R-Squared: 0.5105
F-statistic: 11.99 on 4 and 46 DF,  p-value: 9.33e-07
```

diagonal elements of $(\mathbf{X}'\mathbf{X})^{-1}$. The column marked t-value is the ratio of the estimate to its standard error. The column labelled Pr(>|t|) will be discussed shortly. Below the table are a number of other summary statistics; at this point only the estimate of σ called the residual standard error and its df are familiar.

3.5 THE ANALYSIS OF VARIANCE

For multiple regression, the analysis of variance is a very rich technique that is used to compare mean functions that include different nested sets of terms. In the *overall analysis of variance*, the mean function with all the terms

$$E(Y|X = \mathbf{x}) = \boldsymbol{\beta}'\mathbf{x} \tag{3.18}$$

is compared with the mean function that includes only an intercept:

$$E(Y|X = \mathbf{x}) = \beta_0 \tag{3.19}$$

For simple regression, these correspond to (2.16) and (2.13), respectively. For mean function (3.19), $\hat{\beta}_0 = \overline{y}$ and the residual sum of squares is SYY. For mean function (3.18), the estimate of β is given by (3.9) and RSS is given in (3.11). We must have $RSS < SYY$, and the difference between these two

$$SSreg = SYY - RSS \tag{3.20}$$

corresponds to the sum of squares of Y explained by the larger mean function that is not explained by the smaller mean function. The number of df associated with $SSreg$ is equal to the number of df in SYY minus the number of df in RSS, which equals p, the number of terms in the mean function excluding the intercept.

These results are summarized in the analysis of variance table in Table 3.4. We can judge the importance of the regression on the terms in the larger model by determining if $SSreg$ is sufficiently large by comparing the ratio of the mean square for regression to $\hat{\sigma}^2$ to the $F(p, n - p')$ distribution[2] to get a significance

TABLE 3.4 The Overall Analysis of Variance Table

Source	df	SS	MS	F	p-value
Regression	p	$SSreg$	$SSreg/1$	$MSreg/\hat{\sigma}^2$	
Residual	$n - (p + 1)$	RSS	$\hat{\sigma}^2 = RSS/(n - 2)$		
Total	$n - 1$	SYY			

[2]Reminder: $p' = p$ for mean functions with no intercept, and $p' = p + 1$ for mean functions with an intercept.

level. If the computed significance level is small enough, then we would judge that the mean function (3.18) provides a significantly better fit than does (3.19). The ratio will have an exact F distribution if the errors are normal and (3.19) is true. The hypothesis tested by this F-test is

$$\text{NH:} \quad E(Y|X = \mathbf{x}) = \beta_0$$
$$\text{AH:} \quad E(Y|X = \mathbf{x}) = \mathbf{x}'\boldsymbol{\beta}$$

3.5.1 The Coefficient of Determination

As with simple regression, the ratio

$$R^2 = \frac{SSreg}{SYY} = 1 - \frac{RSS}{SYY} \tag{3.21}$$

gives the proportion of variability in Y explained by regression on the terms. R^2 can also be shown to be the square of the correlation between the observed values Y and the fitted values $\hat{Y}$; we will explore this further in the next chapter. R^2 is also called the *multiple correlation coefficient* because it is the maximum of the correlation between Y and *any* linear combination of the terms in the mean function.

Fuel Consumption Data

The overall analysis of variance table is given by

```
            Df Sum Sq Mean Sq F value    Pr(>F)
Regression 4 201994   50499  11.992   9.33e-07
Residuals 46 193700    4211
Total      50 395694
```

To get a significance level for the test, we would compare $F = 11.992$ with the $F(4, 46)$ distribution. Most computer packages do this automatically, and the result is shown in the column marked Pr(>F) to be about 0.0000009, a very small number, leading to very strong evidence against the null hypothesis that the mean of *Fuel* does not depend on any of the terms. The value of $R^2 = 201994/395694 = 0.5105$ indicates that about half the variation in *Fuel* is explained by the terms. The value of F, its significance level, and the value of R^2 are given in Table 3.3.

3.5.2 Hypotheses Concerning One of the Terms

Obtaining information on one of the terms may be of interest. Can we do as well, understanding the mean function for *Fuel* if we delete the *Tax* variable? This amounts to the following hypothesis test of

$$\text{NH:} \quad \beta_1 = 0, \quad \beta_0, \beta_2, \beta_3, \beta_4 \text{ arbitrary}$$
$$\text{AH:} \quad \beta_1 \neq 0, \quad \beta_0, \beta_2, \beta_3, \beta_4 \text{ arbitrary} \tag{3.22}$$

The following procedure can be used. First, fit the mean function that excludes the term for *Tax* and get the residual sum of squares for this smaller mean function.

Then fit again, this time including *Tax*, and once again get the residual sum of squares. Subtracting the residual sum of squares for the larger mean function from the residual sum of squares for the smaller mean function will give the sum of squares for regression on *Tax* after adjusting for the terms that are in both mean functions, *Dlic*, *Income* and log(*Miles*). Here is a summary of the computations that are needed:

	Df	SS	MS	F	Pr(>F)
Excluding Tax	47	211964			
Including Tax	46	193700			
Difference	1	18264	18264	4.34	0.043

The row marked "Excluding Tax" gives the df and *RSS* for the mean function without *Tax*, and the next line gives these values for the larger mean function including *Tax*. The difference between these two given on the next line is the sum of squares explained by *Tax* after adjusting for the other terms in the mean function. The F-test is given by $F = (18{,}264/1)/\hat{\sigma}^2 = 4.34$, which, when compared to the F distribution with $(1, 46)$ df gives a significance level of about 0.04. We thus have modest evidence that the coefficient for *Tax* is different from zero. This is called a *partial F-test*. Partial F-tests can be generalized to testing *several* coefficients to be zero, but we delay that generalization to Section 5.4.

3.5.3 Relationship to the t-Statistic

Another reasonable procedure for testing the importance of *Tax* is simply to compare the estimate of the coefficient divided by its standard error to the $t(n - p')$ distribution. One can show that the square of this t-statistic is the same number of the F-statistic just computed, so these two procedures are identical. Therefore, the t-statistic tests hypothesis (3.22) concerning the importance of terms adjusted for all the other terms, not ignoring them.

From Table 3.3, the t-statistic for *Tax* is $t = -2.083$, and $t^2 = (-2.083)^2 = 4.34$, the same as the F-statistic we just computed. The significance level for *Tax* given in Table 3.3 also agrees with the significance level we just obtained for the F-test, and so the significance level reported is for the two-sided test. To test the hypothesis that $\beta_1 = 0$ against the one-sided alternative that $\beta_1 < 0$, we could again use the same t-value, but the significance level would be one-half of the value for the two-sided test.

A t-test that β_j has a specific value versus a two-sided or one-sided alternative (with all other coefficients arbitrary) can be carried out as described in Section 2.8.

3.5.4 t-Tests and Added-Variable Plots

In Section 3.1, we discussed adding a term to a simple regression mean function. The same general procedure can be used to add a term to *any* linear regression mean function. For the added-variable plot for a term, say X_1, plot the residuals from the regression of Y on all the other X's versus the residuals for the regression of X_1 on all the other Xs. One can show (Problem 3.2) that (1) the slope of the regression in the added-variable plot is the estimated coefficient for X_1 in the regression with all the terms, and (2) the t-test for testing the slope to the zero in the added-variable plot is essentially the same as the t-test for testing $\beta_1 = 0$ in the fit of the larger mean function, the only difference being a correction for degrees of freedom.

3.5.5   Other Tests of Hypotheses

We have obtained a test of a hypothesis concerning the effect of *Tax* adjusted for all the other terms in the mean function. Equally well, we could obtain tests for the effect of *Tax* adjusting for some of the other terms or for none of the other terms. In general, these tests will not be equivalent, and a variable can be judged useful ignoring variables but useless when adjusted for them. Furthermore, a predictor that is useless by itself may become important when considered in concert with the other variables. The outcome of these tests depends on the sample correlations between the terms.

3.5.6   Sequential Analysis of Variance Tables

By separating *Tax* from the other terms, *SSreg* is divided into two pieces, one for fitting the first three terms, and one for fitting *Tax* after the other three. This subdivision can be continued by dividing *SSreg* into a sum of squares "explained" by each term separately. Unless all the terms are uncorrelated, this breakdown is not unique. For example, we could first fit *Dlic*, then *Tax* adjusted for *Dlic*, then *Income* adjusted for both *Dlic* and *Tax*, and finally log(*Miles*) adjusted for the other three. The resulting table is given in Table 3.5a. Alternatively, we could fit in the order log(*Miles*), *Income*, *Dlic* and then *Tax* as in Table 3.5b. The sums of squares can be quite different in the two tables. For example, the sum of squares for *Dlic* ignoring the other terms is about 25% larger than the sum of squares for *Dlic* adjusting for the other terms. In this problem, the terms are nearly uncorrelated, see Table 3.2, so the effect of ordering is relatively minor. In problems with high sample correlations between terms, order can be very important.

TABLE 3.5   Two Analysis of Variance Tables with Different Orders of Fitting

(a) First analysis

	Df	Sum Sq	Mean Sq
Dlic	1	86854	86854
Tax	1	19159	19159
Income	1	61408	61408
logMiles	1	34573	34573
Residuals	46	193700	4211

(b) Second analysis

	Df	Sum Sq	Mean Sq
logMiles	1	70478	70478
Income	1	49996	49996
Dlic	1	63256	63256
Tax	1	18264	18264
Residuals	46	193700	4211

3.6 PREDICTIONS AND FITTED VALUES

Suppose we have observed, or will in the future observe, a new case with its own set of predictors that result in a vector of terms $\mathbf{x}_*$. We would like to predict the value of the response given $\mathbf{x}_*$. In exactly the same way as was done in simple regression, the point prediction is $\tilde{y}_* = \mathbf{x}_*'\hat{\boldsymbol{\beta}}$, and the standard error of prediction, sepred($\tilde{y}_*|\mathbf{x}_*$), using Appendix A.8, is

$$\text{sepred}(\tilde{y}_*|\mathbf{x}_*) = \hat{\sigma}\sqrt{1 + \mathbf{x}_*'(\mathbf{X}'\mathbf{X})^{-1}\mathbf{x}_*} \qquad (3.23)$$

Similarly, the estimated average of all possible units with a value $\mathbf{x}$ for the terms is given by the estimated mean function at $\mathbf{x}$, $\hat{E}(Y|X = \mathbf{x}) = \hat{y} = \mathbf{x}'\hat{\boldsymbol{\beta}}$ with standard error given by

$$\text{sefit}(\hat{y}|\mathbf{x}) = \hat{\sigma}\sqrt{\mathbf{x}'(\mathbf{X}'\mathbf{X})^{-1}\mathbf{x}} \qquad (3.24)$$

Virtually all software packages will give the user access to the fitted values, but getting the standard error of prediction and of the fitted value may be harder. If a program produces sefit but not sepred, the latter can be computed from the former from the result

$$\text{sepred}(\tilde{y}_*|\mathbf{x}_*) = \sqrt{\hat{\sigma}^2 + \text{sefit}(\tilde{y}_*|\mathbf{x}_*)^2}$$

PROBLEMS

3.1. Berkeley Guidance Study The Berkeley Guidance Study enrolled children born in Berkeley, California, between January 1928 and June 1929, and then measured them periodically until age eighteen (Tuddenham and Snyder, 1954). The data we use is described in Table 3.6, and the data is given in the data files BGSgirls.txt for girls only, BGSboys.txt for boys only, and BGSall.txt for boys and girls combined. For this example, use only the data on the girls.

   3.1.1. For the girls only, draw the scatterplot matrix of all the age two variables, all the age nine variables and *Soma*. Write a summary of the information in this scatterplot matrix. Also obtain the matrix of sample correlations between the height variables.

   3.1.2. Starting with the mean function E(*Soma*|*WT9*) = $\beta_0 + \beta_1 WT9$, use added-variable plots to explore adding *LG9* to get the mean function E(*Soma*|*WT9*, *LG9*) = $\beta_0 + \beta_1 WT9 + \beta_2 LG9$. In particular, obtain the four plots equivalent to Figure 3.1, and summarize the information in the plots.

   3.1.3. Fit the multiple linear regression model with mean function

$$\text{E}(Soma|X) = \beta_0 + \beta_1 HT2 + \beta_2 WT2 + \beta_3 HT9 + \beta_4 WT9 + \beta_5 ST9 \tag{3.25}$$

TABLE 3.6   Variable Definitions for the Berkeley Guidance Study in the Files
`BGSgirls.txt`, `BGSboys.txt`, and `BGSall.txt`

Variable	Description
Sex	0 for males, 1 for females
WT2	Age 2 weight, kg
HT2	Age 2 height, cm
WT9	Age 9 weight, kg
HT9	Age 9 height, cm
LG9	Age 9 leg circumference, cm
ST9	Age 9 strength, kg
WT18	Age 18 weight, kg
HT18	Age 18 height, cm
LG18	Age 18 leg circumference, cm
ST18	Age 18 strength, kg
Soma	Somatotype, a scale from 1, very thin, to 7, obese, of body type

Find $\hat{\sigma}$, R^2, the overall analysis of variance table and overall F-test. Compute the t-statistics to be used to test each of the β_j to be zero against two-sided alternatives. Explicitly state the hypotheses tested and the conclusions.

3.1.4. Obtain the sequential analysis of variance table for fitting the variables in the order they are given in (3.25). State the hypotheses tested and the conclusions for each of the tests.

3.1.5. Obtain analysis of variance again, this time fitting with the five terms in the order given from right to left in (3.25). Explain the differences with the table you obtained in Problem 3.1.4. What graphs could help understand the issues?

3.2. Added-variable plots   This problem uses the United Nations example in Section 3.1 to demonstrate many of the properties of added-variable plots. This problem is based on the mean function

$$E(\log(Fertility)|\log(PPgdp) = x_1, Purban = x_2) = \beta_0 + \beta_1 x_1 + \beta_2 x_2$$

There is nothing special about the two-predictor regression mean function, but we are using this case for simplicity.

3.2.1. Show that the estimated coefficient for $\log(PPgdp)$ is the same as the estimated slope in the added-variable plot for $\log(PPgdp)$ after *Purban*. This correctly suggests that *all the estimates in a multiple linear regression model are adjusted for all the other terms in the mean function.* Also, show that the residuals in the added-variable plot are identical to the residuals from the mean function with both predictors.

3.2.2. Show that the t-test for the coefficient for $\log(PPgdp)$ is not quite the same from the added-variable plot and from the regression with both terms, and explain why they are slightly different.

3.3. The following questions all refer to the mean function

$$E(Y|X_1 = x_1, X_2 = x_2) = \beta_0 + \beta_1 x_1 + \beta_2 x_2 \qquad (3.26)$$

3.3.1. Suppose we fit (3.26) to data for which $x_1 = 2.2x_2$, with no error. For example, x_1 could be a weight in pounds, and x_2 the weight of the same object in kg. Describe the appearance of the added-variable plot for X_2 after X_1.

3.3.2. Again referring to (3.26), suppose now that Y and X_1 are perfectly correlated, so $Y = 3X_1$, without any error. Describe the appearance of the added-variable plot for X_2 after X_1.

3.3.3. Under what conditions will the added-variable plot for X_2 after X_1 have exactly the same shape as the scatterplot of Y versus X_2?

3.3.4. True or false: The vertical variation in an added-variable plot for X_2 after X_1 is always less than or equal to the vertical variation in a plot of Y versus X_2. Explain.

3.4. Suppose we have a regression in which we want to fit the mean function (3.1). Following the outline in Section 3.1, suppose that the two terms X_1 and X_2 have sample correlation equal to zero. This means that, if $x_{ij}, i = 1, \ldots, n$, and $j = 1, 2$ are the observed values of these two terms for the n cases in the data, $\sum_{i=1}^{n}(x_{i1} - \overline{x}_1)(x_{i2} - \overline{x}_2) = 0$.

3.4.1. Give the formula for the slope of the regression for Y on X_1, and for Y on X_2. Give the value of the slope of the regression for X_2 on X_1.

3.4.2. Give formulas for the residuals for the regressions of Y on X_1 and for X_2 on X_1. The plot of these two sets of residuals corresponds to the added-variable plot in Figure 3.1d.

3.4.3. Compute the slope of the regression corresponding to the added-variable plot for the regression of Y on X_2 after X_1, and show that this slope is exactly the same as the slope for the simple regression of Y on X_2 ignoring X_1. Also find the intercept for the added-variable plot.

3.5. Refer to the data described in Problem 1.5, page 18. For this problem, consider the regression problem with response *BSAAM*, and three predictors as terms given by *OPBPC*, *OPRC* and *OPSLAKE*.

3.5.1. Examine the scatterplot matrix drawn for these three terms and the response. What should the correlation matrix look like (that is, which correlations are large and positive, which are large and negative, and which are small)? Compute the correlation matrix to verify your results. Get the regression summary for the regression of *BSAAM* on these three terms. Explain what the "t-values" column of your output means.

3.5.2. Obtain the overall test if the hypothesis that *BSAAM* is independent of the three terms versus the alternative that it is not independent of them, and summarize your results.

3.5.3. Obtain three analysis of variance tables fitting in the order (*OPBPC*, *OPRC* and *OPSLAKE*), then (*OPBPC*, *OPSLAKE* and *OPRC*), and finally (*OPSLAKE*, *OPRC* and *OPBPC*). Explain the resulting tables, and discuss in particular any apparent inconsistencies. Which *F*-tests in the Anova tables are equivalent to *t*-tests in the regression output?

3.5.4. Using the output from the last problem, test the hypothesis that the coefficients for both *OPRC* and *OPBPC* are both zero against the alternative that they are not both zero.

CHAPTER 4

Drawing Conclusions

The computations that are done in multiple linear regression, including drawing graphs, creation of terms, fitting models, and performing tests, will be similar in most problems. Interpreting the results, however, may differ by problem, even if the outline of the analysis is the same. Many issues play into drawing conclusions, and some of them are discussed in this chapter.

4.1 UNDERSTANDING PARAMETER ESTIMATES

Parameters in mean functions have *units* attached to them. For example, the fitted mean function for the fuel consumption data is

$$\mathrm{E}(Fuel|X) = 154.19 - 4.23\,Tax + 0.47\,Dlic - 6.14\,Income + 18.54\log(Miles)$$

Fuel is measured in gallons, and so all the quantities on the right of this equation must also be in gallons. The intercept is 154.19 gallons. Since *Income* is measured in thousands of dollars, the coefficient for *Income* must be in gallons per thousand dollars of income. Similarly, the units for the coefficient for *Tax* is gallons per cent of tax.

4.1.1 Rate of Change

The usual interpretation of an estimated coefficient is as a rate of change: increasing *Tax* rate by one cent should decrease consumption, all other factors being held constant, by about 4.23 gallons per person. This assumes that a predictor can in fact be changed without affecting the other terms in the mean function and that the available data will apply when the predictor is so changed. The fuel data are *observational* since the assignment of values for the predictors was not under the control of the analyst, so whether increasing taxes would *cause*

Applied Linear Regression, Third Edition, by Sanford Weisberg
ISBN 0-471-66379-4   Copyright © 2005 John Wiley & Sons, Inc.

a decrease in fuel consumption cannot be assessed from these data. From these data, we can observe *association* but not cause: states with higher tax rates are *observed* to have lower fuel consumption. To draw conclusions concerning the effects of changing tax rates, the rates must in fact be changed and the results observed.

The coefficient estimate of log(*Miles*) is 18.55, meaning that a change of one unit in log(*Miles*) is associated with an 18.55 gallon per person increase in consumption. States with more roads have higher per capita fuel consumption. Since we used base-two logarithms in this problem, increasing log(*Miles*) by one unit means that the value of *Miles doubles*. If we double the amount of road in a state, we expect to increase fuel consumption by about 18.55 gallons per person. If we had used base-ten logarithms, then the fitted mean function would be

$$E(Fuel|X) = 154.19 - 4.23\,Tax + 0.47\,Dlic - 6.14\,Income + 61.61\log_{10}(Miles)$$

The only change in the fitted model is for the coefficient for the log of Miles, which is now interpreted as a change in expected *Fuel* consumption when $\log_{10}(Miles)$ increases by one unit, or when *Miles* is multiplied by 10.

4.1.2 Signs of Estimates

The sign of a parameter estimate indicates the direction of the relationship between the term and the response. In multiple regression, if the terms are correlated, the sign of a coefficient may change depending on the other terms in the model. While this is mathematically possible and, occasionally, scientifically reasonable, it certainly makes interpretation more difficult. Sometimes this problem can be removed by redefining the terms into new linear combinations that are easier to interpret.

4.1.3 Interpretation Depends on Other Terms in the Mean Function

The value of a parameter estimate not only depends on the other terms in a mean function but it can also change if the other terms are replaced by linear combinations of the terms.

Berkeley Guidance Study

Data from the Berkeley Guidance Study on the growth of boys and girls are given in Problem 3.1. As in Problem 3.1, we will view *Soma* as the response, but consider the three predictors *WT2*, *WT9*, *WT18* for the $n = 70$ girls in the study. The scatterplot matrix for these four variables is given in Figure 4.1. First look at the last row of this figure, giving the marginal response plots of *Soma* versus each of the three potential predictors. For each of these plots, we see that *Soma* is increasing with the potential predictor on the average, although the relationship is strongest at the oldest age and weakest at the youngest age. The two-dimensional plots of each pair of predictors suggest that the predictors are correlated among themselves. Taken together, we have evidence that the regression on all three predictors cannot

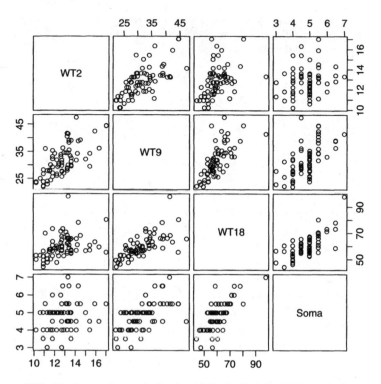

FIG. 4.1   Scatterplot matrix for the girls in the Berkeley Guidance Study.

be viewed as just the sum of the three separate simple regressions because we must account for the correlations between the terms.

We will proceed with this example using the three original predictors as terms and *Soma* as the response. We are encouraged to do this because of the appearance of the scatterplot matrix. Since each of the two-dimensional plots appear to be well summarized by a straight-line mean function, we will see later that this suggests that the regression of the response on the original predictors without transformation is likely to be appropriate.

The parameter estimates for the regression of *Soma* on *WT2*, *WT9*, and *WT18* given in the column marked "Model 1" in Table 4.1 leads to the unexpected conclusion that heavier girls at age two may tend to be thinner, have lower expected somatotype, at age 18. We reach this conclusion because the *t*-statistic for testing the coefficient equal to zero, which is not shown in the table, has a significance level of about 0.06. The sign, and the weak significance, may be due to the correlations between the terms. In place of the preceding variables, consider the following:

$$WT2 = \text{ Weight at age 2}$$

$$DW9 = WT9 - WT2 = \text{ Weight gain from age 2 to 9}$$

$$DW18 = WT18 - WT9 = \text{ Weight gain from age 9 to 18}$$

TABLE 4.1 Regression of *Soma* on Different
Combinations of Three Weight Variables for the $n = 70$
Girls in the Berkeley Guidance Study

Term	Model 1	Model 2	Model 3
(Intercept)	1.5921	1.5921	1.5921
WT2	−0.1156	−0.0111	−0.1156
WT9	0.0562		0.0562
WT18	0.0483		0.0483
DW9		0.1046	NA
DW18		0.0483	NA

Since all three original terms measure weight, combining them in this way is reasonable. If the variables measured different quantities, then combining them could lead to conclusions that are even less useful than those originally obtained. The parameter estimates for *Soma* on *WT2*, *DW9*, and *DW18* are given in the column marked "Model 2" in Table 4.1. Although not shown in the table, summary statistics for the regression like R^2 and $\hat{\sigma}^2$ are identical for all the mean functions in Table 4.1 but coefficient estimates and *t*-tests are not the same. For example, the slope estimate for *WT2* is about -0.12, with $t = -1.87$ in the column "Model 1," while in Model 2, the estimate is about one-tenth the size, and the *t*-value is -0.21. In the former case, the effect of *WT2* appears plausible, while in the latter it does not. Although the estimate is negative in each, we would be led in the latter case to conclude that the effect of *WT2* is negligible. Thus, interpretation of the effect of a variable depends not only on the other variables in a model but also upon which linear transformation of those variables is used.

Another interesting feature of Table 4.1 is that the estimate for *WT18* in Model 1 is identical to the estimate for *DW18* in Model 2. In Model 1, the estimate for *WT18* is the effect on *Soma* of changing *WT18* by one unit, with all other terms held fixed. In Model 2, the estimate for *DW18* is the change in *Soma* when *DW18* changes by one unit, when all other terms are held fixed. *But the only way* DW18 = WT18 − WT9 *can be changed by one unit with the other variables including* WT9 = DW9 − WT2 *held fixed is by changing* WT18 *by one unit.* Consequently, the terms *WT18* in Model 1 and *DW18* in Model 2 play identical roles and therefore we get the same estimates.

The linear transformation of the three weight variables we have used so far could be replaced by other linear combinations, and, depending on the context, others might be preferred. For example, another set might be

$$AVE = (WT2 + WT9 + WT18)/3$$

$$LIN = WT18 - WT2$$

$$QUAD = WT2 - 2WT9 + WT18$$

This transformation focuses on the fact that *WT2*, *WT9* and *WT18* are ordered in time and are more or less equally spaced. Pretending that the weight measurements are equally spaced, *AVE, LIN* and *QUAD* are, respectively, the average, linear, and quadratic time trends in weight gain.

4.1.4   Rank Deficient and Over-Parameterized Mean Functions

In the last example, several combinations of the basic predictors *WT2, WT9,* and *WT18* were studied. One might naturally ask what would happen if more than three combinations of these predictors were used in the same regression model. As long as we use linear combinations of the predictors, as opposed to nonlinear combinations or transformations of them, we cannot use more than three, the number of linearly independent quantities.

To see why this is true, consider adding *DW9* to the mean function including *WT2, WT9* and *WT18.* As in Chapter 3, we can learn about adding *DW9* using an added-variable plot of the residuals from the regression of *Soma* on *WT2, WT9* and *WT18* versus the residuals from the regression of *DW9* on *WT2, WT9* and *WT18.* Since *DW9* can be written as an exact linear combination of the other predictors, $DW9 = WT9 - WT2$, the residuals from this second regression are all exactly zero. A slope coefficient for *DW9* is thus not defined after adjusting for the other three terms. We would say that the four terms *WT2, WT9, WT18,* and *DW9* are *linearly dependent*, since one can be determined exactly from the others. The three variables *WT2, WT9* and *WT18* are *linearly independent* because one of them cannot be determined exactly by a linear combination of the others. The maximum number of linearly independent terms that could be included in a mean function is called the *rank* of the data matrix **X**.

Model 3 in Table 4.1 gives the estimates produced in a computer package when we tried to fit using an intercept and the five terms *WT2, WT9, WT18, DW9,* and *DW18.* Most computer programs, including this one, will select the first three, and the estimated coefficients for them. For the remaining terms, this program sets the estimates to "NA," a code for a missing value; the word *aliased* is sometimes used to indicate a term that is a linear combination of terms already in the mean function, and so a coefficient for it is not estimable.

Mean functions that are over-parameterized occur most often in designed experiments. The simplest example is the one-way design. Suppose that a unit is assigned to one of three treatment groups, and let $X_1 = 1$ if the unit is in group one and zero otherwise, $X_2 = 1$ if the unit is in group two and zero otherwise, and $X_3 = 1$ if the unit is in group three and zero otherwise. For each unit, we must have $X_1 + X_2 + X_3 = 1$ since each unit is in only one of the three groups. We therefore cannot fit the model

$$E(Y|X) = \beta_0 + \beta_1 X_1 + \beta_2 X_2 + \beta_3 X_3$$

because the sum of the X_j is equal to the column of ones, and so, for example, $X_3 = 1 - X_1 - X_2$. To fit a model, we must do something else. The options are: (1) place a constraint like $\beta_1 + \beta_2 + \beta_3 = 0$ on the parameters; (2) exclude one of the X_j from the model, or (3) leave out an explicit intercept. All of these options will in some sense be equivalent, since the same R^2, σ^2 and overall F-test and predictions will result. Of course, some care must be taken in using parameter estimates, since these will surely depend on the parameterization used to get a full rank model. For further reading on matrices and models of less than full rank, see, for example, Searle (1971, 1982).

4.1.5 Tests

Even if the fitted model were correct and errors were normally distributed, tests and confidence statements for parameters are difficult to interpret because correlations among the terms lead to a multiplicity of possible tests. Sometimes, tests of effects adjusted for other variables are clearly desirable, such as in assessing a treatment effect after adjusting for other variables to reduce variability. At other times, the order of fitting is not clear, and the analyst must expect ambiguous results. In most situations, the only true test of significance is repeated experimentation.

4.1.6 Dropping Terms

Suppose we have a sample of n rectangles from which we want to model $\log(Area)$ as a function of $\log(Length)$, perhaps through the simple regression mean function

$$E(\log(Area)|\log(Length)) = \eta_0 + \eta_1\log(Length) \tag{4.1}$$

From elementary geometry, we know that $Area = Length \times Width$, and so the "true" mean function for $\log(Area)$ is

$$E(\log(Area)|\log(Length), \log(Width)) = \beta_0 + \beta_1\log(Length) + \beta_2\log(Width) \tag{4.2}$$

with $\beta_0 = 0$, and $\beta_1 = \beta_2 = 1$. The questions of interest are: (1) can the incorrect mean function specified by (4.1) provide a useful approximation to the true mean function (4.2), and if so, (2) what are the relationships between ηs, in (4.1) and the βs in (4.2)?

The answers to these questions comes from Appendix A.2.4. Suppose that the true mean function were

$$E(Y|X_1 = x_1, X_2 = x_2) = \beta_0 + \beta'_1 x_1 + \beta'_2 x_2 \tag{4.3}$$

but we want to fit a mean function with X_1 only. The mean function for $Y|X_1$ is obtained by averaging (4.3) over X_2,

$$E(Y|X_1 = x_1) = E[E(Y|X_1 = x_1, X_2)|X_1 = x_1]$$
$$= \beta_0 + \beta'_1 x_1 + \beta'_2 E(X_2|X_1 = x_1) \tag{4.4}$$

We cannot, in general, simply drop a set of terms from a correct mean function, but we need to substitute the conditional expectation of the terms dropped given the terms that remain in the mean function.

In the context of the rectangles example, we get

$$E(\log(Area)|\log(Length)) = \eta_0 + \eta_1\log(Length) + \beta_2 E(\log(Width)|\log(Length)) \tag{4.5}$$

The answers to the questions posed depend on the mean function for the regression of log(*Width*) on log(*Length*). This conditional expectation has little to do with the area of rectangles, but much to do with the way we obtain a sample of rectangles to use in our study. We will consider three cases.

In the first case, imagine that each of the rectangles in the study is formed by sampling a log(*Length*) and a log(*Width*) from independent distributions. If the mean of the log(*Width*) distribution is W, then by independence

$$E(\log(Width)|\log(Length)) = E(\log(Width)) = W$$

Substituting into (4.5),

$$E(\log(Area)|\log(Length)) = \beta_0 + \beta_1 \log(Length) + \beta_2 W$$
$$= (\beta_0 + \beta_2 W) + \beta_1 \log(Length)$$
$$= W + \log(Length)$$

where the last equation follows by substituting $\beta_0 = 0$, $\beta_1 = \beta_2 = 1$. For this case, the mean function (4.1) would be appropriate for the regression of log(*Area*) on log(*Width*). The intercept for the mean function (4.1) would be W, and so it depends on the distribution of the widths in the data. The slope for log(*Length*) is the same for fitting (4.1) or (4.2).

In the second case, suppose that

$$E(\log(Width)|\log(Length)) = \gamma_0 + \gamma_1 \log(Length)$$

so the mean function for the regression of log(*Width*) on log(*Length*) is a straight line. This could occur, for example, if the rectangles in our study were obtained by sampling from a family of similar rectangles, so the ratio *Width/Length* is the same for all rectangles in the study. Substituting this into (4.5) and simplifying gives

$$E(\log(Area)|\log(Length)) = \beta_0 + \beta_1 \log(Length) + \beta_2(\gamma_0 + \gamma_1 \log(Length))$$
$$= (\beta_0 + \beta_2 \gamma_0) + (\beta_1 + \beta_2 \gamma_1)\log(Length)$$
$$= \gamma_0 + (1 + \gamma_1)\log(Length)$$

Once again fitting using (4.1) will be appropriate, but the values of $\eta_0 = \gamma_0$ and $\eta_1 = 1 + \gamma_1$ depend on the parameters of the regression of log(*Width*) on log(*Length*). The γs are a characteristic of the sampling plan, not of rectangles. Two experimenters who sample rectangles of different shapes will end up estimating different parameters.

For a final case, suppose that the mean function

$$E(\log(Width)|\log(Length)) = \gamma_0 + \gamma_1 \log(Length) + \gamma_2 \log(Length)^2$$

is quadratic. Substituting into (4.5), setting $\beta_0 = 0, \beta_1 = \beta_2 = 1$ and simplifying gives

$$\mathrm{E}(\log(Area)|\log(Length)) = \beta_0 + \beta_1 \log(Length)$$
$$+ \beta_2 \left(\gamma_0 + \gamma_1 \log(Length) + \gamma_2 \log(Length)^2 \right)$$
$$= \gamma_0 + (1 + \gamma_1)\log(Length) + \gamma_2 \log(Length)^2$$

which is a quadratic function of $\log(Length)$. If the mean function is quadratic, or any other function beyond a straight line, then fitting (4.1) is inappropriate.

From the above three cases, we see that both the mean function and the parameters for the response depend on the mean function for the regression of the removed terms on the remaining terms. If the mean function for the regression of the removed terms on the retained terms is not linear, then a linear mean function will not be appropriate for the regression problem with fewer terms.

Variances are also affected when terms are dropped. Returning to the true mean function given by (4.3), the general result for the regression of Y on X_1 alone is, from Appendix A.2.4,

$$\mathrm{Var}(Y|X_1 = \mathbf{x}_1) = \mathrm{E}\left[\mathrm{Var}(Y|X_1 = \mathbf{x}_1, X_2)|X_1 = \mathbf{x}_1\right]$$
$$+ \mathrm{Var}\left[\mathrm{E}(Y|X_1 = \mathbf{x}_1, X_2)|X_1 = \mathbf{x}_1\right]$$
$$= \sigma^2 + \boldsymbol{\beta}_2' \mathrm{Var}(X_2|X_1 = \mathbf{x}_1)\boldsymbol{\beta}_2 \qquad (4.6)$$

In the context of the rectangles example, $\beta_2 = 1$ and we get

$$\mathrm{Var}(\log(Area)|\log(Length)) = \sigma^2 + \mathrm{Var}(\log(Width)|\log(Length))$$

Although fitting (4.1) can be appropriate if $\log(Width)$ and $\log(Length)$ are linearly related, the errors for this mean function can be much larger than those for (4.2) if $\mathrm{Var}(\log(Width)|\log(Length))$ is large. If $\mathrm{Var}(\log(Width)|\log(Length))$ is small enough, then fitting (4.2) can actually give answers that are nearly as accurate as fitting with the true mean function (4.2).

4.1.7 Logarithms

If we start with the simple regression mean function,

$$\mathrm{E}(Y|X = x) = \beta_0 + \beta_1 x$$

a useful way to interpret the coefficient β_1 is as the first derivative of the mean function with respect to x,

$$\frac{d\mathrm{E}(Y|X = x)}{dx} = \beta_1$$

We recall from elementary geometry that the first derivative is the rate of change, or the slope of the tangent to a curve, at a point. Since the mean function for simple regression is a straight line, the slope of the tangent is the same value β_1 for any value of x, and β_1 completely characterizes the change in the mean when the predictor is changed for any value of x.

When the predictor is replaced by $\log(x)$, the mean function as a function of x

$$E(Y|X = x) = \beta_0 + \beta_1 \log(x)$$

is no longer a straight line, but rather it is a curve. The tangent at the point $x > 0$ is

$$\frac{dE(Y|X = x)}{dx} = \frac{\beta_1}{x}$$

The slope of the tangent is different for each x and the effect of changing x on $E(Y|X = x)$ is largest for small values of x and gets smaller as x is increased.

When the response is in log scale, we can get similar approximate results by exponentiating both sides of the equation:

$$E(\log(Y)|X = x) = \beta_0 + \beta_1 x$$

$$E(Y|X = x) \approx e^{\beta_0} e^{\beta_1 x}$$

Differentiating this second equation gives

$$\frac{dE(Y|X = x)}{dx} = \beta_1 E(Y|X = x)$$

The rate of change at x is thus equal to β_1 times the mean at x. We can also write

$$\frac{dE(Y|X = x)/dx}{E(Y|X = x)} = \beta_1$$

is constant, and so β_1 can be interpreted as the constant rate of change in the response per unit of response.

4.2 EXPERIMENTATION VERSUS OBSERVATION

There are fundamentally two types of predictors that are used in a regression analysis, *experimental* and *observational*. Experimental predictors have values that are under the control of the experimenter, while for observational predictors, the values are observed rather than set. Consider, for example, a hypothetical study of factors determining the yield of a certain crop. Experimental variables might include the amount and type of fertilizers used, the spacing of plants, and the amount of irrigation, since each of these can be assigned by the investigator to the units, which are plots of land. Observational predictors might include characteristics of the plots in the study, such as drainage, exposure, soil fertility, and weather variables. All of these are beyond the control of the experimenter, yet may have important effects on the observed yields.

The primary difference between experimental and observational predictors is in the inferences we can make. From experimental data, we can often infer causation.

If we assign the level of fertilizer to plots, usually on the basis of a randomization scheme, and observe differences due to levels of fertilizer, we can infer that the fertilizer is causing the differences. Observational predictors allow weaker inferences. We might say that weather variables are associated with yield, but the causal link is not available for variables that are not under the experimenter's control. Some experimental designs, including those that use randomization, are constructed so that the effects of observational factors can be ignored or used in analysis of covariance (see, e.g., Cox, 1958; Oehlert, 2000).

Purely observational studies that are not under the control of the analyst can only be used to predict or model the events that were observed in the data, as in the fuel consumption example. To apply observational results to predict future values, additional assumptions about the behavior of future values compared to the behavior of the existing data must be made. From a purely observational study, we cannot infer a causal relationship without additional information external to the observational study.

Feedlots

A *feedlot* is a farming operation that includes large number of cattle, swine or poultry in a small area. Feedlots are efficient producers of animal products, and can provide high-paying skilled jobs in rural areas. They can also cause environmental problems, particularly with odors, ground water pollution, and noise.

Taff, Tiffany, and Weisberg (1996) report a study on the effect of feedlots on property values. This study was based on all 292 rural residential property sales in two southern Minnesota counties in 1993–94. Regression analysis was used. The response was sale price. Predictors included house characteristics such as size, number of bedrooms, age of the property, and so on. Additional predictors described the relationship of the property to existing feedlots, such as distance to the nearest feedlot, number of nearby feedlots, and related features of the feedlots such as their size. The "feedlot effect" could be inferred from the coefficients for the feedlot variables.

In the analysis, the coefficient estimates for feedlot effects were generally positive and judged to be nonzero, meaning that close proximity to feedlots was associated with an *increase* in sale prices. While association of the opposite sign was expected, the positive sign is plausible if the positive economic impact of the feedlot outweighs the negative environmental impact. The positive effect is estimated to be small, however, and equal to 5% or less of the sale price of the homes in the study.

These data are purely observational, with no experimental predictors. The data collectors had no control over the houses that actually sold, or siting of feedlots. Consequently, any inference that nearby feedlots *cause* increases in sale price is unwarranted from this study. Given that we are limited to association, rather than causation, we might next turn to whether we can generalize the results. Can we infer the same association to houses that were *not* sold in these counties during this period? We have no way of knowing from the data if the same relationship would hold for homes that did not sell. For example, some home-owners may have perceived that they could not get a reasonable price and may have decided not to sell. This would create a bias in favor of a positive effect of feedlots.

Can we generalize geographically, to other Minnesota counties or to other places in the Midwest United States? The answer to this question depends on the characteristics of the two counties studied. Both are rural counties with populations of about 17,000. Both have very low property values with median sale price in this period of less than \$50,000. Each had different regulations for operators of feedlots, and these regulations could impact pollution problems. Applying the results to a county with different demographics or regulations cannot be justified by these data alone, and additional information and assumptions are required.

Joiner (1981) coined the picturesque phrase *lurking variable* to describe a predictor variable not included in a mean function that is correlated with terms in the mean function. Suppose we have a regression with predictors X that are included in the regression and a lurking variable L not included in the study, and that the true regression mean function is

$$E(Y|X = \mathbf{x}, L = \ell) = \beta_0 + \sum_{j-1}^{p} \beta_j x_j + \delta \ell \qquad (4.7)$$

with $\delta \neq 0$. We assume that X and L are correlated and for simplicity we assume further that $E(L|X = \mathbf{x}) = \gamma_0 + \sum \gamma_j x_j$. When we fit the incorrect mean function that ignores the lurking variable, we get, from Section 4.1.6,

$$E(Y|X = \mathbf{x}) = \beta_0 + \sum_{j-1}^{p} \beta_j x_j + \delta E(L|X = \mathbf{x})$$

$$= (\beta_0 + \delta \gamma_0) + \sum_{j-1}^{p} (\beta_j + \delta \gamma_j) x_j \qquad (4.8)$$

Suppose we are particularly interested in inferences about the coefficient for X_1, and, unknown to us, β_1 in (4.7) is equal to zero. If we were able to fit with the lurking variable included, we would probably conclude that X_1 is not an important predictor. If we fit the incorrect mean function (4.8), the coefficient for X_1 becomes $(\beta_1 + \delta \gamma_1)$, which will be non zero if $\gamma_1 \neq 0$. The lurking variable masquerades as the variable of interest to give an incorrect inference. A lurking variable can also hide the effect of an important variable if, for example, $\beta_1 \neq 0$ but $\beta_1 + \delta \gamma_1 = 0$.

All large observational studies like this feedlot study potentially have lurking variables. For this study, a casino had recently opened near these counties, creating many jobs and a demand for housing that might well have overshadowed any effect of feedlots. In experimental data with random assignment, the potential effects of lurking variables are greatly decreased, since the random assignment guarantees that the correlation between the terms in the mean function and any lurking variable is small or zero.

The interpretation of results from a regression analysis depend on the details of the data design and collection. The feedlot study has extremely limited scope, and is but one element to be considered in trying to understand the effect of feedlots on property values. Studies like this feedlot study are easily misused. As recently as spring 2004, the study was cited in an application for a permit to build a feedlot in Starke county, Indiana, claiming that the study supports the positive effect of feedlots on property values, confusing association with causation, and inferring generalizability to other locations without any logical foundation for doing so.

4.3 SAMPLING FROM A NORMAL POPULATION

Much of the intuition for the use of least squares estimation is based on the assumption that the observed data are a sample from a multivariate normal population. While the assumption of multivariate normality is almost never tenable in practical regression problems, it is worthwhile to explore the relevant results for normal data, first assuming random sampling and then removing that assumption.

Suppose that all of the observed variables are normal random variables, and the observations on each case are independent of the observations on each other case. In a two-variable problem, for the ith case observe (x_i, y_i), and suppose that

$$\begin{pmatrix} x_i \\ y_i \end{pmatrix} \sim \mathrm{N}\left(\begin{pmatrix} \mu_x \\ \mu_y \end{pmatrix}, \begin{pmatrix} \sigma_x^2 & \rho_{xy}\sigma_x\sigma_y \\ \rho_{xy}\sigma_x\sigma_y & \sigma_y^2 \end{pmatrix} \right) \qquad (4.9)$$

Equation (4.9) says that x_i and y_i are each realizations of normal random variables with means μ_x and μ_y, variances σ_x^2 and σ_y^2 and correlation ρ_{xy}. Now, suppose we consider the conditional distribution of y_i given that we have already observed the value of x_i. It can be shown (see e.g., Lindgren, 1993; Casella and Berger, 1990) that the conditional distribution of y_i given x_i, is normal and,

$$y_i|x_i \sim \mathrm{N}\left(\mu_y + \rho_{xy}\frac{\sigma_y}{\sigma_x}(x_i - \mu_x), \sigma_y^2(1 - \rho_{xy}^2) \right) \qquad (4.10)$$

If we define

$$\beta_0 = \mu_y - \beta_1\mu_x \qquad \beta_1 = \rho_{xy}\frac{\sigma_y}{\sigma_x} \qquad \sigma^2 = \sigma_y^2(1 - \rho_{xy}^2) \qquad (4.11)$$

then the conditional distribution of y_i given x_i is simply

$$y_i|x_i \sim \mathrm{N}(\beta_0 + \beta_1 x_i, \sigma^2) \qquad (4.12)$$

which is essentially the same as the simple regression model with the added assumption of normality.

Given random sampling, the five parameters in (4.9) are estimated, using the notation of Table 2.1, by

$$\hat{\mu}_x = \bar{x} \quad \hat{\sigma}_x^2 = SD_x^2 \quad \hat{\rho}_{xy} = r_{xy}$$
$$\hat{\mu}_y = \bar{y} \quad \hat{\sigma}_y^2 = SD_y^2 \tag{4.13}$$

Estimates of β_0 and β_1 are obtained by substituting estimates from (4.13) for parameters in (4.11), so that $\hat{\beta}_1 = r_{xy} SD_y / SD_x$, and so on, as derived in Chapter 2. However, $\hat{\sigma}^2 = [(n-1)/(n-2)]SD_y^2(1 - r_{xy}^2)$ to correct for degrees of freedom.

If the observations on the ith case are y_i and a $p \times 1$ vector $\mathbf{x}_i$ not including a constant, multivariate normality is shown symbolically by

$$\begin{pmatrix} \mathbf{x}_i \\ y_i \end{pmatrix} \sim N\left(\begin{pmatrix} \mu_x \\ \mu_y \end{pmatrix}, \begin{pmatrix} \Sigma_{xx} & \Sigma_{xy} \\ \Sigma_{xy} & \sigma_y^2 \end{pmatrix} \right) \tag{4.14}$$

where Σ_{xx} is a $p \times p$ matrix of variances and covariances between the elements of $\mathbf{x}_i$ and Σ_{xy} is a $p \times 1$ vector of covariances between $\mathbf{x}_i$ and y_i. The conditional distribution of y_i given x_i is then

$$y_i | \mathbf{x}_i \sim N\left((\mu_y - \boldsymbol{\beta}^{*\prime}\boldsymbol{\mu}_x) + \boldsymbol{\beta}^{*\prime}\mathbf{x}_i, \sigma^2 \right) \tag{4.15}$$

If $\mathcal{R}^2$ is the population multiple correlation,

$$\boldsymbol{\beta}^* = \Sigma_{xx}^{-1}\Sigma_{xy}; \quad \sigma^2 = \sigma_y^2 \Sigma_{xy}^{\prime} \Sigma_{xx}^{-1} \Sigma_{xy} = \sigma_y^2 (1 - \mathcal{R}^2)$$

The formulas for $\hat{\boldsymbol{\beta}}^*$ and σ^2 and the formulas for their least squares estimators differ only by the substitution of estimates for parameters, with $(n-1)^{-1}(\mathcal{X}'\mathcal{X})$ estimating Σ_{xx}, and $(n-1)^{-1}(\mathcal{X}'\mathcal{Y})$ estimating Σ_{xy}.

4.4 MORE ON R^2

The conditional distribution in (4.10) or (4.15) does not depend on random sampling, but only on normal distributions, so whenever multivariate normality seems reasonable, a linear regression model is suggested for the conditional distribution of one variable, given the others. However, if random sampling is not used, some of the usual summary statistics, including R^2, lose their connection to population parameters.

Figure 4.2a repeats Figure 1.1, the scatterplot of *Dheight* versus *Mheight* for the heights data. These data closely resemble a bivariate normal sample, and so $R^2 = 0.24$ estimates the population $\mathcal{R}^2$ for this problem. Figure 4.2b repeats this last figure, except that all cases with *Mheight* between 61 and 64 inches—the lower and upper quartile of the mother's heights rounded to the nearest inch—have been removed form the data. The OLS regression line appears similar, but the value of $R^2 = 0.37$ is about 50% larger. By removing the middle of the data, we have made

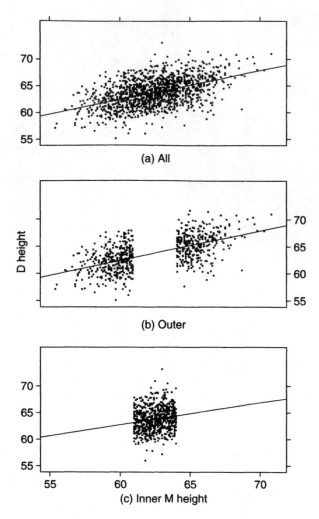

FIG. 4.2   Three views of the heights data.

R^2 larger, and it no longer estimates a population value. Similarly, in Figure 4.2c, we exclude all the cases with *Mheight* outside the quartiles, and get $R^2 = 0.027$, and the relationship between *Dheight* and *Mheight* virtually disappears.

This example points out that even in the unusual event of analyzing data drawn from a multivariate normal population, if sampling of the population is not random, the interpretation of R^2 may be completely misleading, as this statistic will be strongly influenced by the method of sampling. In particular, a few cases with unusual values for the predictors can largely determine the observed value of this statistic.

We have seen that we can manipulate the value of R^2 merely by changing our sampling plan for collecting data: if the values of the terms are widely dispersed, then R^2 will tend to be too large, while if the values are over a very small range, then R^2 will tend to be too small. Because the notion of proportion of variability explained is so useful, a diagnostic method is needed to decide if it is a useful concept in any particular problem.

4.4.1 Simple Linear Regression and R^2

In simple regression linear problems, we can always determine the appropriateness of R^2 as a summary by examining the summary graph of the response versus the predictor. If the plot looks like a sample from a bivariate normal population, as in Figure 4.2a, then R^2 is a useful measure. The less the graph looks like this figure, the less useful is R^2 as a summary measure.

Figure 4.3 shows six summary graphs. Only for the first three of them is R^2 a useful summary of the regression problem. In Figure 4.3e, the mean function appears curved rather than straight so correlation is a poor measure of dependence. In Figure 4.3d the value of R^2 is virtually determined by one point, making R^2 necessarily unreliable. The regular appearance of the remaining plot suggests a different type of problem. We may have several identifiable groups of points caused by a lurking variable not included in the mean function, such that the mean function for each group has a negative slope, but when groups are combined the slope becomes positive. Once again R^2 is not a useful summary of this graph.

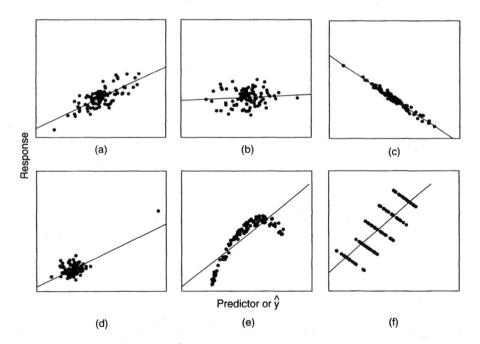

FIG. 4.3 Six summary graphs. R^2 is an appropriate measure for a–c, but inappropriate for d–f.

4.4.2 Multiple Linear Regression

In multiple linear regression, R^2 can also be interpreted as the square of the correlation in a summary graph, this time of Y versus fitted values $\hat{Y}$. This plot can be interpreted exactly the same way as the plot of the response versus the single term in simple linear regression to decide on the usefulness of R^2 as a summary measure.

For other regression methods such as nonlinear regression, we can define R^2 to be the square of the correlation between the response and the fitted values, and use this summary graph to decide if R^2 is a useful summary.

4.4.3 Regression through the Origin

With regression through the origin, the proportion of variability explained is given by $1 - SSreg/\sum y_i^2$, using uncorrected sums of squares. This quantity is *not invariant under location change*, so, for example, if units are changed from Fahrenheit to Celsius, you will get a different value for the proportion of variability explained. For this reason, use of an R^2-like measure for regression through the origin is not recommended.

4.5 MISSING DATA

In many problems, some variables will be unrecorded for some cases. The methods we study in this book generally assume and require complete data, without any missing values. The literature on analyzing incomplete data problems is very large, and our goal here is more to point out the issues than to provide solutions. Two recent books on this topic are Little and Rubin (1987) and Schafer (1997).

4.5.1 Missing at Random

The most common solution to missing data problems is to delete either cases or variables so the resulting data set is complete. Many software packages delete partially missing cases by default, and fit regression models to the remaining, complete, cases. This is a reasonable approach as long as the fraction of cases deleted is small enough, and the cause of values being unobserved is unrelated to the relationships under study. This would include data lost through an accident like dropping a test tube, or making an illegible entry in a logbook. If the reason for not observing values depends on the values that would have been observed, then the analysis of data may require modeling the cause of the failure to observe values. For example, if values of a measurement are unrecorded if the value is less than the minimum detection limit of an instrument, then the value is missing because the value that should have been observed is too small. A simple expedient in this case that is sometimes helpful is to substitute a value less than or equal to the detection limit for the unobserved values. This expedient is not always entirely satisfactory because substituting, or imputing, a fixed value for the unobserved quantity can reduce the variation on the filled-in variable, and yield misleading inferences.

As a second example, suppose we have a clinical trial that enrolls subjects with a particular medical condition, assigns each subject a treatment, and then the subjects are followed for a period of time to observe their response, which may be time until a particular landmark occurs, such as improvement of the medical condition. Subjects who do not respond well to the treatment may drop out of the study early, while subjects who do well may be more likely to remain in the study. Since the probability of observing a value depends on the value that would have been observed, simply deleting subjects who drop out early can easily lead to incorrect inferences because the successful subjects will be overrepresented among those who complete the study.

In many clinical trials, the response variable is not observed because the study ends, not because of patient characteristics. In this case, we call the response times *censored*; so for each patient, we know either the time to the landmark or the time to censoring. This is a different type of missing data problem, and analysis needs to include both the uncensored and censored observations. Book-length treatments of censored survival data are given by Kalbfleisch and Prentice (1980) and Cox and Oakes (1984), among others.

As a final example, consider a cross-cultural demographic study. Some demographic variables are harder to measure than others, and some variables, such as the rate of employment for women over the age of 15, may not be available for less-developed countries. Deleting countries that do not have this variable measured could change the population that is studied by excluding less-developed countries.

Rubin (1976) defined data to be *missing at random* (MAR) if the failure to observe a value does not depend on the value that would have been observed. With MAR data, case deletion can be a useful option. Determining whether an assumption of MAR is appropriate for a particular data set is an important step in the analysis of incomplete data.

4.5.2 Alternatives

All the alternatives we briefly outline here require strong assumptions concerning the data that may be impossible to check in practice.

Suppose first that we combine the response and predictors into a single vector Z. We assume that the distribution of Z is fully known, apart from unknown parameters. The simplest assumption is that $Z \sim N(\mu, \Sigma)$. If we had reasonable estimates of μ and Σ, then we could use (4.15) to estimate parameters for the regression of the response on the other terms. The *EM algorithm* (Dempster, Laird, and Rubin , 1977) is a computational method that is used to estimate the parameters of the known joint distribution based on data with missing values.

Alternatively, given a model for the data like multivariate normality, one could impute values for the missing data and then analyze the completed data as if it were fully observed. *Multiple imputation* carries this one step further by creating several imputed data sets that, according to the model used, are plausible, filled-in data sets, and then "average" the analyses of the filled-in data sets. Software for both imputation and the EM algorithm for maximum likelihood estimate is available in several standard statistical packages, including the "missing" package in S-plus and the "MI" procedure in SAS.

The third approach is more comprehensive, as it requires building a model for the process of interest and the missing data process simultaneously. Examples of this approach are given by Ibrahim, Lipsitz, and Horton (2001), and Tang, Little, and Raghunathan (2003).

The data described in Table 4.2 provides an example. Allison and Cicchetti (1976) presented data on sleep patterns of 62 mammal species along with several other possible predictors of sleep. The data were in turn compiled from several other sources, and not all values are measured for all species. For example, *PS*, the number of hours of paradoxical sleep, was measured for only 50 of the 62 species in the data set, and *GP*, the gestation period, was measured for only 58 of the species. If we are interested in the dependence of hours of sleep on the other predictors, then we have at least three possible responses, *PS*, *SWS*, and *TS*, all observed on only a subset of the species. To use case deletion and then standard methods to analyze the conditional distributions of interest, we need to assume that the chance of a value being missing does not depend on the value. For example, the four missing values of *GP* are missing because no one had (as of 1976) published this value for these species. Using the imputation or the maximum likelihood methods are alternatives for these data, but they require making assumptions like normality, which might be palatable for many of the variables if transformed to logarithmic scale. Some of the variables, like *P* and *SE* are categorical, so other assumptions beyond multivariate normality might be needed.

TABLE 4.2   The Sleep Data[a]

Variable	Type	Number Observed	Percent Missing	Description
BodyWt	Variate	62	0	Body weight in kg
BrainWt	Variate	62	0	Brain weight in g
D	Factor	62	0	Danger index, 1 = least danger, ..., 5 = most
GP	Variate	58	6	Gestation time, days
Life	Variate	58	6	Maximum life span, years
P	Factor	62	0	Predation index, 1 = lowest , ... , 5 = highest
SE	Factor	62	0	Sleep exposure index, 1 = more exposed, ... , 5 = most protected
PS	Response	50	19	Paradoxical dreaming sleep, hrs/day
SWS	Response	48	23	Slow wave nondreaming sleep, hrs/day
TS	Response	58	6	Total sleep, hrs/day
Species	Labels	62	0	Species of mammal

[a] 10 variables, 62 observations, 8 patterns of missing values; 5 variables (50%) have at least one missing value; 20 observations (32%) have at least one missing value.

4.6 COMPUTATIONALLY INTENSIVE METHODS

Suppose we have a sample $y_1, \ldots, y_n$ from a particular distribution G, for example a standard normal distribution. What is a confidence interval for the population median?

We can obtain an approximate answer to this question by computer simulation, set up as follows:

1. Obtain a simulated random sample $y_1^*, \ldots, y_n^*$ from the known distribution G. Most statistical computing languages include functions for simulating random deviates (see Thisted, 1988 for computational methods).
2. Compute and save the median of the sample in step 1.
3. Repeat steps 1 and 2 a large number of times, say B times. The larger the value of B, the more precise the ultimate answer.
4. If we take $B = 999$, a simple *percentile-based* 95% confidence interval for the median is the interval between the 25th smallest value and the 975th largest value, which are the sample 2.5 and 97.5 percentiles, respectively.

In most interesting problems, we will not actually know G and so this simulation is not available. Efron (1979) pointed out that the observed data can be used to estimate G, and then we can sample from the estimate $\hat{G}$. The algorithm becomes:

1. Obtain a random sample $y_1^*, \ldots, y_n^*$ from $\hat{G}$ by sampling *with replacement* from the observed values $y_1, \ldots, y_n$. In particular, the i-th element of the sample y_i^* is equally likely to be any of the original $y_1, \ldots, y_n$. Some of the y_i will appear several times in the random sample, while others will not appear at all.
2. Continue with steps 2–4 of the first algorithm. A test at the 5% level concerning the population median can be rejected if the hypothesized value of the median does not fall in the confidence interval computed at step 4.

Efron called this method the *bootstrap*, and we call B the number of bootstrap samples. Excellent references for the bootstrap are the books by Efron and Tibshirani (1993), and Davison and Hinkley (1997).

4.6.1 Regression Inference without Normality

Bootstrap methods can be applied in more complex problems like regression. Inferences and accurate standard errors for parameters and mean functions require either normality of regression errors or large sample sizes. In small samples without normality, standard inference methods can be misleading, and in these cases a bootstrap can be used for inference.

Transactions Data

The data in this example consists of a sample of branches of a large Australian bank (Cunningham and Heathcote, 1989). Each branch makes transactions of two types, and for each of the branches we have recorded the number of transactions T_1 and T_2, as well as *Time*, the total number of minutes of labor used by the branch in type 1 and type 2 transactions. If β_j is the average number of minutes for a transaction of type j, $j = 1, 2$, then the total number of minutes in a branch for transaction type j is $\beta_j T_j$, and the total number of minutes is expected to be

$$\text{E}(Time|T_1, T_2) = \beta_0 + \beta_1 T_1 + \beta_2 T_2 \tag{4.16}$$

possibly with $\beta_0 = 0$ because zero transactions should imply zero time spent. The data are displayed in Figure 4.4, and are given in the data file `transact.txt`. The key features of the scatterplot matrix are: (1) the marginal response plots in the last row appear to have reasonably linear mean functions; (2) there appear to be a number of branches with no T_1 transactions but many T_2 transactions; and (3) in the plot of *Time* versus T_2, variability appears to increase from left to right.

The errors in this problem probably have a skewed distribution. Occasional transactions take a very long time, but since transaction time is bounded below by

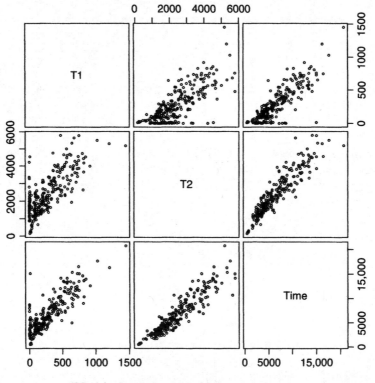

FIG. 4.4 Scatterplot matrix for the transactions data.

TABLE 4.3 Summary for $B = 999$ Case Bootstraps for the Transactions Data, Giving 95% Confidence Intervals, Lower to Upper, Based on Standard Normal Theory and on the Percentile Bootstrap

	Normal Theory			Bootstrap		
	Estimate	Lower	Upper	Estimate	Lower	Upper
Intercept	144.37	−191.47	480.21	136.09	−254.73	523.36
T_1	5.46	4.61	6.32	5.48	4.08	6.77
T_2	2.03	1.85	2.22	2.04	1.74	2.36

zero, there can not be any really extreme "quick" transactions. Inferences based on normal theory are therefore questionable.

Following the suggestion of Pardoe and Weisberg (2001) for this example, a bootstrap is computed as follows:

1. Number the cases in the data set from 1 to n. Take a random sample *with replacement* of size n from these case numbers. Thus, the i-th case number in the sample is equally likely to be any of the n cases in the original data.

2. Create a data set from the original data, but repeating each row in the data set the number of times that row was selected in the random sample in step 1. Some cases will appear several times and others will not appear at all. Compute the regression using this data set, and save the values of the coefficient estimates.

3. Repeat steps 1 and 2 a large number of times, say, B times.

4. Estimate a 95% confidence interval for each of the estimates by the 2.5 and 97.5 percentiles of the sample of B bootstrap samples.

Table 4.3 summarizes the percentile bootstrap for the transactions data. The column marked Estimate gives the OLS estimate under "Normal theory" and the average of the B bootstrap simulations under "Bootstrap." The difference between these two is called the *bootstrap bias*, which is quite small for all three terms relative to the size of the confidence intervals. The 95% bootstrap intervals are consistently wider than the corresponding normal intervals, indicating that the normal-theory confidence intervals are probably overly optimistic. The bootstrap intervals given in Table 4.3 are random, since if the bootstrap is repeated, the answers will be a little different. The variability in the end-points of the interval can be decreased by increasing the number B of bootstrap samples.

4.6.2 Nonlinear Functions of Parameters

One of the important uses of the bootstrap is to get estimates of error variability in problems where standard theory is either missing, or, equally often, unknown to the analyst. Suppose, for example, we wanted to get a confidence interval for the ratio β_1/β_2 in the transactions data. This is the ratio of the time for a type 1 transaction to the time for a type 2 transaction. The point estimate for this ratio is just $\hat{\beta}_1/\hat{\beta}_2$, but we will not learn how to get a normal-theory confidence interval for a nonlinear function of parameters like this until Section 6.1.2. Using the bootstrap, this computation is easy: just compute the ratio in each of the bootstrap samples and then use the percentiles of the bootstrap distribution to get the confidence interval. For these data, the point estimate is 2.68 with 95% bootstrap confidence interval from 1.76 to 3.86, so with 95% confidence, type 1 transactions take on average from about 1.76 to 3.86 times as long as do type 2 transactions.

4.6.3 Predictors Measured with Error

Predictors and the response are often measured with error. While we might have a theory that tells us the mean function for the response, given the true values of the predictors, we must fit with the response, given the imperfectly measured values of the predictors. We can sometimes use simulation to understand how the measurement error affects our answers.

Here is the basic setup. We have a true response Y^* and a set of terms X^* and a true mean function

$$E(Y^*|X^* = \mathbf{x}^*) = \boldsymbol{\beta}'\mathbf{x}^*$$

In place of Y^* and X^* we observe $Y = Y^* + \delta$ and $X = X^* + \eta$, where δ and η are measurement errors. If we fit the mean function

$$E(Y|X = \mathbf{x}) = \boldsymbol{\gamma}'\mathbf{x}$$

what can we say about the relationship between $\boldsymbol{\beta}$ and $\boldsymbol{\gamma}$? While there is a substantial theoretical literature on this problem (for example, Fuller, 1987), we shall attempt to get an answer to this question using simulation. To do so, we need to know something about δ and η.

Catchability of Northern Pike

One of the questions of interest to fisheries managers is the difficulty of catching a fish. A useful concept is the idea of *catchability*. Suppose that Y^* is the catch for an angler for a fixed amount of effort, and X^* is the abundance of fish available in the population that the angler is fishing. Suppose further that

$$E(Y^*|X^* = x^*) = \beta_1 x^* \tag{4.17}$$

If this mean function were to hold, then we could define β_1 to be the catchability of this particular fish species.

The data we use comes from a study of Northern Pike, a popular game fish in inland lakes in the United States. Data were collected on 16 lakes by Rob Pierce of the Minnesota Department of Natural Resources. On each lake we have a measurement called *CPUE* or catch per unit effort, which is the catch for a specific amount of fishing effort. Abundance on the lake is measured using the fish *Density* that is defined to be the number of fish in the lake divided by the surface area of the lake. While surface area can be determined with reasonable accuracy, the number of fish in the lake is estimated using a capture–recapture experiment (Seber, 2002). Since both *CPUE* and *Density* are experimentally estimated, they both have standard errors attached to them. In terms of (4.17), we have observed $CPUE = Y^* + \delta$ and $Density = x^* + \eta$. In addition, we can obtain estimates of the standard deviations of the δs and ηs from the properties of the methods used to find *CPUE* and *Density*. The data file `npdata.txt` includes both the *CPUE* and *Density* and their standard errors *SECPUE* and *SEdens*.

Figure 4.5 is the plot of the estimated *CPUE* and *Density*. Ignoring the lines on the graph, a key characteristic of this graph is the large variability in the points. A straight line mean function seems plausible for these data, but many other curves are equally plausible. We continue under the assumption that a straight-line mean function is sensible.

The two lines on Figure 4.5 are the OLS simple regression fits through the origin (solid line) and not through the origin (dashed line). The *F*-test comparing them has a *p*-value of about 0.13, so we are encouraged to use the simpler through-the-origin model that will allow us to interpret the slope as the catchability. The estimate is $\hat{\beta}_1 = 0.34$ with standard error 0.035, so a 95% confidence interval for β_1 ignoring measurement errors is $(0.250, 0.0399)$.

To assess the effect of measurement error on the estimate and on the confidence interval, we first make some assumptions. First, we suppose that the estimated standard errors of the measurements are the actual standard errors of the measurements. Second, we assume that the measurement errors are independently and normally

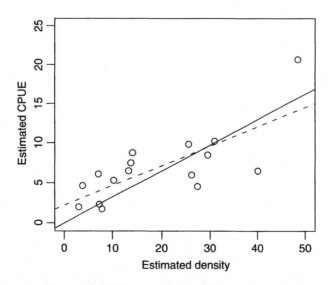

FIG. 4.5 Scatterplot of estimated *CPUE* versus *Density* for the northern pike data. Solid line is the OLS mean function through the origin, and the dashed line is the OLS line allowing an intercept.

TABLE 4.4 Simulation Summary for the Northern Pike Data

	Point Estimate	95% ConfidenceInterval
Normal theory	0.324	(0.250, 0.399)
Simulation	0.309	(0.230, 0.387)

distributed. Neither of these assumptions are checkable from these data, but for the purposes of a simulation these seem like reasonable assumptions.

The simulation proceeds as follows:

1. Generate a pseudo-response vector given by $\tilde{Y} = CPUE + \tilde{\delta}$, where the i-th element of $\tilde{\delta}$ is a normal random number with mean zero and variance given by the square of the estimated standard error for the i-th $CPUE$ value. In this problem, each observation has its own estimated error variance, but in other problems there may be a common estimate for all elements of the response.

2. Repeat step 1, but for the predictor to get $\tilde{x} = Density + \tilde{\eta}$.

3. Fit the simple regression model of $\tilde{Y}$ on $\tilde{x}$ and save the estimated slope.

4. Repeat steps 1–3 B times. The average of the B values of the slope estimate is an estimate of the slope in the problem with no measurement error. A confidence interval for the slope is found using the percentile method discussed with the bootstrap.

The samples generated in steps 1–2 are not quite from the right distribution, as they are centered at the observed values of $CPUE$ and $Density$ rather than the unobserved values of Y^* and x^*, but the observed values estimate the unobserved true values, so this substitution adds variability to the results, but does not affect the validity of the methodology.

The results for $B = 999$ simulations are summarized in Table 4.4. The results of the normal theory and the simulation that allows for measurement error are remarkably similar. In this problem, we judge the measurement error to be unimportant.

PROBLEMS

4.1. Fit the regression of *Soma* on *AVE*, *LIN* and *QUAD* as defined in Section 4.1 for the girls in the Berkeley Guidance Study data, and compare to the results in Section 4.1.

4.2.

    4.2.1. Starting with (4.10), we can write

$$y_i = \mu_y + \rho_{xy}\frac{\sigma_y}{\sigma_x}(x_i - \mu_x) + \varepsilon_i$$

Ignoring the error term ε_i, solve this equation for x_i as a function of y_i and the parameters.

4.2.2. Find the conditional distribution of $x_i | y_i$. Under what conditions is the equation you obtained in Problem 4.2.1, which is computed by inverting the regression of y on x, the same as the regression of x on y?

4.3. For the transactions data described in Section 4.6.1, define $A = (T_1 + T_2)/2$ to be the average transaction time, and $D = T_1 - T_2$, and fit the following four mean functions

$$M1 : E(Y|T_1, T_2) = \beta_{01} + \beta_{11}T_1 + \beta_{21}T_2$$

$$M2 : E(Y|T_1, T_2) = \beta_{02} + \beta_{32}A + \beta_{42}D$$

$$M3 : E(Y|T_1, T_2) = \beta_{03} + \beta_{23}T_2 + \boxed{\beta_{43}D}$$

$$M4 : E(Y|T_1, T_2) = \beta_{04} + \beta_{14}T_1 + \beta_{24}T_2 + \beta_{34}A + \beta_{44}D$$

4.3.1. In the fit of M4, some of the coefficients estimates are labelled as either "aliased" or as missing. Explain what this means.

4.3.2. What aspects of the fitted regressions are the same? What is different?

4.3.3. Why is the estimate for T_2 different in M1 and M3?

4.4. Interpreting coefficients with logarithms

4.4.1. For the simple regression with mean function $E(\log(Y)|X = x) = \beta_0 + \beta_1 \log(x)$, provide an interpretation for β_1 as a rate of change in Y for a small change in x.

4.4.2. Show that the results of Section 4.1.7 do not depend on the base of the logarithms.

4.5. Use the bootstrap to estimate confidence intervals of the coefficients in the fuel data.

4.6. Windmill data For the windmill data in the data file wm1.txt discussed in Problem 2.13, page 45, use $B = 999$ replications of the bootstrap to estimate a 95% confidence interval for the long-term average wind speed at the candidate site and compare this to the prediction interval in Problem 2.13.5. See the comment at the end of Problem 2.13.4 to justify using a bootstrap confidence interval for the mean as a prediction interval for the long-term mean.

4.7. Suppose we fit a regression with the true mean function

$$E(Y|X_1 = x_1, X_2 = x_2) = 3 + 4x_1 + 2x_2$$

Provide conditions under which the mean function for $E(Y|X_1 = x_1)$ is linear but has a negative coefficient for x_1.

4.8. In a study of faculty salaries in a small college in the Midwest, a linear regression model was fit, giving the fitted mean function

$$E(\widehat{Salary|Sex}) = 24697 - 3340Sex \qquad (4.18)$$

where *Sex* equals one if the faculty member was female and zero if male. The response *Salary* is measured in dollars (the data are from the 1970s).

4.8.1. Give a sentence that describes the meaning of the two estimated coefficients.

4.8.2. An alternative mean function fit to these data with an additional term, *Years*, the number of years employed at this college, gives the estimated mean function

$$E(\widehat{Salary|Sex}, Years) = 18065 + 201Sex + 759Years \qquad (4.19)$$

The important difference between these two mean functions is that the coefficient for *Sex* has changed signs. Using the results of this chapter, explain how this could happen. (Data consistent with these equations are presented in Problem 6.13).

4.9. Sleep data

4.9.1. For the sleep data described in Section 4.5, describe conditions under which the missing at random assumption is reasonable. In this case, deleting the partially observed species and analyzing the complete data can make sense.

4.9.2. Describe conditions under which the missing at random assumption for the sleep data is not reasonable. In this case, deleting partially observed species can change the inferences by changing the definition of the sampled population.

4.9.3. Suppose that the sleep data were fully observed, meaning that values for all the variables were available for all 62 species. Assuming that there are more than 62 species of mammals, provide a situation where examining the missing at random assumption could still be important.

4.10. The data given in `longley.txt` were first given by Longley (1967) to demonstrate inadequacies of regression computer programs then available. The variables are:

Def = GNP price deflator, in percent

GNP = GNP, in millions of dollars

Unemployed = Unemployment, in thousands of persons

Armed.Forces = Size of armed forces, in thousands

Population = Population 14 years of age and over, in thousands

Employed = Total derived employment in thousands the response

Year = Year

4.10.1. Draw the scatterplot matrix for these data excluding *Year*, and explain from the plot why this might be a good example to illustrate numerical problems of regression programs. (*Hint*: Numerical problems arise through rounding errors, and these are most likely to occur when terms in the regression model are very highly correlated.)

4.10.2. Fit the regression of *Employed* on the others excluding *Year*.

4.10.3. Suppose that the values given in this example were only accurate to three significant figures (two figures for *Def*). The effects of measurement errors can be assessed using a simulation study in which we add uniform random values to the observed values, and recompute estimates for each simulation. For example, *Unemp* for 1947 is given as 2356, which corresponds to 2,356,000. If we assume only three significant figures, we only believe the first three digits. In the simulation we would replace 2356 by 2356 + *u*, where *u* is a uniform random number between −5 and +5. Repeat the simulation 1000 times, and on each simulation compute the coefficient estimates. Compare the standard deviation of the coefficient estimates from the simulation to the coefficient standard errors from the regression on the unperturbed data. If the standard deviations in the simulation are as large or larger than the standard errors, we would have evidence that rounding would have important impact on results.

CHAPTER 5

Weights, Lack of Fit, and More

This chapter introduces a number of additional basic tools for fitting and using multiple linear regression models.

5.1 WEIGHTED LEAST SQUARES

The assumption that the variance function $\text{Var}(Y|X)$ is the same for all values of the terms X can be relaxed in a number of ways. In the simplest case, suppose we have the multiple regression mean function given for the ith case by

$$E(Y|X = \mathbf{x}_i) = \boldsymbol{\beta}'\mathbf{x}_i \tag{5.1}$$

but rather than assume that errors are constant, we assume that

$$\text{Var}(Y|X = \mathbf{x}_i) = \text{Var}(e_i) = \sigma^2/w_i \tag{5.2}$$

where $w_1, \ldots, w_n$ are *known positive numbers*. The variance function is still characterized by only one unknown positive number σ^2, but the variances can be different for each case. This will lead to the use of *weighted least squares*, or WLS, in place of OLS, to get estimates.

In matrix terms, let $\mathbf{W}$ be an $n \times n$ diagonal matrix with the w_i on the diagonal. The model we now use is

$$\mathbf{Y} = \mathbf{X}\boldsymbol{\beta} + \mathbf{e} \qquad \text{Var}(\mathbf{e}) = \sigma^2\mathbf{W}^{-1} \tag{5.3}$$

We will continue to use the symbol $\hat{\boldsymbol{\beta}}$ for the estimator of $\boldsymbol{\beta}$, even though the estimate will be obtained via weighted, not ordinary, least squares. The estimator $\hat{\boldsymbol{\beta}}$ is chosen to minimize the weighted residual sum of squares function,

$$RSS(\boldsymbol{\beta}) = (\mathbf{Y} - \mathbf{X}\boldsymbol{\beta})'\mathbf{W}(\mathbf{Y} - \mathbf{X}\boldsymbol{\beta}) \tag{5.4}$$

$$= \sum w_i(y_i - \mathbf{x}_i'\boldsymbol{\beta})^2 \tag{5.5}$$

Applied Linear Regression, Third Edition, by Sanford Weisberg
ISBN 0-471-66379-4   Copyright © 2005 John Wiley & Sons, Inc.

The use of the weighted residual sum of squares recognizes that some of the errors are more variable than others since cases with large values of w_i will have small variance and will therefore be given more weight in the weighted *RSS*. The WLS estimator is given by

$$\hat{\beta} = (\mathbf{X'WX})^{-1}\mathbf{X'WY} \tag{5.6}$$

While this last equation can be found directly, it is convenient to transform the problem specified by (5.3) to one that can be solved by OLS. Then, all of the results for OLS can be applied to WLS problems.

Let $\mathbf{W}^{1/2}$ be the $n \times n$ diagonal matrix with ith diagonal element $\sqrt{w_i}$, and so $\mathbf{W}^{-1/2}$ is a diagonal matrix with $1/\sqrt{w_i}$ on the diagonal, and $\mathbf{W}^{1/2}\mathbf{W}^{-1/2} = \mathbf{I}$. Using Appendix A.7 on random vectors, the covariance matrix of $\mathbf{W}^{1/2}\mathbf{e}$ is

$$
\begin{aligned}
\mathrm{Var}(\mathbf{W}^{1/2}\mathbf{e}) &= \mathbf{W}^{1/2}\mathrm{Var}(\mathbf{e})\mathbf{W}^{1/2} \\
&= \mathbf{W}^{1/2}(\sigma^2\mathbf{W}^{-1})\mathbf{W}^{1/2} \\
&= \mathbf{W}^{1/2}(\sigma^2\mathbf{W}^{-1/2}\mathbf{W}^{-1/2})\mathbf{W}^{1/2} \\
&= \sigma^2(\mathbf{W}^{1/2}\mathbf{W}^{-1/2})(\mathbf{W}^{-1/2}\mathbf{W}^{1/2}) \\
&= \sigma^2\mathbf{I}
\end{aligned}
\tag{5.7}
$$

This means that the vector $\mathbf{W}^{1/2}\mathbf{e}$ is a random vector but with covariance matrix equal to σ^2 times the identity matrix. Multiplying both sides of equation (5.3) by $\mathbf{W}^{1/2}$ gives

$$\mathbf{W}^{1/2}\mathbf{Y} = \mathbf{W}^{1/2}\mathbf{X}\beta + \mathbf{W}^{1/2}\mathbf{e} \tag{5.8}$$

Define $\mathbf{Z} = \mathbf{W}^{1/2}\mathbf{Y}$, $\mathbf{M} = \mathbf{W}^{1/2}\mathbf{X}$, and $\mathbf{d} = \mathbf{W}^{1/2}\mathbf{e}$, and (5.8) becomes

$$\mathbf{Z} = \mathbf{M}\beta + \mathbf{d} \tag{5.9}$$

From (5.7), $\mathrm{Var}(\mathbf{d}) = \sigma^2\mathbf{I}$, and in (5.9) β is exactly the same as β in (5.3). Model (5.9) can be solved using OLS,

$$
\begin{aligned}
\hat{\beta} &= (\mathbf{M'M})^{-1}\mathbf{M'Z} \\
&= \left((\mathbf{W}^{1/2}\mathbf{X})'(\mathbf{W}^{1/2}\mathbf{X})\right)^{-1}(\mathbf{W}^{1/2}\mathbf{X})'(\mathbf{W}^{1/2}\mathbf{Y}) \\
&= \left(\mathbf{X'W}^{1/2}\mathbf{W}^{1/2}\mathbf{X}\right)^{-1}(\mathbf{X'W}^{1/2}\mathbf{W}^{1/2}\mathbf{Y}) \\
&= (\mathbf{X'WX})^{-1}(\mathbf{X'WY})
\end{aligned}
$$

which is the estimator given at (5.6).

To summarize, the WLS regression of **Y** on **X** with weights given by the diagonal elements of **W** is the same as the OLS regression of **Z** on **M**, where

$$
\mathbf{M} = \begin{pmatrix} \sqrt{w_1} & \sqrt{w_1}x_{11} & \cdots & \sqrt{w_1}x_{1p} \\ \sqrt{w_2} & \sqrt{w_2}x_{21} & \cdots & \sqrt{w_2}x_{2p} \\ \vdots & \vdots & \vdots & \vdots \\ \sqrt{w_n} & \sqrt{w_n}x_{n1} & \cdots & \sqrt{w_n}x_{np} \end{pmatrix} \quad \mathbf{Z} = \begin{pmatrix} \sqrt{w_1}y_1 \\ \sqrt{w_2}y_2 \\ \vdots \\ \sqrt{w_n}y_n \end{pmatrix}
$$

Even the column of ones gets multiplied by the $\sqrt{w_i}$. The regression problem is then solved using **M** and **Z** in place of **X** and **Y**.

5.1.1 Applications of Weighted Least Squares

Known weights w_i can occur in many ways. If the ith response is an average of n_i equally variable observations, then $\text{Var}(y_i) = \sigma^2/n_i$, and $w_i = n_i$. If y_i is a total of n_i observations, $\text{Var}(y_i) = n_i\sigma^2$, and $w_i = 1/n_i$. If variance is proportional to some predictor x_i, $\text{Var}(y_i) = x_i\sigma^2$, then $w_i = 1/x_i$.

Strong Interaction

The purpose of the experiment described here is to study the interactions of unstable elementary particles in collision with proton targets (Weisberg *et al.*, 1978). These particles interact via the so-called strong interaction force that holds nuclei together. Although the electromagnetic force is well understood, the strong interaction is somewhat mysterious, and this experiment was designed to test certain theories of the nature of the strong interaction.

The experiment was carried out with beam having various values of incident momentum, or equivalently for various values of s, the square of the total energy in the center-of-mass frame of reference system. For each value of s, we observe the *scattering cross-section y*, measured in millibarns (μb). A theoretical model of the strong interaction force predicts that

$$
E(y|s) = \beta_0 + \beta_1 s^{-1/2} + \text{relatively small terms} \tag{5.10}
$$

The theory makes quantitative predictions about β_0 and β_1 and their dependence on particular input and output particle type. Of interest, therefore, are: (1) estimation of β_0 and β_1, given (5.10) and (2) assessment of whether (5.10) provides an accurate description of the observed data.

The data given in Table 5.1 and in the file `physics.txt` summarize the results of experiments when both the input and output particle was the π^- meson. At each value of s, a very large number of particles was counted, and as a result the values of $\text{Var}(y|s = s_i) = \sigma^2/w_i$ are known almost exactly; the square roots of these values are given in the third column of Table 5.1, labelled SD_i. The variables in the file are labelled as x, y, and SD, corresponding to Table 5.1.

Ignoring the smaller terms, mean function (5.10) is a simple linear regression mean function with terms for an intercept and $x = s^{-1/2}$. We will need to use WLS,

TABLE 5.1   The Strong Interaction Data

$x = s^{-1/2}$	y (μb)	SD_i
0.345	367	17
0.287	311	9
0.251	295	9
0.225	268	7
0.207	253	7
0.186	239	6
0.161	220	6
0.132	213	6
0.084	193	5
0.060	192	5

TABLE 5.2   WLS **Estimates for the Strong Interaction Data**

```
Coefficients:
            Estimate Std. Error t value Pr(>|t|)
(Intercept) 148.473      8.079    18.38 7.91e-08
x           530.835     47.550    11.16 3.71e-06

Residual standard error: 1.657 on 8 degrees of freedom
Multiple R-Squared: 0.9397

Analysis of Variance Table
          Df Sum Sq Mean Sq F value    Pr(>F)
x          1 341.99  341.99  124.63 3.710e-06
Residuals  8  21.95    2.74
```

because the variances are not constant but are different for each value of s. In this problem, we are in the unusual situation that we not only know the weights *but also know the value of σ^2/w_i for each value of i.* There are 11 quantities $w_1, \ldots, w_{10}$ and σ^2 that describe the values of only 10 variances, so we have too many parameters, and we are free to specify one of the 11 parameters to be any nonzero value we like. The simplest approach is to set $\sigma^2 = 1$. If $\sigma^2 = 1$, then the last column of Table 5.1 gives $1/\sqrt{w_i}$, $i = 1, 2, \ldots, n$, and so the weights are just the inverse squares of the last column of this table.

The fit of the simple regression model via WLS is summarized in Table 5.2. R^2 is large, and the parameter estimates are well determined. The next question is whether (5.10) does in fact fit the data. This question of fit or lack of fit of a model is the subject of the next section.

5.1.2 Additional Comments

Many statistical models, including mixed effects, variance components, time series, and some econometric models, will specify that Var($\mathbf{e}$) = $\mathbf{\Sigma}$, where $\mathbf{\Sigma}$ is an $n \times n$ symmetric matrix that depends on a small number of parameters. Estimates for the *generalized least squares* problem minimize (5.4), with $\mathbf{W}$ replaced by $\mathbf{\Sigma}^{-1}$. Pinheiro and Bates (2000) is one recent source for discussion of these models.

If many observations are taken at each value of $\mathbf{x}$, the inverse of the sample variance of y given $\mathbf{x}$ can provide useful estimated weights. This method was used to get weights in the strong interaction data, where the number of cases per value of x was extremely large. Problem 5.6 provides another example using estimated weights as if they were true weights. The usefulness of this method depends on having a large sample size at each value of $\mathbf{x}$.

In some problems, $\text{Var}(Y|X)$ will depend on the mean $\text{E}(Y|X)$. For example, if the response is a count that follows a Poisson distribution, then $\text{Var}(Y|X) = \text{E}(Y|X)$, while if the response follows a gamma distribution, $\text{Var}(Y|X) = \sigma^2(\text{E}(Y|X))^2$. The traditional approach to fitting when the variance depends on the mean is to use a variance stabilizing transformation, as will be described in Section 8.3, in which the response is replaced by a transformation of it so that the variance function is approximately constant in the transformed scale.

Nelder and Wedderburn (1972) introduced *generalized linear models* that extend linear regression methodology to problems in which the variance function depends on the mean. One particular generalized linear model when the response is a binomial count leads to *logistic regression* and is discussed in Chapter 12. The other most important example of a generalized linear model is Poisson regression and log-linear models and is discussed by Agresti (1996). McCullagh and Nelder (1989) provide a general introduction to generalized linear models.

5.2 TESTING FOR LACK OF FIT, VARIANCE KNOWN

When the mean function used in fitting is correct, then the residual mean square $\hat{\sigma}^2$ provides an unbiased estimate of σ^2. If the mean function is wrong, then $\hat{\sigma}^2$ will estimate a quantity larger than σ^2, since its size will depend both on the errors and on systematic bias from fitting the wrong mean function. If σ^2 is known, or if an estimate of it is available that does not depend on the fitted mean function, a test for lack of fit of the model can be obtained by comparing $\hat{\sigma}^2$ to the model-free value. If $\hat{\sigma}^2$ is too large, we may have evidence that the mean function is wrong.

In the strong interaction data, we want to know if the straight-line mean function (5.10) is correct. As outlined in Section 5.1, the inverses of the squares of the values in column 3 of Table 5.1 are used as weights when we set $\sigma^2 = 1$, a known value. From Table 5.2, $\hat{\sigma}^2 = 2.744$. Evidence against the simple regression model will be obtained if we judge $\hat{\sigma}^2 = 2.744$ large when compared with the known value of $\sigma^2 = 1$. To assign a p-value to this comparison, we use the following result.

If the $e_i \sim \text{NID}(0, \sigma^2/w_i), i = 1, 2, \ldots, n$, with the w_i and σ^2 known, and parameters in the mean function are estimated using WLS, and the mean function is correct, then

$$X^2 = \frac{RSS}{\sigma^2} = \frac{(n - p')\hat{\sigma}^2}{\sigma^2} \tag{5.11}$$

is distributed as a chi-squared random variable with $n - p'$ df. As usual, RSS is the residual sum of squares. For the example, from Table 5.4,

$$X^2 = \frac{21.953}{1} = 21.953$$

Using a table or a computer program that computes quantiles of the χ^2 distribution, $\chi^2(0.01, 8) = 20.09$, so the p-value associated with the test is less than 0.01, which suggests that the straight-line mean function may not be adequate.

When this test indicates the lack of fit, it is usual to fit alternative mean functions either by transforming some of the terms or by adding polynomial terms in the predictors. The available physical theory suggests this latter approach, and the quadratic mean function

$$E(y|s) = \beta_0 + \beta_1 s^{-1/2} + \beta_2(s^{-1/2})^2 + \text{smaller terms}$$

$$\approx \beta_0 + \beta_1 x + \beta_2 x^2 \tag{5.12}$$

with $x = s^{-1/2}$ should be fit to the data. This mean function has three terms, an intercept, $x = s^{-1/2}$, and $x^2 = s^{-1}$. Fitting must use WLS with the same weights as before, as given in Table 5.3. The fitted curve for the quadratic fit is shown in Figure 5.1. The curve matches the data very closely. We can test for lack of fit of this model by computing

$$X^2 = \frac{RSS}{\sigma^2} = \frac{3.226}{1} = 3.226$$

Comparing this value with the percentage points of $\chi^2(7)$ gives a p-value of about 0.86, indicating no evidence of lack of fit for mean function (5.12).

TABLE 5.3 WLS **Estimates for the Quadratic Mean Function for the Strong Interaction Data**

```
Coefficients:
             Estimate Std. Error t value Pr(>|t|)
(Intercept)   183.830      6.459   28.46  1.7e-08
x               0.971     85.369    0.01  0.99124
x^2          1597.505    250.587    6.38  0.00038

Residual standard error: 0.679 on 7 degrees of freedom
Multiple R-Squared: 0.991

Analysis of Variance Table
Response: y
            Df  Sum Sq Mean Sq F value     Pr(>F)
x            1 341.991 341.991 742.1846 2.3030e-08
x^2          1  18.727  18.727  40.6413 0.00037612
Residuals    7   3.226   0.461
```

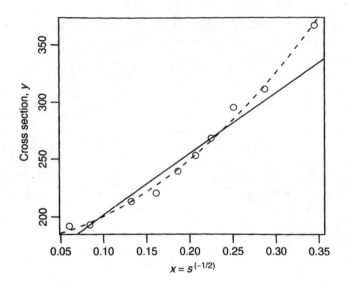

FIG. 5.1 Scatterplot for the strong interaction data. Solid line: simple linear regression mean function. Dashed line: quadratic mean function.

Although (5.10) does not describe the data, (5.12) does result in an adequate fit. Judgment of the success or failure of the model for the strong interaction force requires analysis of data for other choices of incidence and product particles, as well as the data analyzed here. On the basis of this further analysis, Weisberg et al. (1978) concluded that the theoretical model for the strong interaction force is consistent with the observed data.

5.3   TESTING FOR LACK OF FIT, VARIANCE UNKNOWN

When σ^2 is unknown, a test for lack of fit requires a model-free estimate of the variance. The most common model-free estimate makes use of the variation between cases with the same values on all of the predictors. For example, consider the artificial data with $n = 10$ given in Table 5.4. The data were generated by first choosing the values of x_i and then computing $y_i = 2.0 + 0.5x_i + e_i, i = 1, 2, \ldots, 10$, where the e_i are standard normal random deviates. If we consider only the values of y_i corresponding to $x = 1$, we can compute the average response $\bar{y}$ and the standard deviation with $3 - 1 = 2$ df, as shown in the table. If we assume that variance is the same for all values of x, a pooled estimate of the common variance is obtained by pooling the individual standard deviations into a single estimate. If n is the number of cases at a value of x and SD is the standard deviation for that value of x, then the sum of squares for pure error, symbolically SS_{pe}, is given by

$$SS_{pe} = \sum (n - 1)SD^2 \tag{5.13}$$

TABLE 5.4 An Illustration of the Computation of Pure Error

X	Y	$\bar{y}$	$(n-1)SD^2$	SD	df
1	2.55 ⎫				
1	2.75 ⎬	2.6233	0.0243	0.1102	2
1	2.57 ⎭				
2	2.40	2.4000	0	0	0
3	4.19 ⎫	4.4450	0.1301	0.3606	1
3	4.70 ⎭				
4	3.81 ⎫				
4	4.87 ⎬	4.0325	2.2041	0.8571	3
4	2.93 ⎪				
4	4.52 ⎭				
			2.3585		6

where the sum is over all groups of cases. For example, SS_{pe} is simply the sum of the numbers in the fourth column of Table 5.4,

$$SS_{pe} = 0.0243 + 0.0000 + 0.1301 + 2.2041 = 2.3585$$

Associated with SS_{pe} is its df, $df_{pe} = \sum(n-1) = 2 + 0 + 1 + 3 = 6$. The pooled, or *pure error* estimate of variance is $\hat{\sigma}_{pe} = SS_{pe}/df_{pe} = 0.3931$. This is the same estimate that would be obtained for the residual variance if the data were analyzed using a one-way analysis of variance, grouping according to the value of x. The pure error estimate of variance makes no reference to the linear regression mean function. It only uses the assumption that the residual variance is the same for each x.

Now suppose we fit a linear regression mean function to the data. The analysis of variance is given in Table 5.5. The residual mean square in Table 5.5 provides an estimate of σ^2, but this estimate depends on the mean function. Thus we have two estimates of σ^2, and if the latter is much larger than the former, the model is inadequate.

We can obtain an F-test if the residual sum of squares in Table 5.5 is divided into two parts, the sum of squares for pure error, as given in Table 5.4, and the remainder, called the *sum of squares for lack of fit*, or $SS_{lof} = RSS - SS_{pe} = 4.2166 - 2.3585 = 1.8581$ with $df = n - p' - df_{pe}$. The F-test is the ratio of the

TABLE 5.5 Analysis of Variance for the Data in Table 5.4

```
Analysis of Variance Table
              Df Sum Sq Mean Sq F value  Pr(>F)
Regression     1 4.5693  4.5693 11.6247 0.01433
Residuals      8 4.2166  0.5271
  Lack of fit  2 1.8582  0.9291  2.3638 0.17496
  Pure error   6 2.3584  0.3931
```

mean square for lack of fit to the mean square for pure error. The observed $F = 2.36$ is considerably smaller than $F(0.05; 2, 6) = 5.14$, suggesting no lack of fit of the model to these data.

Although all the examples in this section have a single predictor, the ideas used to get a model-free estimate of σ^2 are perfectly general. The pure-error estimate of variance is based on the sum of squares between the values of the response for all cases with the same values on all of the predictors.

Apple Shoots

Many types of trees produce two types of morphologically different shoots. Some branches remain vegetative year after year and contribute considerably to the size of the tree. Called long shoots or leaders, they may grow as much as 15 or 20 cm in one growing season. Short shoots, rarely exceeding 1 cm in total length, produce fruit. To complicate the issue further, long shoots occasionally change to short in a new growing season and vice versa. The mechanism that the tree uses to control the long and short shoots is not well understood.

Bland (1978) has done a descriptive study of the difference between long and short shoots of McIntosh apple trees. Using healthy trees of clonal stock planted in 1933 and 1934, he took samples of long and short shoots from the trees every few days throughout the 1971 growing season of about 106 days. The shoots sampled are presumed to be a sample of available shoots at the sampling dates. The sampled shoots were removed from the tree, marked and taken to the laboratory for analysis.

Among the many measurements taken, Bland counted the number of stem units in each shoot. The long and the short shoots could differ because of the number of stem units, the average size of stem units, or both. Bland's data is given in the data files longshoots.txt for data on long shoots, shortshoots.txt for data on short shoots, and allshoots.txt for both long and short shoots. We will consider only the long shoots, leaving the short shoots to the problems section.

Our goal is to find an equation that can adequately describe the relationship between Day = days from dormancy and Y = number of stem units. Lacking a theoretical form for this equation, we first examine Figure 5.2, a scatterplot of average number of stem units versus Day. The apparent linearity of this plot should encourage us to fit a straight-line mean function,

$$E(Y|Day) = \beta_0 + \beta_1 Day \tag{5.14}$$

If this mean function were adequate, we would have the interesting result that the observed rate of production of stem units per day is constant over the growing season.

For each sampled day, Table 5.6 reports n = number of shoots sampled, $\bar{y}$ = average number of stem units on that day, and SD = within-day standard deviation. Assuming that residual variance is constant from day to day, we can do the regression in two ways.

On the basis of the summaries given, since $Var(\bar{y}|Day) = \sigma^2/n$, we must compute the WLS regression of $\bar{y}$ on Day with weights given by the values of n. This is

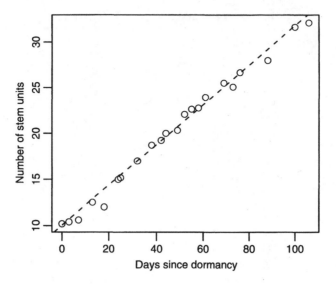

FIG. 5.2 Scatterplot for long shoots in the apple shoot data.

summarized in Table 5.7a. Alternatively, if the original 189 data points were available, we could compute the unweighted regression of the original data on *Day*. This is summarized in Table 5.7b. Both methods give identical intercept, slope, and regression sum of squares. They differ on any calculation that uses the residual sum of squares, because in Table 5.7b the residual sum of squares is the sum of SS_{pe} and SS_{lof}. For example, the standard errors of the coefficients in the two tables differ because in Table 5.7a the apparent estimate of variance is 3.7196 with 20 d.f., while in Table 5.7b it is 1.7621 with 187 df. Using pure error alone to estimate $\hat{\sigma}^2$ may be appropriate, especially if the model is doubtful; this would lead to a third set of standard errors. The SS_{pe} can be computed directly from Table 5.6 using (5.13), $SS_{pe} = \sum(n-1)SD^2 = 255.11$ with $\sum(n-1) = 167$ df. The *F*-test for lack of fit is $F = 2.43$. Since $F(0.01; 20, 167) = 1.99$, the *p*-value for this test is less than 0.01, indicating that the straight-line mean function (5.14) does not appear to be adequate. However, an *F*-test with this many df is very powerful and will detect very small deviations from the null hypothesis. Thus, while the result here is statistically significant, it may not be scientifically important, and for purposes of describing the growth of apple shoots, the mean function (5.14) may be adequate.

5.4 GENERAL *F* TESTING

We have encountered several situations that lead to computation of a statistic that has an *F* distribution when a null hypothesis (NH) and normality hold. The theory for the *F*-tests is quite general. In the basic structure, a smaller mean function of the null hypothesis is compared with a larger mean function of the alternative

TABLE 5.6　Bland's Data for Long and Short Apple Shoots[a]

	Long Shoots					Short Shoots			
Day	n	$\overline{y}$	SD	Len	Day	n	$\overline{y}$	SD	Len
0	5	10.200	0.830	1	0	5	10.000	0.000	0
3	5	10.400	0.540	1	6	5	11.000	0.720	0
7	5	10.600	0.540	1	9	5	10.000	0.720	0
13	6	12.500	0.830	1	19	11	13.360	1.030	0
18	5	12.000	1.410	1	27	7	14.290	0.950	0
24	4	15.000	0.820	1	30	8	14.500	1.190	0
25	6	15.170	0.760	1	32	8	15.380	0.510	0
32	5	17.000	0.720	1	34	5	16.600	0.890	0
38	7	18.710	0.740	1	36	6	15.500	0.540	0
42	9	19.220	0.840	1	38	7	16.860	1.350	0
44	10	20.000	1.260	1	40	4	17.500	0.580	0
49	19	20.320	1.000	1	42	3	17.330	1.520	0
52	14	22.070	1.200	1	44	8	18.000	0.760	0
55	11	22.640	1.760	1	48	22	18.460	0.750	0
58	9	22.780	0.840	1	50	7	17.710	0.950	0
61	14	23.930	1.160	1	55	24	19.420	0.780	0
69	10	25.500	0.980	1	58	15	20.600	0.620	0
73	12	25.080	1.940	1	61	12	21.000	0.730	0
76	9	26.670	1.230	1	64	15	22.330	0.890	0
88	7	28.000	1.010	1	67	10	22.200	0.790	0
100	10	31.670	1.420	1	75	14	23.860	1.090	0
106	7	32.140	2.280	1	79	12	24.420	1.000	0
					82	19	24.790	0.520	0
					85	5	25.000	1.010	0
					88	27	26.040	0.990	0
					91	5	26.600	0.540	0
					94	16	27.120	1.160	0
					97	12	26.830	0.590	0
					100	10	28.700	0.470	0
					106	15	29.130	1.740	0

[a] Len = 0 for short shoots and 1 for long shoots.

hypothesis (AH), and the smaller mean function can be obtained from the larger by setting some parameters in the larger mean function equal to zero, equal to each other, or equal to some specific value. One example previously encountered is testing to see if the last q terms in a mean function are needed after fitting the first $p' - q$. In matrix notation, partition $\mathbf{X} = (\mathbf{X}_1, \mathbf{X}_2)$, where $\mathbf{X}_1$ is $n \times (p' - q)$, $\mathbf{X}_2$ is $n \times q$, and partition $\boldsymbol{\beta}' = (\boldsymbol{\beta}_1', \boldsymbol{\beta}_2')$, where $\boldsymbol{\beta}_1$ is $(p' - q) \times 1$, $\boldsymbol{\beta}_2$ is $q \times 1$, so the two hypotheses in matrix terms are

$$\text{NH:} \quad \mathbf{Y} = \mathbf{X}_1 \boldsymbol{\beta}_1 + \mathbf{e}$$
$$\text{AH:} \quad \mathbf{Y} = \mathbf{X}_1 \boldsymbol{\beta}_1 + \mathbf{X}_2 \boldsymbol{\beta}_2 + \mathbf{e} \tag{5.15}$$

TABLE 5.7   Regression for Long Shoots in the Apple Data

(a) WLS regression using day means

```
            Estimate Std. Error t value Pr(>|t|)
(Intercept) 9.973754   0.314272   31.74   <2e-16
Day         0.217330   0.005339   40.71   <2e-16
```

Residual standard error: 1.929 on 20 degrees of freedom
Multiple R-Squared: 0.988

Analysis of Variance Table

	Df	Sum Sq	Mean Sq	F value	Pr(>F)
Day	1	6164.3	6164.3	1657.2	< 2.2e 16
Residuals	20	74.4	3.7		

(b) OLS regression of **y** on *Day*

```
            Estimate Std. Error t value Pr(>|t|)
(Intercept) 9.973754   0.21630    56.11   <2e-16
Day         0.217330   0.00367    59.12   <2e-16
```

Residual standard error: 1.762 on 187 degrees of freedom
Multiple R Squared: 0.949

Analysis of Variance Table

	Df	Sum Sq	Mean Sq	F value	Pr(>F)
Regression	1	6164.3	6164.3	1657.2	< 2.2e-16
Residual	187	329.5	1.8		
Lack of fit	20	74.4	3.7	2.43	0.0011
Pure error	167	255.1	1.5		

The smaller model is obtained from the larger by setting $\beta_2 = 0$.

The most general approach to computing the F-test is to fit two regressions. After fitting NH, find the residual sum of squares and its degrees of freedom RSS_{NH} and df_{NH}. Similarly, under the alternative mean function compute RSS_{AH} and df_{AH}. We must have $df_{NH} > df_{AH}$, since the alternative mean function has more parameters. Also, $RSS_{NH} - RSS_{AH} > 0$, since the fit of the AH must be at least as good as the fit of the NH. The F-test then gives evidence against NH if

$$F = \frac{(RSS_{NH} - RSS_{AH})/(df_{NH} - df_{AH})}{RSS_{AH}/df_{AH}} \quad (5.16)$$

is large when compared with the $F(df_{NH} - df_{AH}, df_{AH})$ distribution.

5.4.1   Non-null   Distributions

The numerator and denominator of (5.16) are independent of each other. Assuming normality, apart from the degrees of freedom, the denominator is distributed as σ^2 times a χ^2 random variable under both NH and AH. Ignoring the degrees of freedom, *when NH is true* the numerator is also distributed as σ^2 times a χ^2, so the ratio (5.16) has an F-distribution under NH because the F is defined to be the ratio of two independent χ^2 random variables, each divided by their degrees of freedom. When AH is true, apart from degrees of freedom the numerator is distributed as a σ^2 times a *noncentral* χ^2. In particular, the expected value of the numerator of (5.16) will be

$$E(\text{numerator of } (5.16)) = \sigma^2(1 + \text{noncentrality parameter}) \qquad (5.17)$$

For hypothesis (5.15), the noncentrality parameter is given by the expression

$$\frac{\beta_2'X_2'(I - X_1(X_1'X_1)^{-1}X_1')X_2\beta_2}{q\sigma^2} \qquad (5.18)$$

To help understand this, consider the special case of $X_2'X_2 = I$ and $X_1'X_2 = 0$ so the terms in X_2 are uncorrelated with each other and with the terms in X_1. Then (5.17) becomes

$$E(\text{numerator}) = \sigma^2 + \beta_2'\beta_2 \qquad (5.19)$$

For this special case, the expected value of the numerator of (5.16), and the power of the F-test, will be large if β_2 is large. In the general case where $X_1'X_2 \neq 0$, the results are more complicated, and the size of the noncentrality parameter, and power of the F-test, depend not only on σ^2 but also on the sample correlations between the variables in X_1 and those in X_2. If these correlations are large, then the power of F may be small even if $\beta_2'\beta_2$ is large. More general results on F-tests are presented in advanced linear model texts such as Seber (1977).

5.4.2 Additional Comments

The F distribution for (5.16) is exact if the errors are normally distributed, and in this case it is the *likelihood ratio test* for (5.15). The F-test is generally *robust* to departures from normality of the errors, meaning that estimates, tests, and confidence procedures are only modestly affected by modest departures from normality. In any case, when normality is in doubt, the bootstrap described in Section 4.6 can be used to get significance levels for (5.16).

5.5 JOINT CONFIDENCE REGIONS

Just as confidence intervals for a single parameter are based on the t distribution, confidence regions for several parameters will require use of an F distribution. The regions are elliptical.

The $(1 - \alpha) \times 100\%$ confidence region for β is the set of vectors β such that

$$\frac{(\beta - \hat{\beta})'(X'X)(\beta - \hat{\beta})}{p'\hat{\sigma}^2} \leq F(\alpha; p', n - p') \qquad (5.20)$$

The confidence region for $\boldsymbol{\beta}^*$, the parameter vector excluding β_0, is, using the notation of Chapter 3, the set of vectors $\boldsymbol{\beta}^*$ such that

$$\frac{(\boldsymbol{\beta}^* - \hat{\boldsymbol{\beta}}^*)'(\mathcal{X}'\mathcal{X})(\boldsymbol{\beta}^* - \hat{\boldsymbol{\beta}}^*)}{p\hat{\sigma}^2} \leq F(\alpha; p, n - p) \qquad (5.21)$$

The region (5.20) is a p'-dimensional ellipsoid centered at $\boldsymbol{\beta}$, while (5.21) is a p-dimensional ellipsoid centered at $\boldsymbol{\beta}^*$.

For example, the 95% confidence region for (β_1, β_2) in the regression of log(*Fertility*) on log(*PPgdp*) and *Purban* in the UN data is given in Figure 5.3. This ellipse is centered at $(-0.13, -0.0035)$. The orientation of the ellipse, the direction of the major axis, is negative, reflecting the negative correlation between the estimates of these two coefficients. The horizontal and vertical lines shown of the plot are the marginal 95% confidence intervals for each of the two coefficient estimates. From the graph, it is apparent that there are values of the coefficients that are in the 95% joint confidence region that would be viewed as implausible if we examined only the marginal intervals.

A slight generalization is needed to get a confidence ellipsoid for an arbitrary subset of $\boldsymbol{\beta}$. Suppose that $\boldsymbol{\beta}_1$ is a subvector of $\boldsymbol{\beta}$ with q elements. Let $\mathbf{S}$ be the $q \times q$ submatrix of $(\mathbf{X}'\mathbf{X})^{-1}$ corresponding to the q elements of $\boldsymbol{\beta}_1$. Then the 95% confidence region is the set of points $\boldsymbol{\beta}_l$ such that

$$\frac{(\boldsymbol{\beta}_1 - \hat{\boldsymbol{\beta}}_1)'\mathbf{S}^{-1}(\boldsymbol{\beta}_1 - \hat{\boldsymbol{\beta}}_1)}{q\hat{\sigma}^2} \leq F(\alpha; q, n - p') \qquad (5.22)$$

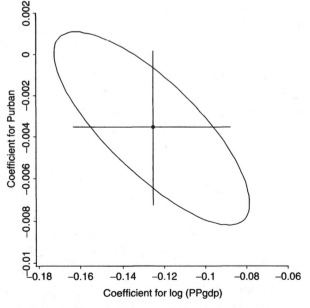

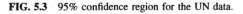

FIG. 5.3 95% confidence region for the UN data.

The bootstrap can also be used to get joint confidence regions by generalization of the method outlined for confidence intervals in Section 4.6, but the number of bootstrap replications B required is likely to be much larger than the number needed for intervals. The bootstrap confidence region would be the smallest set that includes $(1 - \alpha) \times 100\%$ of the bootstrap replications.

PROBLEMS

5.1. Galton's sweet peas Many of the ideas of regression first appeared in the work of Sir Francis Galton on the inheritance of characteristics from one generation to the next. In a paper on "Typical Laws of Heredity," delivered to the Royal Institution on February 9, 1877, Galton discussed some experiments on sweet peas. By comparing the sweet peas produced by parent plants to those produced by offspring plants, he could observe inheritance from one generation to the next. Galton categorized parent plants according to the typical diameter of the peas they produced. For seven size classes from 0.15 to 0.21 inches, he arranged for each of nine of his friends to grow 10 plants from seed in each size class; however, two of the crops were total failures. A summary of Galton's data was later published by Karl Pearson (1930) (see Table 5.8 and the data file `galtonpeas.txt`). Only average diameter and standard deviation of the offspring peas are given by Pearson; sample sizes are unknown.

5.1.1. Draw the scatterplot of *Progeny* versus *Parent*.

5.1.2. Assuming that the standard deviations given are population values, compute the weighted regression of *Progeny* on *Parent*. Draw the fitted mean function on your scatterplot.

5.1.3. Galton wanted to know if characteristics of the parent plant such as size were passed on to the offspring plants. In fitting the regression, a parameter value of $\beta_1 = 1$ would correspond to perfect inheritance, while $\beta_1 < 1$ would suggest that the offspring are "reverting" toward "what may be roughly and perhaps fairly described as the average ancestral type" (The substitution of "regression" for "reversion" was

TABLE 5.8   Galton's Peas Data

Parent Diameter (.01 in)	Progeny Diameter (.01 in)	SD
21	17.26	1.988
20	17.07	1.938
19	16.37	1.896
18	16.40	2.037
17	16.13	1.654
16	16.17	1.594
15	15.98	1.763

probably due to Galton in 1885). Test the hypothesis that $\beta_1 = 1$ versus the alternative that $\beta_1 < 1$.

5.1.4. In his experiments, Galton took the average size of all peas produced by a plant to determine the size class of the parental plant. Yet for seeds to represent that plant and produce offspring, Galton chose seeds that were as close to the overall average size as possible. Thus, for a small plant, the exceptional large seed was chosen as a representative, while larger, more robust plants were represented by relatively smaller seeds. What effects would you expect these experimental biases to have on (1) estimation of the intercept and slope and (2) estimates of error?

5.2. Apple shoots Apply the analysis of Section 5.3 to the data on short shoots in Table 5.6.

5.3. Nonparametric lack of fit The lack-of-fit tests in Sections 5.2–5.3 require either a known value for σ^2 or repeated observations for a given value of the predictor that can be used to obtain a model-free, or pure-error, estimate of σ^2. Loader (2004, Sec. 4.3) describes a lack-of-fit test that can be used without repeated observations or prior knowledge of σ^2 based on comparing the fit of the parametric model to the fit of a smoother. For illustration, consider Figure 5.4, which uses data that will be described later in this problem. For each data point, we can find the fitted value $\hat{y}_i$ from the parametric fit, which is just a point on the line, and $\tilde{y}_i$, the fitted value from the smoother, which is

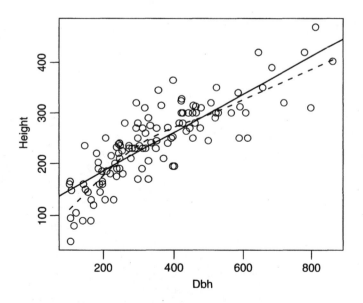

FIG. 5.4 *Height* versus *Dbh* for the Upper Flat Creek grand fir data. The solid line is the OLS fit. The dashed line is the *loess* fit with smoothing parameter 2/3.

a point on the dashed line. If the parametric model is appropriate for the data, then the differences $(\hat{y}_i - \tilde{y}_i)$ should all be relatively small. A suggested test statistic is based on looking at the squared differences and then dividing by an estimate of σ^2,

$$G = \frac{\sum_{i=1}^{n}(\hat{y}_i - \tilde{y}_i)^2}{\hat{\sigma}^2} \tag{5.23}$$

where $\hat{\sigma}^2$ is the estimate of variance from the parametric fit. Large values of G provide evidence against the NH that the parametric mean function matches the data. Loader (2004) provides an approximation to the distribution of G and also a bootstrap for computing an approximate significance level for a test based on G. In this problem, we will present the bootstrap.

5.3.1. The appropriate bootstrap algorithm is a little different from what we have seen before, and uses a *parametric bootstrap*. It works as follows:

1. Fit the parametric and smooth regression to the data, and compute G from (5.23). Save the residuals, $\hat{e}_i = y_i - \hat{y}_i$ from the parametric fit.

2. Obtain a bootstrap sample $\hat{e}_1^*, \ldots, \hat{e}_n^*$ by sampling with replacement from $\hat{e}_1, \ldots, \hat{e}_n$. Some residuals will appear in the sample many times, some not at all.

3. Given the bootstrap residuals, compute a bootstrap response $\mathbf{Y}^*$ with elements $y_i^* = \hat{y}_i + \hat{e}_i^*$. Use the original predictors unchanged in every bootstrap sample. Obtain the parametric and nonparametric fitted values with the response $\mathbf{Y}^*$, and then compute G from (5.23).

4. Repeat steps 2–3 B times, perhaps $B = 999$.

5. The significance level of the test is estimated to be the fraction of bootstrap samples that give a value of (5.23) that exceed the observed G.

The important problem of selecting a smoothing parameter for the smoother has been ignored. If the *loess* smoother is used, selecting the smoothing parameter to be 2/3 is a reasonable default, and statistical packages may include methods to choose a smoothing parameter. See Simonoff (1996), Bowman, and Azzalini (1997), and Loader (2004) for more discussion of this issue.

Write a computer program that implements this algorithm for regression with one predictor.

5.3.2. The data file ufcgf.txt gives the diameter *Dbh* in millimeters at 137 cm perpendicular to the bole, and the *Height* of the tree in decimeters for a sample of grand fir trees at Upper Flat Creek, Idaho, in 1991, courtesy of Andrew Robinson. Also included in the file are the *Plot* number, the *Tree* number in a plot, and the *Species* that is always "GF" for these data. Use the computer program you wrote in the last subproblem to test for lack of fit of the simple linear regression mean function for the regression of *Height* on *Dbh*.

5.4. An F-test In simple regression, derive an explicit formula for the F-test of

$$\text{NH:} \quad E(Y|X = x) = x \qquad (\beta_0 = 0, \beta_1 = 1)$$
$$\text{AH:} \quad E(Y|X = x) = \beta_0 + \beta_1 x$$

5.5. Snow geese Aerial surveys sometimes rely on visual methods to estimate the number of animals in an area. For example, to study snow geese in their summer range areas west of Hudson Bay in Canada, small aircraft were used to fly over the range, and when a flock of geese was spotted, an experienced person estimated the number of geese in the flock.

To investigate the reliability of this method of counting, an experiment was conducted in which an airplane carrying two observers flew over $n = 45$ flocks, and each observer made an independent estimate of the number of birds in each flock. Also, a photograph of the flock was taken so that a more or less exact count of the number of birds in the flock could be made. The resulting data are given in the data file `snowgeese.txt` (Cook and Jacobson, 1978). The three variables in the data set are *Photo* = photo count, *Obs1* = aerial count by observer one and *Obs2* = aerial count by observer 2.

5.5.1. Draw scatterplot matrix of three variables. Do these graphs suggest that a linear regression model might be appropriate for the regression of *Photo* on either of the observer counts, or on both of the observer counts? Why or why not? For the simple regression model of *Photo* on *Obs1*, what do the error terms measure? Why is it appropriate to fit the regression of *Photo* on *Obs1* rather than the regression of *Obs1* on *Photo*?

5.5.2. Compute the regression of *Photo* on *Obs1* using OLS, and test the hypothesis of Problem 5.4. State in words the meaning of this hypothesis and the result of the test. Is the observer reliable (you must define reliable)? Summarize your results.

5.5.3. Repeat Problem 5.5.2, except fit the regression of $Photo^{1/2}$ on $Obs1^{1/2}$. The square-root scale is used to stabilize the error variance.

5.5.4. Repeat Problem 5.5.2, except assume that the variance of an error is $Obs1 \times \sigma^2$.

5.5.5. Do both observers combined do a better job at predicting *Photo* than either predictor separately? To answer this question, you may wish to look at the regression of *Photo* on both *Obs1* and *Obs2*. Since from the scatterplot matrix the two terms are highly correlated, interpretation of results might be a bit hard. An alternative is to replace *Obs1* and *Obs2* by $Average = (Obs1 + Obs2)/2$ and $Diff = Obs1 - Obs2$. The new terms have the same information as the observer counts, but they are much less correlated. You might also need to consider using WLS.

As a result of this experiment, the practice of using visual counts of flock size to determine population estimates was discontinued in favor of using photographs.

TABLE 5.9 Jevons Gold Coinage Data

Age, Decades	Sample Size n	Average Weight	SD	Minimum Weight	Maximum Weight
1	123	7.9725	0.01409	7.900	7.999
2	78	7.9503	0.02272	7.892	7.993
3	32	7.9276	0.03426	7.848	7.984
4	17	7.8962	0.04057	7.827	7.965
5	24	7.873	0.05353	7.757	7.961

5.6. Jevons' gold coins The data in this example are deduced from a diagram in a paper written by W. Stanley Jevons (1868) and provided by Stephen M. Stigler. In a study of coinage, Jevons weighed 274 gold sovereigns that he had collected from circulation in Manchester, England. For each coin, he recorded the weight after cleaning to the nearest 0.001 g, and the date of issue. Table 5.9 lists the average, minimum, and maximum weight for each age class. The age classes are coded 1 to 5, roughly corresponding to the age of the coin in decades. The standard weight of a gold sovereign was supposed to be 7.9876 g; the minimum legal weight was 7.9379 g. The data are given the file `jevons.txt`.

5.6.1. Draw a scatterplot of *Weight* versus *Age*, and comment on the applicability of the usual assumptions of the linear regression model. Also draw a scatterplot of SD versus *Age*, and summarize the information in this plot.

5.6.2. Since the numbers of coins n in each age class are all fairly large, it is reasonable to pretend that the variance of coin weight for each *Age* is well approximated by SD^2, and hence Var(*Weight*) is given by SD^2/n. Compute the implied WLS regression.

5.6.3. Compute a lack-of-fit test for the linear regression model, and summarize results.

5.6.4. Is the fitted regression consistent with the known standard weight for a new coin?

5.6.5. For previously unsampled coins of $Age = 1, 2, 3, 4, 5$, estimate the probability that the weight of the coin is less than the legal minimum. (*Hints:* The standard error of prediction is a sum of two terms, the known variance of an unsampled coin of known *Age* and the estimated variance of the fitted value for that *Age*. You should use the normal distribution rather than a t to get the probabilities.)

5.7. The data file `physics1.txt` gives the results of the experiment described in Section 5.1.1, except in this case the input is the π^- meson as before, but the output is the π^+ meson.

Analyze these data following the analysis done in the text, and summarize your results.

CHAPTER 6

Polynomials and Factors

6.1 POLYNOMIAL REGRESSION

If a mean function with one predictor X is smooth but not straight, integer powers of the predictors can be used to approximate $E(Y|X)$. The simplest example of this is *quadratic regression*, in which the mean function is

$$E(Y|X = x) = \beta_0 + \beta_1 x + \beta_2 x^2 \tag{6.1}$$

Depending on the signs of the βs, a quadratic mean function can look like either of curves shown in Figure 6.1. Quadratic mean functions can therefore be used when the mean is expected to have a minimum or maximum in the range of the predictor. The minimum or maximum will occur for the value of X for which the derivative $dE(Y|X = x)/dx = 0$, which occurs at

$$x_M = -\beta_1/(2\beta_2) \tag{6.2}$$

x_M is estimated by substituting estimates for the βs into (6.2).

Quadratics can also be used when the mean function is curved but does not have a minimum or maximum within the range of the predictor. Referring to Figure 6.1a, if the range of X is between the dashed lines, then the mean function is everywhere increasing but not linear, while in Figure 6.1b it is decreasing but not linear.

Quadratic regression is an important special case of *polynomial regression*. With one predictor, the polynomial mean function of degree d is

$$E(Y|X) = \beta_0 + \beta_1 X + \beta_2 X^2 + \cdots + \beta_d X^d \tag{6.3}$$

If $d = 2$, the model is quadratic, $d = 3$ is cubic, and so on. Any smooth function can be estimated by a polynomial of high-enough degree, and polynomial mean functions are generally used as approximations and rarely represent a physical model.

Applied Linear Regression, Third Edition, by Sanford Weisberg
ISBN 0-471-66379-4   Copyright © 2005 John Wiley & Sons, Inc.

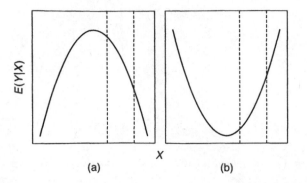

FIG. 6.1 Generic quadratic curves. A quadratic is the simplest curve that can approximate a mean function with a minimum or maximum within the range of possible values of the predictor. It can also be used to approximate some nonlinear functions without a minimum or maximum in the range of interest, possibly using the part of the curve between the dashed lines.

The mean function (6.3) can be fit via OLS with $p' = d + 1$ terms given by an intercept and $X, X^2, \ldots, X^d$. Any regression program can be used for fitting polynomials, but if d is larger than three, serious numerical problems may arise with some computer packages, and direct fitting of (6.3) can be unreliable. Some numerical accuracy can be retained by centering, using terms such as $Z_k = (X - \overline{x})^k, k = 1, \ldots, d$. Better methods using orthogonal polynomials are surveyed by Seber (1977, Chapter 8).

An example of quadratic regression has already been given with the physics data in Section 5.1.1. From Figure 5.1, page 102, a maximum value of the mean function does not occur within the range of the data, and we are in a situation like Figure 6.1b with the range of x between the dashed lines. The approximating mean function may be very accurate within the range of X observed in the data, but it may be very poor outside this range; see Problem 6.15.

In the physics example, a test for lack of fit that uses extra information about variances indicated that a straight-line mean function was not adequate for the data, while the test for lack of fit after the quadratic model indicated that this model was adequate. When a test for lack of fit is not available, comparison of the quadratic model

$$E(Y|X) = \beta_0 + \beta_1 X + \beta_2 X^2$$

to the simple linear regression model

$$E(Y|X) = \beta_0 + \beta_1 X$$

is usually based on a t-test of $\beta_2 = 0$. A simple strategy for choosing d is to continue adding terms to the mean function until the t-test for the highest-order term is nonsignificant. An elimination scheme can also be used, in which a maximum value of d is fixed, and terms are deleted from the mean function one at a time, starting with the highest-order term, until the highest-order remaining term has a significant t-value. Kennedy and Bancroft (1971) suggest using a significance level of about 0.10 for this procedure. In most applications of polynomial regression, only $d = 1$ or $d = 2$ are considered. For larger values of d, the fitted polynomial curves become wiggly, providing an increasingly better fit by matching the variation in the observed data more and more closely. The curve is then modeling the random variation rather than the overall shape of the relationship between variables.

6.1.1 Polynomials with Several Predictors

With more than one predictor, we can contemplate having integer powers and products of all the predictors as terms in the mean function. For example, for the important special case of two predictors the *second-order mean function* is given by

$$E(Y|X_1 = x_1, X_2 = x_2) = \beta_0 + \beta_1 x_1 + \beta_2 x_2 + \beta_{11} x_1^2 + \beta_{22} x_2^2 + \beta_{12} x_1 x_2 \quad (6.4)$$

The new term in (6.4) is the multiplicative term $x_1 x_2$ called an *interaction*. With k predictors, the second-order model includes an intercept, k linear terms, k quadratic terms, and $k(k + 1)/2$ interaction terms. If $k = 5$, the second-order mean function has 26 terms, and with $k = 10$, it has 76 terms. A usual strategy is to view the second-order model as consisting of too many terms and use testing or other selection strategies such as those to be outlined in Section 10.2.1 to delete terms for unneeded quadratics and interactions. We will provide an alternative approach in Section 6.4.

The most important new feature of the second-order model is the interaction. Return to the $k = 2$ predictor mean function (6.4). If x_1 is changed to $x_1 + \delta$, then the value of the mean function is

$$E(Y|X_1 = x_1 + \delta, X_2 = x_2) = \beta_0 + \beta_1 (x_1 + \delta) + \beta_2 x_2 + \beta_{11}(x_1 + \delta)^2$$
$$+ \beta_{22} x_2^2 + \beta_{12}(x_1 + \delta) x_2 \quad (6.5)$$

The change in the expected response is the difference between (6.5) and (6.4),

$$E(Y|X_1 = x_1 + \delta, X_2 = x_2) - E(Y|X_1 = x_1, X_2 = x_2)$$
$$= (\beta_{11}\delta^2 + \beta_1\delta) + 2\beta_{11}\delta x_1 + \beta_{12}\delta x_2 \quad (6.6)$$

If $\beta_{12} = 0$, the expected change is the same for every value of x_2. If $\beta_{12} \neq 0$, then $\beta_{12}\delta x_2$ will be different for each value of x_2, and so the effect of changing x_1 will depend on the value of x_2. Without the interaction, the effect of changing one predictor is the same for every value of the other predictor.

Cakes

Oehlert (2000, Example 19.3) provides data from a small experiment on baking packaged cake mixes. Two factors, X_1 = baking time in minutes and X_2 = baking temperature in degrees F, were varied in the experiment. The response Y was the average palatability score of four cakes baked at a given combination of (X_1, X_2), with higher values desirable. Figure 6.2 is a graphical representation of the experimental design, from which we see that the center point at $(35, 350)$ was replicated

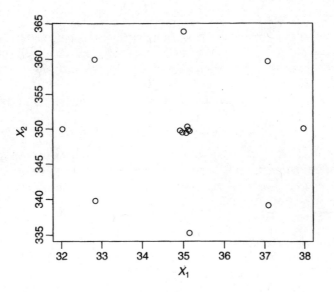

FIG. 6.2 Central composite design for the cake example. The center points have been slightly jittered to avoid overprinting.

six times. Replication allows for estimation of pure error and tests for lack of fit. The experiment consisted of $n = 14$ runs.

The estimated mean function based on (6.4) and using the data in the file cakes.txt is

$$E(Y|X_1, X_2) = -2204.4850 + 25.9176X_1 + 9.9183X_2$$
$$-0.1569X_1^2 - 0.0120X_2^2 - 0.0416X_1X_2 \qquad (6.7)$$

Each of the coefficient estimates, including both quadratics and the interaction, has significance level of 0.005 or less, so all terms are useful in the mean function (see Problem 6.1). Since each of X_1 and X_2 appears in three of the terms in (6.7), interpreting this mean function is virtually impossible without the aid of graphs. Figure 6.3 presents a useful way of summarizing the fitted mean function. In Figure 6.3a, the horizontal axis is the baking time X_1, and the vertical axis is the response Y. The three curves shown on the graph are obtained by fixing the value of temperature X_2 at either 340, 350, or 360, and substituting into (6.7). For example, when $X_2 = 350$, substitute 350 for X_2 in (6.7), and simplify to get

$$E(Y|X_1, \widehat{X_2} = 350) = \hat{\beta}_0 + \hat{\beta}_2(350) + \hat{\beta}_{22}(350)^2$$
$$+\hat{\beta}_1 X_1 + \hat{\beta}_{12}(350)X_1$$
$$+\hat{\beta}_{11}X_1^2 \qquad (6.8)$$
$$= -196.9664 + 11.3488X_1 - 0.1569X_1^2$$

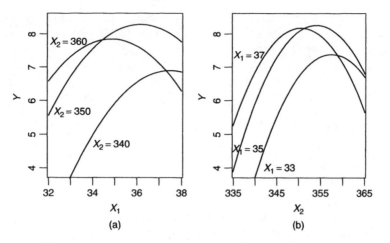

FIG. 6.3 Estimated response curves for the cakes data, based on (6.7).

Each of the lines within a plot is a quadratic curve, because both the X_1^2 and X_2^2 terms are in the mean function. Each of the curves has a somewhat different shape. For example, in Figure 6.3a, the baking time X_1 that maximizes the response is lower at $X_2 = 360$ degrees than it is at $X_2 = 340$ degrees. Similarly, we see from Figure 6.3b that the response curves are about the same for baking time of 35 or 37 minutes, but the response is lower at the shorter baking time. The palatability score is perhaps surprisingly sensitive to changes in temperature of 10 or 15 degrees and baking times of just a few minutes.

If we had fit the mean function (6.4), but with $\beta_{12} = 0$ so the interaction is absent, we would get the fitted response curves shown in Figure 6.4. Without the

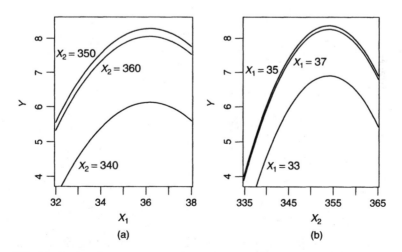

FIG. 6.4 Estimated response curves for the cakes data, based on fitting with $\beta_{12} = 0$.

interaction, all the curves within a plot have the same shape, and all are maximized at the same point. Without the interaction, we could say, for example, that for all temperatures the response is maximized for baking time of around 36 min, and for all baking times, the response is maximized for temperature around 355 degrees. While this mean function is simpler, F-testing would show that *it does not adequately match the data*, and so (6.8) and Figure 6.3 give appropriate summaries for these data.

6.1.2   Using the Delta Method to Estimate a Minimum or a Maximum

We have seen at (6.2) that the value of the predictor that will maximize or minimize a quadratic, depending on the signs of the βs, is $x_M = -\beta_1/(2\beta_2)$. This is a *nonlinear* combination of the βs, and so its estimate, $\hat{x}_M = -\hat{\beta}_1/(2\hat{\beta}_2)$ is a nonlinear combination of estimates. The *delta method* provides an approximate standard error of a nonlinear combination of estimates that is accurate in large samples. The derivation of the delta method, and possibly its use, requires elementary calculus.

We will use different notation for this derivation to emphasize that the results are much more general than just for ratios of coefficient estimates in multiple linear regression. Let θ be a $k \times 1$ parameter vector, with estimator $\hat{\theta}$ such that

$$\hat{\theta} \sim N(\theta, \sigma^2 \mathbf{D}) \tag{6.9}$$

where $\mathbf{D}$ is a known, positive definite, matrix. Equation (6.9) can be exact, as it is for the multiple linear regression model with normal errors, or asymptotically valid, as in nonlinear or generalized linear models. In some problems, σ^2 may be known, but in the multiple linear regression problem it is usually unknown and also estimated from data.

Suppose $g(\theta)$ is a nonlinear continuous function of θ that we would like to estimate. Suppose that θ^* is the true value of θ. To approximate $g(\hat{\theta})$, we can use a Taylor series expansion (see Section 11.1) about $g(\theta^*)$,

$$g(\hat{\theta}) = g(\theta^*) + \sum_{j=1}^{k} \frac{\partial g}{\partial \theta_j}(\hat{\theta}_j - \theta_j^*) + \text{small terms}$$

$$\approx g(\theta^*) + \dot{\mathbf{g}}(\theta^*)'(\hat{\theta} - \theta^*) \tag{6.10}$$

where we have defined

$$\dot{\mathbf{g}}(\theta^*) = \frac{\partial g}{\partial \theta} = \left(\frac{\partial g}{\partial \theta_1}, \dots, \frac{\partial g}{\partial \theta_k}\right)'$$

evaluated at θ^*. The vector $\dot{\mathbf{g}}$ has dimension $k \times 1$. We have expressed in (6.10) our estimate $g(\hat{\theta})$ as approximately a constant $g(\theta^*)$ plus a linear combination of data. The variance of a constant is zero, as is the covariance between a constant and a function of data. We can therefore approximate the variance of $g(\hat{\theta})$ by

$$
\begin{aligned}
\text{Var}(g(\hat{\theta})) &= \text{Var}(g(\theta^*)) + \text{Var}\left[\dot{\mathbf{g}}(\theta^*)'(\hat{\theta} - \theta^*)\right] \\
&= \dot{\mathbf{g}}(\theta^*)'\text{Var}(\hat{\theta})\dot{\mathbf{g}}(\theta^*) \\
&= \sigma^2\dot{\mathbf{g}}(\theta^*)'\mathbf{D}\dot{\mathbf{g}}(\theta^*)
\end{aligned} \tag{6.11}
$$

This equation is the heart of the delta method, so we will write it out again as a scalar equation. Let $\dot{g}_i$ be the i-th element of $\dot{\mathbf{g}}(\hat{\theta})$, so $\dot{g}_i$ is the partial derivative of $g(\theta)$ with respect to θ_i, and let d_{ij} be the (i, j)-element of the matrix $\mathbf{D}$. Then the estimated variance of $g(\hat{\theta})$ is

$$
\text{Var}(g(\hat{\theta})) = \sigma^2 \sum_{i=1}^{k}\sum_{j=1}^{k} \dot{g}_i\dot{g}_j d_{ij} \tag{6.12}
$$

In practice, all derivatives are evaluated at $\hat{\theta}$, and σ^2 is replaced by its estimate.

In large samples and under regularity conditions, $g(\hat{\theta})$ will be normally distributed with mean $g(\theta^*)$ and variance (6.11). In small samples, the normal approximation may be poor, and inference based on the bootstrap, Problem 6.16, might be preferable.

For quadratic regression (6.1), the minimum or maximum occurs at $g(\boldsymbol{\beta}) = -\beta_1/(2\beta_2)$, which is estimated by $g(\hat{\boldsymbol{\beta}})$. To apply the delta method, we need the partial derivative, evaluated at $\hat{\boldsymbol{\beta}}$,

$$
\left(\frac{\partial g}{\partial \boldsymbol{\beta}}\right)' = \left(0, -\frac{1}{2\hat{\beta}_2}, \frac{\hat{\beta}_1}{2\hat{\beta}_2^2}\right)
$$

Using (6.12), straightforward calculation gives

$$
\text{Var}(g(\hat{\boldsymbol{\beta}})) = \frac{1}{4\hat{\beta}_2^2}\left(\text{Var}(\hat{\beta}_1) + \frac{\hat{\beta}_1^2}{\hat{\beta}_2^2}\text{Var}(\hat{\beta}_2) - \frac{2\hat{\beta}_1}{\hat{\beta}_2}\text{Cov}(\hat{\beta}_1, \hat{\beta}_2)\right) \tag{6.13}
$$

The variances and covariances in (6.13) are elements of the matrix $\sigma^2(\mathbf{X'X})^{-1}$, and so the estimated variance is obtained from $\hat{\sigma}^2\mathbf{D} = \hat{\sigma}^2(\mathbf{X'X})^{-1}$.

As a modestly more complicated example, the estimated mean function for palatability for the cake data when the temperature is 350 degrees is given by (6.8). The estimated maximum palatability occurs when the baking time is

$$
\hat{x}_M = -\frac{\hat{\beta}_1 + \hat{\beta}_{12}(350)}{2\hat{\beta}_{11}} = 36.2 \text{ min}
$$

which depends on the estimate $\hat{\beta}_{12}$ for the interaction as well as on the linear and quadratic terms for X_1. The standard error from the delta method can be computed to be 0.4 minutes. If we can believe the normal approximation, a 95% confidence interval for x_M is $36.2 \pm 1.96 \times 0.4$ or about 35.4 to 37 min.

Writing a function for computing the delta method is not particularly hard using a language such as Maple, Mathematica, MatLab, R or S-plus that can do symbolic differentiation to get $\dot{g}$. If your package will not do the differentiation for you, then you can still compute the derivatives by hand and use (6.12) to get the estimated standard error. The estimated variance matrix $\hat{\sigma}^2(\mathbf{X'X})^{-1}$ is computed by all standard regression programs, although getting access to it may not be easy in all programs.

6.1.3 Fractional Polynomials

Most problems that use polynomials use only integer powers of the predictors as terms. Royston and Altman (1994) considered using fractional powers of predictors in addition to integer powers. This provides a wider class of mean functions that can be approximated using only a few terms and gives results similar to the results we will get in choosing a transformation in Section 7.1.1.

6.2 FACTORS

Factors allow the inclusion of qualitative or categorical predictors in the mean function of a multiple linear regression model. Factors can have two levels, such as male or female, treated or untreated, and so on, or they can have many levels, such as eye color, location, or many others.

To include factors in a multiple linear regression mean function, we need a way to indicate which particular level of the factor is present for each case in the data. For a factor with two levels, a *dummy variable*, which is a term that takes the value 1 for one of the categories and 0 for the other category, can be used. Assignment of labels to the values is generally arbitrary, and will not change the outcome of the analysis. Dummy variables can alternatively be defined with a different set of values, perhaps -1 and 1, or possibly 1 and 2. The only important point is the term has only two values.

As an example, we return to the sleep data described in Section 4.5. This is an observational study of the sleeping patterns of 62 mammal species. One of the response variables in the study is *TS*, the total hours of sleep per day. Consider here as an initial predictor the variable D, which is a categorical index to measure the overall *danger* of that species. D has five categories, with $D = 1$ indicating species facing the least danger from other animals, to $D = 5$ for species facing the most danger. Category labels here are the numbers 1, 2, 3, 4, and 5, but D is not a measured variable. We could have just as easily used names such as "lowest," "low," "middle," "high," and "highest" for these five category names. The data are in the file sleep1.txt. *TS* was not given for three of the species, so this analysis is based on the 59 species for which data are provided.

6.2.1 No Other Predictors

We begin this discussion by asking how the mean of TS varies as D changes from category to category. We would like to be able to write down a mean function that allows each level of D to have its own mean, and we do that with a set of dummy variables. Since D has five levels, the j-th dummy variable U_j for the factor, $j = 1, \ldots, 5$ has ith value u_{ij}, for $i = 1, \ldots, n$, given by

$$u_{ij} = \begin{cases} 1 & \text{if } D_i = j\text{th category of } D \\ 0 & \text{otherwise} \end{cases} \qquad (6.14)$$

If the factor had three levels rather than five, and the sample size $n = 7$ with cases 1, 2, and 7 at the first level of the factor, cases 4 and 5 at the second level, and cases 3 and 6 at the third level, then the three dummy variables would be

U_1	U_2	U_3
1	0	0
1	0	0
0	0	1
0	1	0
0	1	0
0	0	1
1	0	0

If these three dummy variables are added together, we will get a column of ones. This is an important characteristic of a set of dummy variables for a factor: their sum always adds up to the same value, usually one, for each case because each case has one and only one level of the factor.

Returning to the sleep data, we can write the mean function as

$$E(TS|D) = \beta_1 U_1 + \beta_2 U_2 + \beta_3 U_3 + \beta_4 U_4 + \beta_5 U_5 \qquad (6.15)$$

and we can interpret β_j as the population mean for all species with danger index equal to j. Mean function (6.15) *does not appear to include an intercept*. Since the sum of the U_j is just a column of ones, the intercept is implicit in (6.15). Since it is usual to have an intercept included explicitly, we can leave out one of the dummy variables, leading to the factor rule:

The factor rule A factor with d levels can be represented by at most d dummy variables. If the intercept is in the mean function, at most $d - 1$ of the dummy variables can be used in the mean function.

One common choice is to delete the first dummy variable, and get the mean function

$$E(TS|D) = \eta_0 + \eta_2 U_2 + \eta_3 U_3 + \eta_4 U_4 + \eta_5 U_5 \qquad (6.16)$$

where we have changed the names of the parameters because they now have different meanings. The means for the five groups are now $\eta_0 + \eta_j$ for levels

$j = 2, 3, 4, 5$ of D, and η_0 for $D = 1$. Although the parameters have different meanings in (6.15) and (6.16), both are fitting a separate mean for each level of D, and so both are really the same mean function. The mean function (6.16) is a usual mean function for *one-way analysis of variance* models, which simply fits a separate mean for each level of the classification factor.

Most computer programs allow the user to use a factor[1] in a mean function without actually computing the dummy variables. For example, in the packages S-plus and R, D would first be declared to be a factor, and then the mean function (6.16) would be specified by

$$TS \sim 1 + D \qquad (6.17)$$

where the "1" specifies fitting the intercept, and the D specifies fitting the *terms that are created for the factor* D. As is common in the specification of mean functions in linear regression computer programs, (6.17) specifies the terms in the mean function but not the parameters. This will work for linear models because each term has one corresponding parameter.

Since most mean functions include an intercept, the specification

$$TS \sim D$$

is equivalent to (6.17)[2]; to fit (6.15) without an explicit intercept, use

$$TS \sim D - 1$$

Sets of dummy variables are not the only way to convert a factor into a set of terms, and each computer package can have its own rules for getting the terms that will represent the factor, and it is important to know what your package is doing if you need to interpret and use coefficient estimates. Some packages, like R and S-plus, allow the user to choose the way that a factor will be represented.

Figure 6.5 provides a scatterplot of TS versus D for the sleep data. Some comments about this plot are in order. First, D is a categorical variable, but since the categories are ordered, it is reasonable to plot them in the order from one to five. However, the spacings on the horizontal axis between the categories are arbitrary. Figure 6.5 appears to have an approximately linear mean function, but we are not using this discovery in the one-way analysis of variance model (but see Problem 6.3). If D had unordered categories, the graph could be drawn with the categories on the horizontal axis so that the average response within group is increasing from left to right. Also from Figure 6.5, variability seems to be more or less the same for each group, suggesting that fitting with constant variance is appropriate.

Table 6.1 summarizes the fit of the one-way analysis of variance, first using (6.15), then using (6.16). In Table 6.1a, the coefficient for each U_j is the corresponding estimated mean for level j of D, and the t-value is for testing the

---

[1] A factor is called a *class variable* in SAS.

[2] In SAS, the equivalent model specification would be TS=D.

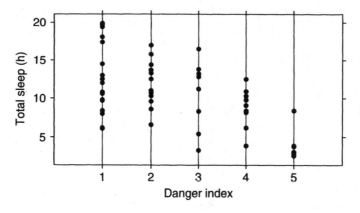

FIG. 6.5 Total sleep versus danger index for the sleep data.

TABLE 6.1 One-Way Mean Function for the Sleep Data Using Two Parameterizations

	Estimate	Std. Error	t-value	$\Pr(>|t|)$
(a) Mean function (6.15)				
U_1	13.0833	0.8881	14.73	0.0000
U_2	11.7500	1.0070	11.67	0.0000
U_3	10.3100	1.1915	8.65	0.0000
U_4	8.8111	1.2559	7.02	0.0000
U_5	4.0714	1.4241	2.86	0.0061

	Df	Sum Sq	Mean Sq	F-value	$\Pr(>F)$
D	5	6891.72	1378.34	97.09	0.0000
Residuals	53	752.41	14.20		

	Estimate	Std. Error	t-value	$\Pr(>|t|)$
(b) Mean function (6.16)				
Intercept	13.0833	0.8881	14.73	0.0000
U_2	−1.3333	1.3427	−0.99	0.3252
U_3	−2.7733	1.4860	−1.87	0.0675
U_4	−4.2722	1.5382	−2.78	0.0076
U_5	−9.0119	1.6783	−5.37	0.0000

	Df	Sum Sq	Mean Sq	F-value	$\Pr(>F)$
D	4	457.26	114.31	8.05	0.0000
Residuals	53	752.41	14.20		

hypothesis that the mean for level j is zero versus the alternative that the mean is not zero. In Table 6.1b, the estimate for the intercept is the mean for level one of D, and the other estimates are the differences between the mean for level one and the jth level. Similarly, the t-test that the coefficient for U_j is zero for $j > 1$ is really testing the difference between the mean for the jth level of D and the first level of D.

There are also differences in the analysis of variance tables. The analysis of variance in Table 6.1a corresponding to (6.15) is testing the null hypothesis that all the βs equal zero or that $E(TS|D) = 0$, against the alternative (6.15). This is not the usual hypothesis that one wishes to test using the overall analysis of variance. The analysis of variance in Table 6.1b is the usual table, with null hypothesis $E(TS|D) = \beta_0$ versus (6.16).

In summary, both the analyst and a computer package have considerable flexibility in the way that dummy variables for a factor are defined. Different choices have both advantages and disadvantages, and the analyst should be aware of the choice made by a particular computer program.

6.2.2 Adding a Predictor: Comparing Regression Lines

To the sleep data, suppose we add $\log(BodyWt)$, the base-two logarithm of the species average body weight, as a predictor. We now have two predictors, the danger index D, a factor with five levels, and $\log(BodyWt)$. We assume for now that for a fixed value of D,

$$E(TS|\log(BodyWt) = x, D = j) = \beta_{0j} + \beta_{1j}x \qquad (6.18)$$

We distinguish four different situations:

Model 1: Most general Every level of D has a different slope and intercept, corresponding to Figure 6.6a. We can write this most general mean function in several ways. If we include a dummy variable for each level of D, we can write

$$E(TS|\log(BodyWt) = x, D = j) = \sum_{j=1}^{d} \left(\beta_{0j}U_j + \beta_{1j}U_jx \right) \qquad (6.19)$$

Mean function (6.19) has $2d$ terms, the d dummy variables for d intercepts and d interactions formed by multiplying each dummy variable by the continuous predictor for d slope parameters. In the computer packages R and S-plus, if D has been declared to be a factor, this mean function can be fit by the statement

$$TS \sim -1 + D + D{:}\log(BodyWt)$$

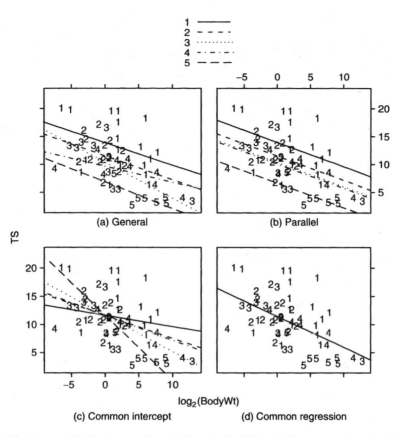

FIG. 6.6 Four models for the regression of *TS* on log(*BodyWt*) with five groups determined by *D*.

where the "-1" explicitly deletes the intercept, the term "*D*" fits a separate intercept for each level of *D*, and the term "*D*:log(*BodyWt*)" specifies interactions between each of the dummy variables for *D* and log(*BodyWt*)[3].

Using a different letter for the parameters, this mean function can also be written as

$$E(TS|\log(BodyWt) = x, D = j) = \eta_0 + \eta_1 x + \sum_{j=2}^{d} \left(\eta_{0j} U_j + \eta_{1j} U_j x \right)$$

(6.20)

Comparing the two parameterizations, we have $\eta_0 = \beta_{01}$, $\eta_1 = \beta_{11}$, and for $j > 1$, $\eta_{0j} = \beta_{0j} - \beta_{01}$ and $\eta_{1j} = \beta_{1j} - \beta_{11}$. The parameterization (6.19) is more convenient for getting interpretable parameters, while (6.20) is useful for comparing mean functions. Mean function (6.20) can be specified in R

---

[3]Other programs such as SAS may replace the ":" with a "*".

and S-plus by

$$\log(TS) \sim \log(BodyWt) + D + D{:}\log(BodyWt)$$

where again the overall intercept is implicit in the mean function and need not be specified.

In Figure 6.6a, the estimated mean functions for each level of D appear to be nearly parallel, so we should expect a simpler mean function might be appropriate for these data.

Model 2: Parallel regressions For this mean function, all the within-group mean functions are parallel as in Figure 6.6b, so $\beta_{11} = \beta_{12} = \cdots = \beta_{1d}$ in (6.19), or $\eta_{12} = \eta_{12} = \cdots = \eta_{1d} = 0$ in (6.20). Each level of D can have its own intercept. This mean function can be specified as

$$\log(TS) \sim D + \log(BodyWt)$$

The difference between levels of D is the same for every value of the continuous predictor because no dummy variable by predictor interactions is included in the mean function. *This mean function should only be used if it is in fact appropriate for the data.* This mean function is fit with terms for the intercept, $\log(BodyWt)$ and D. The number of parameters estimated is $d + 1$. We can see from Figure 6.6b that the fitted mean function for $D = 5$ has the smallest intercept, for $D = 1$ the intercept is largest, and for the three intermediate categories, the mean functions are very nearly the same. This might suggest that the three middle categories could be combined; see Problem 6.3.

Model 3: Common intercept In this mean function, the intercepts are all equal, $\beta_{01} = \cdots = \beta_{0d}$ in (6.19) or $\eta_{02} = \cdots = \eta_{0d} = 0$ in (6.20), but slopes are arbitrary, as illustrated in Figure 6.6c. This particular mean function is probably inappropriate for the sleep data, as it requires that the expected hours of TS given that a species is of 1-kg body weight, so $\log(BodyWt) = 0$, is the same for all levels of danger, and this seems to be totally arbitrary. The mean function would change if we used different units, like grams or pounds.

This mean function is fit with terms for the intercept, $\log(BodyWt)$ and the $\log(BodyWt) \times D$ interaction, for a total of $d + 1$ parameters. The R or S-plus specification of this mean function is

$$TS \sim 1 + D{:}\log(BodyWt)$$

Model 4: Coincident regression lines Here, all lines are the same, $\beta_{01} = \cdots = \beta_{0m}$ and $\beta_{11} = \cdots = \beta_{1m}$ in (6.19) or $\eta_{02} = \cdots = \eta_{0d} = \eta_{12} = \cdots = \eta_{1d} = 0$ in (6.20). This is the most stringent model, as illustrated in Figure 6.6d. This mean function requires only a term for the intercept and for $\log(BodyWt)$, for a total of 2 parameters and is given by

$$TS \sim \log(BodyWt)$$

TABLE 6.2   Residual Sum of Squares and df for the Four Mean Functions for the Sleep Data

	df	RSS	F	P(>F)
Model 1, most general	48	565.46		
Model 2, parallel	52	581.22	0.33	0.853
Model 3, common intercept	52	709.49	3.06	0.025
Model 4, all the same	56	866.23	3.19	0.006

It is usually of interest to test the plausibility of models 4 or 2 against a different, less stringent model as an alternative. The form of these tests is from the formulation of the general F-test given in Section 5.4.

Table 6.2 gives the RSS and df for each of the four models for the sleep data. Most tests concerning the slopes and intercepts of different regression lines will use the general mean function of Model 1 as the alternative model. The usual F-test for testing mean functions 2, 3, and 4 is then given for $\ell = 2, 3, 4$ by

$$F_\ell = \frac{(RSS_\ell - RSS_1)(df_\ell - df_1)}{RSS_1/df_1} \tag{6.21}$$

If the hypothesis provides as good a model as does the alternative, then F will be small. If the model is not adequate when compared with the general model, then F will be large when compared with the percentage points of the $F(df_\ell - df_1, df_1)$ distribution. The F-values for comparison to the mean function for Model 1 are given in Table 6.2. Both the common intercept mean function and the coincident mean function are clearly worse than Model 1, since the p-values are quite small. However, the p-value for the parallel regression model is very large, suggesting that the parallel regression model is appropriate for these data. The analysis is completed in Problem 6.3.

6.2.3   Additional Comments

Probably the most common problem in comparing groups is testing for parallel slopes in simple regression with two groups. Since this F-test has 1 df in the numerator, it is equivalent to a t-test. Let $\hat{\beta}_j, \hat{\sigma}_j^2, n_j$ and SXX_j be, respectively, the estimated slope, residual mean square, sample size, and corrected sum of squares for the fit of the mean function in group j, $j = 1, 2$. Then a pooled estimate of σ^2 is

$$\hat{\sigma}^2 = \left(\frac{(n_1 - 2)\hat{\sigma}_1^2 + (n_2 - 2)\hat{\sigma}_2^2}{n_1 + n_2 - 4} \right) \tag{6.22}$$

and the t-test for equality of slopes is

$$t = \frac{\hat{\beta}_1 - \hat{\beta}_2}{\hat{\sigma}(1/SXX_1 + 1/SXX_2)^{1/2}} \tag{6.23}$$

with $n_1 + n_2 - 4$ df. The square of this t-statistic is numerically identical to the corresponding F-statistic.

The model for common intercept, Model 3 of Section 6.2.2, can be easily extended to the case where the regression lines are assumed common at any fixed point, $X = c$. In the sleep data, suppose we wished to test for concurrence at $c = 2$, close to the average log body weight in the data. Simply replace $\log(BrainWt)$ by $z = \log(BrainWt) - 2$ in all the models. Another generalization of this method is to allow the regression lines to be concurrent at some arbitrary and unknown point, so the point of equality must be estimated from the data. This turns out to be a nonlinear regression problem (Saw, 1966).

6.3   MANY FACTORS

Increasing the number of factors or the number of continuous predictors in a mean function can add considerably to complexity but does not really raise any new fundamental issues. Consider first a problem with many factors but no continuous predictors. The data in the file `wool.txt` are from a small experiment to understand the strength of wool as a function of three factors that were under the control of the experimenter (Box and Cox, 1964). The variables are summarized in Table 6.3. Each of the three factors was set to one of three levels, and all $3^3 = 27$ possible combinations of the three factors were used exactly once in the experiment, so we have a single replication of a 3^3 design. The response variable $\log(Cycles)$ is the logarithm of the number of loading cycles to failure of worsted yarn. We will treat each of the three predictors as a factor with three levels.

A *main-effects* mean function for these data would include only an intercept and two dummy variables for each of the three factors, for a total of seven parameters. A full second-order mean function would add all the two-factor interactions to the mean function; each two-factor interaction would require $2 \times 2 = 4$ dummy variables, so the second-order model will have $7 + 3 \times 4 = 19$ parameters. The third-order model includes the three-factor interaction with $2 \times 2 \times 2 = 8$ dummy variables for a total of $19 + 8 = 27$ parameters. This latter mean function will fit the data exactly because it has as many parameters as data points. R and S-plus specification of these three mean functions are, assuming that *Len*, *Amp* and *Load*

TABLE 6.3   The Wool Data

Variable	Definition
Len	Length of test specimen (250, 300, 350 mm)
Amp	Amplitude of loading cycle (8, 9, 10 mm)
Load	Load put on the specimen (40, 45, 50 g)
$\log(Cycles)$	Logarithm of the number of cycles until the specimen fails

have all been declared as factors,

$$\log(Cycles) \sim Len + Amp + Load$$

$$\log(Cycles) \sim Len + Amp + Load$$
$$+ Len{:}Amp + Len{:}Load + Amp{:}Load$$

$$\log(Cycles) \sim Len + Amp + Load$$
$$+ Len{:}Amp + Len{:}Load + Amp{:}Load$$
$$+ Len{:}Amp{:}Load$$

Other mean functions can be obtained by dropping some of the two-factor interactions. Problems with many factors can be neatly handled using analysis of variance methods given, for example, by Oehlert (2000), and in many other books. The analysis of variance models are the same as multiple linear regression models, but the notation is a little different. Analysis of the wool data is continued in Problems 6.20 and 7.6.

6.4 PARTIAL ONE-DIMENSIONAL MEAN FUNCTIONS

A problem with several continuous predictors and factors requires generalization of the four mean functions given in Section 6.2.2. For example, suppose we have two continuous predictors X_1 and X_2 and a single factor F. All of the following are generalizations of the parallel regression mean functions, using a generic response Y and the computer notation of showing the terms but not the parameters:

$$Y \sim 1 + F + X_1$$

$$Y \sim 1 + F + X_2$$

$$Y \sim 1 + F + X_1 + X_2$$

$$Y \sim 1 + F + X_1 + X_2 + X_1 X_2$$

These mean functions differ only with respect to the complexity of the dependence of Y on the continuous predictors. With more continuous predictors, interpreting mean functions such as these can be difficult. In particular, how to summarize these fitted functions using a graph is not obvious.

Cook and Weisberg (2004) have provided a different way to look that problems such as this one can be very useful in practice and can be easily summarized graphically. In the basic setup, suppose we have terms $X = (X_1, \ldots, X_p)$ created from the continuous terms and a factor F with d levels. We suppose that

1. The mean function depends on the X-terms through a single linear combination, so if $X = \mathbf{x}$, the linear combination has the value $\mathbf{x}'\boldsymbol{\beta}^*$ for some unknown $\boldsymbol{\beta}^*$. A term for the intercept is not included in X.

2. For an observation at level j of the factor F, the mean function is

$$E(Y|X = \mathbf{x}, F = j) = \eta_{0j} + \eta_{1j}(\mathbf{x}'\boldsymbol{\beta}^*) \qquad (6.24)$$

This is equivalent to the most general Model 1 given previously, since each level of the factor has its own intercept and slope. We can then summarize the regression problem with a graph like one of the frames in Figure 6.6, with $\mathbf{x}'\boldsymbol{\beta}^*$ or an estimate of it on the horizontal axis. The generalization of the parallel mean functions is obtained by setting all the η_{1j} equal, while the generalization of the common intercepts mean function sets all the η_{0j} to be equal.

There is an immediate complication: the mean function (6.24) is *not a linear mean function* because the unknown parameter η_{1j} multiplies the unknown parameter $\boldsymbol{\beta}^*$, and so the parameters cannot be fit in the usual way using linear least squares software. Even so, estimating parameters is not particularly hard. In Problem 6.21, we suggest a simple computer program that can be written that will use standard linear regression software to estimate parameters, and in Problem 11.6, we show how the parameters can be estimated using a nonlinear least squares program.

Australian Athletes

As an example, we will use data provided by Richard Telford and Ross Cunningham collected on a sample of 202 elite athletes who were in training at the Australian Institute of Sport. The data are in the file `ais.txt`. For this example, we are interested in the conditional distribution of the variable *LBM*, the lean body mass, given three terms, *Ht*, height in cm, *Wt*, weight in kg, and *RCC*, the red cell count, separately for each sex. The data are displayed in Figure 6.7, with "m" for males and "f" for females. With the exception of *RCC*, the variables are all approximately linearly related; *RCC* appears to be at best weakly related to the others.

We begin by computing an F-test to compare the null hypothesis given by

$$E(LBM|Sex, Ht, Wt, RCC) = \beta_0 + \beta_1 Sex + \beta_2 Ht + \beta_3 Wt + \beta_4 RCC$$

to the alternative mean function

$$E(LBM|Sex, Ht, Wt, RCC) = \beta_0 + \beta_1 Sex + \beta_2 Ht + \beta_3 Wt + \beta_4 RCC + \beta_{12}(Sex \times Ht)$$
$$+ \beta_{13}(Sex \times Wt) + \beta_{14}(Sex \times RCC) \qquad (6.25)$$

that has a separate mean function for each of the two sexes. For the first of these two, we find $RSS = 4043.6$ with $202 - 4 = 198$ df. For the second mean function, $RSS = 1136.8$ with $202 - 8 = 194$ df. The value of the test statistic is $F = 104.02$ with $(4, 194)$ df, for a corresponding p-value that is zero to three decimal places. We have strong evidence that the smaller of these mean functions is inadequate.

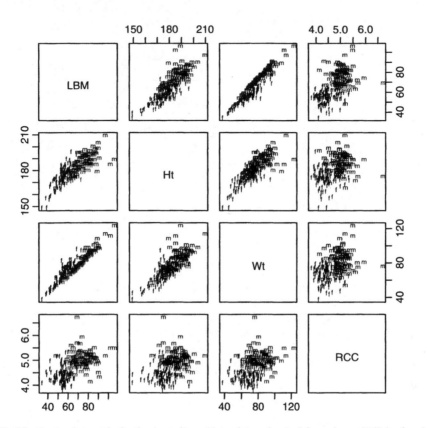

FIG. 6.7 Scatterplot matrix for the Australian athletes data, using "m" for males and "f" for females.

Interpretation of the larger mean function is difficult because we cannot draw simple graphs to summarize the results.

The partial one-dimensional (POD) mean function for these data is given by

$$E(LBM|Sex, Ht, Wt, RCC) = \beta_0 + \beta_1 Sex + \beta_2 Ht + \beta_3 Wt + \beta_4 RCC$$

$$+ \eta_0 Sex + \eta_1 Sex \times (\beta_2 Ht + \beta_3 Wt + \beta_4 RCC) \quad (6.26)$$

which is a modest reparameterization of (6.24). We can fit the mean function (6.26) using either the algorithm outlined in Problem 6.21 or the nonlinear least squares method outlined in Problem 11.6. The residual sum of squares is 1144.2 with $202 - 6 = 196$ df. The F-test for comparing this mean function to (6.25) has value $F = 0.63$ with $(2, 194)$ df, with a p-value of about 0.53. The conclusion is the POD mean function matches the data as well as the more complicated (6.25).

The major advantage of the POD mean function is that we can draw the summarizing graph given in Figure 6.8. The horizontal axis in the graph is the single linear combination of the predictors $\hat{\beta}_2 Ht + \hat{\beta}_3 Wt + \hat{\beta}_4 RCC$ from the fit of (6.26). The vertical axis is the response *LBM*, once again with "m" for males and "f" for

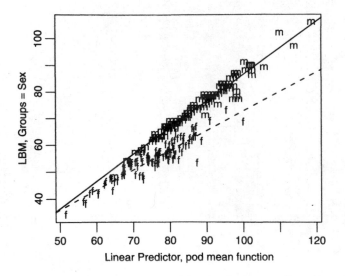

FIG. 6.8   Summary graph for the POD mean function for the Australian athletes data.

females. The two lines shown on the graph are the fitted values for males in the solid line and females in the dashed line. We get the interesting summary that *LBM* depends on the same linear combination of the terms for each of the two sexes, but the fitted regression has a larger slope for males than for females.

6.5   RANDOM COEFFICIENT MODELS

We conclude this chapter with a brief introduction to problems in which the methodology of this chapter *seems* appropriate, but for which different methodology is to be preferred.

Little is known about wetland contamination by road salt, primarily NaCl. An exploratory study examined chloride concentrations in five roadside marshes and four marshes isolated from roads to evaluate potential differences in chloride concentration between marshes receiving road runoff and those isolated from road runoff, and to explore trends in chloride concentrations across an agricultural growing season, from about April to October.

The data in the file `chloride.txt`, provided by Stefanie Miklovic and Susan Galatowitsch, summarize results. Repeated measurements of chloride level were taken in April, June, August, and October, during 2001. Two of the marshes were dry by August 2001, so only April and June measurements were taken on those two. The variables in the file are *Cl*, the measured chloride level in mg/liter, *Month*, the month of measurement, with April as 4, June as 6, and so on; *Marsh*, the marsh number, and *Type*, either isolated or roadside.

Following the methodology of this chapter, we might contemplate fitting multiple linear regression models with a separate intercept and slope for each level of

Type, following the outline in Section 6.2.2. This mean function ignores the possibility of each marsh having a different intercept and slope. Were we to include a factor *Marsh* in the mean function, we would end up fitting up to 18 parameters, but the data include only 32 observations. Furthermore, fitting a separate regression for each marsh does not directly answer the questions of interest that average over marshes.

The data are shown in Figure 6.9. A separate graph is given for the two types, and the points within a wetland are joined in each graph. While it is clear that the overall chloride level is different in the two types of wetlands, the lines within a graph do not tell a coherent story; we cannot tell if there is a dependence on *Month*, or if the dependence is the same for the two types.

To examine data such as these, we use a *random coefficients model*, which in this problem assumes that the marshes within a type are a random sample of marshes that could have been studied. Using a generic notation, suppose that y_{ijk} is the value of the response for the jth marsh of type i at time x_k. The random coefficients model specifies that

$$y_{ijk} = \beta_{0i} + \beta_{1i}x_k + b_{0ij} + b_{1ij}x_k + e_{ijk} \tag{6.27}$$

As in other models, the βs are fixed, unknown parameters that specify separate linear regression for each of the two types, $\beta_{01} + \beta_{11}x_k$ for *Type* = isolated, and $\beta_{02} + \beta_{12}x_k$ for *Type* = roadside. The errors e_{ijk} will be taken to be independent and identically distributed with variance σ^2. Mean function (6.27) is different from other mean functions we have seen because of the inclusion of the bs. We assume

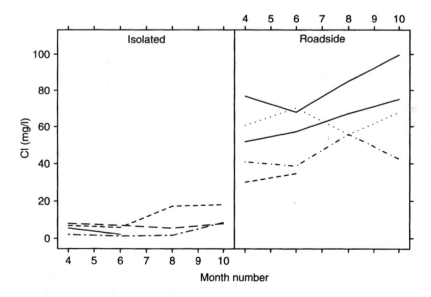

FIG. 6.9 The chloride concentration data.

that the bs are random variables, independent of the e_{ijk}, and that

$$
\begin{pmatrix} b_{0ij} \\ b_{1ij} \end{pmatrix} \sim N \left(\begin{pmatrix} 0 \\ 0 \end{pmatrix} , \begin{pmatrix} \tau_0^2 & \tau_{01} \\ \tau_{01} & \tau_1^2 \end{pmatrix} \right)
$$

According to this model, a particular marsh has intercept $\beta_{0i} + b_{0ij}$ and slope $\beta_{1i} + b_{1ij}$, while the average intercept and slope for type i are β_{0i} and β_{1i}, respectively. The two variances τ_0^2 and τ_1^2 model the variation in intercepts and slopes between marshes within a type, and τ_{01} allows the bs to be correlated within marsh. Rather than estimate all the bs, we will instead estimate the τs that characterize the variability in the bs.

One of the effects of a random coefficient model is that the y_{ijk} are no longer independent as they are in the linear regression models we have considered so far. Using Appendix A.2.2:

$$
\mathrm{Cov}(y_{ijk}, y_{i'j'k'}) = \begin{cases} 0 & i \neq i' \\ 0 & i = i', j \neq j' \\ \tau_0^2 + x_k x_{k'} \tau_1^2 + (x_k + x_{k'}) \tau_{01} & i = i', j = j', k \neq k' \\ \sigma^2 + \tau_0^2 + x_k^2 \tau_1^2 + 2 x_k \tau_{01} & i = i', j = j', k = k' \end{cases}
$$

$$(6.28)$$

The important point here is that *repeated observations on the same marsh are correlated*, while observations on different marshes are not correlated. If we consider the simpler *random intercepts model* given by

$$
y_{ijk} = \beta_{0i} + \beta_{1i} x_k + b_{0ij} + e_{ijk} \tag{6.29}
$$

we will have $b_{1ij} = \tau_1^2 = \tau_{01} = 0$, and the correlation between observations in the same marsh will be $\tau_0^2 / (\sigma^2 + \tau_0^2)$, which is known as the *intra-class correlation*. Since this is always positive, the variation within one marsh will always be smaller than the variation between marshes, as often makes very good sense.

Methods for fitting with a model such as (6.27) is beyond the scope of this book. Software is widely available, however, and books by Pinheiro and Bates (2000), Littell, Milliken, Stroup, and Wolfinger (1996), and Verbeke and Molenberghs (2000) describe the methodology and the software. Without going into detail, generalization of the testing methods discussed in this book suggest that the random intercepts model (6.29) is appropriate for these data with $\beta_{11} = \beta_{12}$, as there is no evidence that either the slope is different for the two types or that the slope varies from marsh to marsh. When we fit (6.29) using the software described by Pinheiro and Bates, we get the summaries shown in Table 6.4. The difference between the types is estimated to be between about 28 and 72 mg/liter, while the level of Cl appears to be increasing over the year by about 1.85 mg/liter per month. The estimates of the variances τ_0 and σ are shown in the standard deviation scale. Since these two are of comparable size, therefore a substantial gain in precision in the analysis by accounting for, and removing, the between marsh variation.

The random coefficients models are but one instance of a general class of *linear mixed models*. These are described in Pinheiro and Bates (2000), Littell, Milliken,

TABLE 6.4 **Approximate 95% Confidence Intervals for Parameters of the Random Coefficients Model Using R for the Chloride Data**

```
Fixed effects:
                lower      est.    upper
(Intercept) -21.50837  -5.5038  10.5007
Month         0.75487   1.8538   2.9527
Type         28.69165  50.5719  72.4521

Random Effects:
                             lower    est.   upper
sd((Intercept))             7.5761  13.335  23.473
Within-group standard error: 4.7501  6.3872  8.5885
```

Stroup, and Wolfinger (1996), and Diggle, Heagerty, Liang, and Zeger (2002). This is a very rich and important class of models that allow fitting with a wide variety of correlation and mean structures.

PROBLEMS

6.1. Cake data The data for this example are in the data file cakes.txt.

6.1.1. Fit (6.4) and verify that the significance levels are all less than 0.005.

6.1.2. Estimate the optimal (X_1, X_2) combination $(\tilde{X}_1, \tilde{X}_2)$ and find the standard errors of $\tilde{X}_1$ and $\tilde{X}_2$.

6.1.3. The cake experiment was carried out in two blocks of seven observations each. It is possible that the response might differ by block. For example, if the blocks were different days, then differences in air temperature or humidity when the cakes were mixed might have some effect on Y. We can allow for block effects by adding a factor for Block to the mean function, and possibly allowing for Block by term interactions. Add block effects to the mean function fit in Section 6.1.1 and summarize results. The blocking is indicated by the variable *Block* in the data file.

6.2. The data in the file lathe1.txt are the results of an experiment on characterizing the life of a drill bit in cutting steel on a lathe. Two factors were varied in the experiment, *Speed* and *Feed* rate. The response is *Life*, the total time until the drill bit fails, in minutes. The values of *Speed* in the data have been coded by computing

$$Speed = \frac{(\text{Actual speed in feet per minute} - 900)}{300}$$

$$Feed = \frac{(\text{Actual feed rate in thousandths of an inch per revolution} - 13)}{6}$$

The coded variables are centered at zero. Coding has no material effect on the analysis but can be convenient in interpreting coefficient estimates.

6.2.1. Draw a scatterplot matrix of *Speed*, *Feed*, *Life*, and log(*Life*), the base-two logarithm of tool life. Add a little jittering to *Speed* and *Feed* to reveal over plotting. The plot of *Speed* versus *Feed* gives a picture of the experimental design, which is called a *central composite design*. It is useful when we are trying to find a value of the factors that maximizes or minimizes the response. Also, several of the experimental conditions were replicated, allowing for a pure-error estimate of variance and lack of fit testing. Comment on the scatterplot matrix.

6.2.2. For experiments in which the response is a time to failure or time to event, the response often needs to be transformed to a more useful scale, typically by taking the log of the response, or sometimes by taking the inverse. For this experiment, log scale can be shown to be appropriate (Problem 9.7).

Fit the full second-order mean function (6.4) to these data using log(*Life*) as the response, and summarize results.

6.2.3. Test for the necessity of the *Speed* × *Feed* interaction, and summarize your results. Draw appropriate summary graphs equivalent to Figure 6.3 or Figure 6.4, depending on the outcome of your test.

6.2.4. For *Speed* = 0.5, estimate the value of *Feed* that minimizes log(*Life*), and obtain a 95% confidence interval for this value using the delta method.

6.3. In the sleep data, do a lack of fit test for *D* linear against the one-way Anova model. Summarize results.

6.4. The data in the file `twins.txt` give the IQ scores of identical twins, one raised in a foster home, *IQf*, and the other raised by birth parents, *IQb*. The data were published by Burt (1966), and their authenticity has been questioned. For purposes of this example, the twin pairs can be divided into three social classes *C*, low, middle or high, coded in the data file 1, 2, and 3, respectively, according to the social class of the birth parents. Treat *IQf* as the response and *IQb* as the predictor, with *C* as a factor.

Perform an appropriate analysis of these data. Be sure to draw and discuss a relevant graph. Are the within-class mean functions straight lines? Are there class differences? If there are differences, what are they?

6.5. Referring to the data in Problem 2.2, compare the regression lines for Forbes' data and Hooker's data, for the mean function $E(\log(Pressure)|Temp) = \beta_0 + \beta_1 Temp$.

6.6. Refer to the Berkeley Guidance study described in Problem 3.1. Using the data file `BGSall.txt`, consider the regression of *HT18* on *HT9* and the grouping factor *Sex*.

6.6.1. Draw the scatterplot of *HT18* versus *HT9*, using a different symbol for males and females. Comment on the information in the graph about an appropriate mean function for these data.

6.6.2. Fit the four mean function suggested in Section 6.2.2, perform the appropriate tests, and summarize your findings.

6.7. In the Berkeley Guidance Study data, Problem 6.6, consider the response *HT18* and predictors *HT2* and *HT9*. Model 1 in Section 6.2.2 allows each level of the grouping variable, in this example the variable *Sex*, to have its own mean function. Write down at least two generalizations of this model for this problem with two continuous predictors rather than one.

6.8. Continuing with Problem 6.7 and assuming no interaction between *HT2* and *HT9*, obtain a test for the null hypothesis that the regression planes are parallel for boys and girls versus the alternative that separate planes are required for each sex.

6.9. Refer to the apple shoot data, Section 5.3, using the data file `all-shoots.txt`, giving information on both long and short shoots.

6.9.1. Compute a mean square for pure error separately for long and short shoots, and show that the pure-error estimate of variance for long shoots is about twice the size of the estimate for short shoots. Since these two estimates are based on completely different observations, they are independent, and so their ratio will have an F distribution under the null hypothesis that the variance is the same for the two types of shoots. Obtain the appropriate test, and summarize results. (Hint: the alternative hypothesis is that the two variances are unequal, meaning that you need to compute a two-tailed significance level, not one-tailed as is usually done with F-tests). Under the assumption that the variance for short shoots is σ^2 and the variance for long shoots is $2\sigma^2$, obtain a pooled pure-error estimate of σ^2.

6.9.2. Draw the scatterplot of *ybar* versus *Day*, with a separate symbol for each of the two types of shoots, and comment on the graph. Are straight-line mean functions plausible? Are the two types of shoots different?

6.9.3. Fit models 1, 3 and 4 from Section 6.2.2 to these data. You will need to use weighted least squares, since each of the responses is an average of *n* values. Also, in light of Problem 6.9.1, assume that the variance for short shoots is σ^2, but the variance for long shoots is $2\sigma^2$.

6.10. Gothic and Romanesque cathedrals The data in the data file `cathedral.txt` gives *Height* = nave height and *Length* = total length, both in feet, for medieval English cathedrals. The cathedrals can be classified according to their architectural style, either Romanesque or, later, Gothic. Some cathedrals have both a Gothic and a Romanesque part, each of differing height; these cathedrals are included twice. Names of the cathedrals are also provided in the file. The data were provided by Gould S.J. based on plans given by Clapham (1934).

6.10.1. For these data, it is useful to draw *separate* plots of *Length* versus *Height* for each architectural style. Summarize the differences apparent in the graphs in the regressions of *Length* on *Height* for the two styles.

6.10.2. Use the data and the plots to fit regression models that summarize the relationship between the response *Length* and the predictor *Height* for the two architectural styles.

6.11. Windmill data In Problem 2.13, page 45, we considered data to predict wind speed *CSpd* at a candidate site based on wind speed *RSpd* at a nearby reference site where long-term data is available. In addition to *RSpd*, we also have available the wind direction, *RDir*, measured in degrees. A standard method to include the direction data in the prediction is to divide the directions into several bins and then fit a separate mean function for of *CSpd* on *RSpd* in each bin. In the wind farm literature, this is called the *measure, correlate, predict* method, Derrick (1992). The data file wm2.txt contains values of *CSpd*, *RSpd*, *RDir*, and *Bin* for 2002 for the same candidate and reference sites considered in Problem 2.13. Sixteen bins are used, the first bin for cases with *RDir* between 0 and 22.5 degrees, the second for cases with *RDir* between 22.5 and 45 degrees, ... , and the last bin between 337.5 and 360 degrees. Both the number of bins and their starting points are arbitrary.

6.11.1. Obtain tests that compare fitting the four mean functions discussed in Section 6.2.2 to the 16 bins. How many parameters are in each of the mean functions?

6.11.2. Do not attempt this problem unless your computer package has a programming language.

Table 6.5 gives the number of observations in each of the 16 bins along with the average wind speed in that bin for the reference site for the period January 1, 1948 to July 31, 2003, excluding the year 2002; the table is also given in the data file wm3.txt. Assuming the most general model of a separate regression in each bin is appropriate, predict the average wind speed at the candidate site for each of the 16 bins, and find the standard error. This will give you 16 predictions and 16 independent standard errors. Finally, combine these 16 estimates into one overall estimate (you should weight according to the number of cases in a bin), and then compare your answer to the prediction and standard error from Problem 4.6.

6.12. Land valuation Taxes on farmland enrolled in a "Green Acres" program in metropolitan Minneapolis–St. Paul are valued only with respect to the land's value as productive farmland; the fact that a shopping center or industrial park

TABLE 6.5   Bin Counts and Means for the Windmill Data[a]

Bin	Bin.count	RSpd	Bin	Bin.count	RSpd
0	2676	6.3185	8	4522	7.7517
1	2073	5.6808	9	32077	6.4943
2	1710	5.4584	10	2694	6.1619
3	1851	5.4385	11	2945	6.5947
4	2194	5.8763	12	4580	7.6865
5	3427	6.6539	13	6528	8.8078
6	5201	7.8756	14	6705	8.5664
7	6392	8.4281	15	4218	7.5656

[a]These data are also given in the file wm3.txt.

has been built nearby cannot enter into the valuation. This creates difficulties because almost all sales, which are the basis for setting assessed values, are priced according to the development potential of the land, not its value as farmland. A method of equalizing valuation of land of comparable quality was needed.

One method of equalization is based on a soil productivity score P, a number between 1, for very poor land, and 100, for the highest quality agricultural land. The data in the file prodscore.txt, provided by Doug Tiffany, gives P along with *Value*, the average assessed value, the *Year*, either 1981 or 1982 and the *County* name for four counties in Minnesota, Le Sueur, Meeker, McLeod, and Sibley, where development pressures had little effect on assessed value of land in 1981–82. The unit of analysis is a township, roughly six miles square.

The goal of analysis is to decide if soil productivity score is a good predictor of assessed value of farmland. Be sure to examine county and year differences, and write a short summary that would be of use to decision makers who need to determine if this method can be used to set property taxes.

6.13. Sex discrimination   The data in the file salary.txt concern salary and other characteristics of all faculty in a small Midwestern college collected in the early 1980s for presentation in legal proceedings for which discrimination against women in salary was at issue. All persons in the data hold tenured or tenure track positions; temporary faculty are not included. The data were collected from personnel files and consist of the quantities described in Table 6.6.

6.13.1. Draw an appropriate graphical summary of the data, and comment of the graph.

6.13.2. Test the hypothesis that the mean salary for men and women is the same. What alternative hypothesis do you think is appropriate?

TABLE 6.6 The Salary Data

Variable	Description
Sex	Sex, 1 for female and 0 for male
Rank	Rank, 1 for Assistant Professor, 2 for Associate Professor, and 3 for Full Professor
Year	Number of years in current rank
Degree	Highest degree, 1 if Doctorate, 0 if Masters
YSdeg	Number of years since highest degree was earned
Salary	Academic year salary in dollars

6.13.3. Obtain a test of the hypothesis that salary adjusted for years in current rank, highest degree, and years since highest degree is the same for each of the three ranks, versus the alternative that the salaries are not the same. Test to see if the sex differential in salary is the same in each rank.

6.13.4. Finkelstein (1980), in a discussion of the use of regression in discrimination cases, wrote, "... [a] variable may reflect a position or status bestowed by the employer, in which case if there is discrimination in the award of the position or status, the variable may be 'tainted'." Thus, for example, if discrimination is at work in promotion of faculty to higher ranks, using rank to adjust salaries before comparing the sexes may not be acceptable to the courts.

Fit two mean functions, one including *Sex, Year, YSdeg* and *Degree*, and the second adding *Rank*. Summarize and compare the results of leaving out rank effects on inferences concerning differential in pay by sex.

6.14. Using the salary data in Problem 6.13, one fitted mean function is

$$\text{E}(Salary|Sex, Year) = 18223 - 571Sex + 741Year + 169Sex \times Year$$

6.14.1. Give the coefficients in the estimated mean function if *Sex* were coded so males had the value 2 and females had the value 1 (the coding given to get the above mean function was 0 for males and 1 for females).

6.14.2. Give the coefficients if *Sex* were codes as -1 for males and $+1$ for females.

6.15. Pens of turkeys were grown with an identical diet, except that each pen was supplemented with an amount A of an amino acid methionine as a percentage of the total diet of the birds. The data in the file turk0.txt give the response average weight *Gain* in grams of all the turkeys in the pen for 35 pens of turkeys receiving various levels of A.

6.15.1. Draw the scatterplot of *Gain* versus *A* and summarize. In particular, does simple linear regression appear plausible?

6.15.2. Obtain a lack of fit test for the simple linear regression mean function, and summarize results. Repeat for the quadratic regression mean function.

6.15.3. To the graph drawn in Problem 6.15.1, add the fitted mean functions based on both the simple linear regression mean function and the quadratic mean function, for values of *A* in the range from 0 to 0.60, and comment.

6.16. For the quadratic regression mean function for the turkey data discussed in Problem 6.15, use the bootstrap to estimate the standard error of the value of *D* that maximizes gain. Compare this estimated standard error with the answer obtained using the delta method.

6.17. Refer to Jevons' coin data, Problem 5.6. Determine the *Age* at which the predicted weight of coins is equal to the legal minimum, and use the delta method to get a standard error for the estimated age. This problem is called *inverse regression*, and is discussed by Brown (1994).

6.18. The data in the file `mile.txt` give the world record times for the one-mile run. For males, the records are for the period from 1861–2003, and for females, for the period 1967–2003. The variables in the file are *Year*, year of the record, *Time*, the record time, in seconds, *Name*, the name of the runner, *Country*, the runner's home country, *Place*, the place where the record was run (missing for many of the early records), and *Gender*, either Male or Female. The data were taken from http://www.saunalahti.fi/~sut/eng/.

6.18.1. Draw a scatterplot of *Time* versus *Year*, using a different symbol for men and women. Comment on the graph.

6.18.2. Fit separate simple linear regression mean functions to each sex, and show that separate slopes and intercepts are required. Provide an interpretation of the slope parameters for each sex.

6.18.3. Find the year in which the female record is expected to be 240 seconds, or four minutes. This will require inverting the fitted regression equation. Use the delta method to estimate the standard error of this estimate.

6.18.4. Using the model fit in Problem 6.18.2, estimate the year in which the female record will match the male record, and use the delta method to estimate the standard error of the year in which they will agree. Comment on whether you think using the point at which the fitted regression lines cross as a reasonable estimator of the crossing time.

6.19. Use the delta method to get a 95% confidence interval for the ratio β_1/β_2 for the transactions data, and compare with the bootstrap interval obtained at the end of Section 4.6.1.

6.20. Refer to the wool data discussed in Section 6.3.

    6.20.1. Write out in full the main-effects and the second-order mean functions, assuming that the three predictors will be turned into factors, each with three levels. This will require you to define appropriate dummy variables and parameters.

    6.20.2. For the two mean functions in Problem 6.20.1, write out the expected change in the response when *Len* and *Amp* are fixed at their middle levels, but *Load* is increased from its middle level to its high level.

6.21. A POD model for a problem with p predictors $X = (X_1 \ldots, X_p)$ and a factor F with d levels is specified, for the jth level of F, by

$$E(Y|X = \mathbf{x}, F = j) = \eta_{0j} + \eta_{1j}(\mathbf{x}'\boldsymbol{\beta}^*) \qquad (6.30)$$

This is a nonlinear model because η_{ij} multiplies the parameter $\boldsymbol{\beta}^*$. Estimation of parameters can use the following two-step algorithm:

1. Assume that the η_{1j}, $j = 1, \ldots, d$ are known. At the first step of the algorithm, set $\eta_{1j} = 1$, $j = 1, \ldots, d$. Define a new term $\mathbf{z}_j = \eta_{1j}\mathbf{x}$, and substituting into (6.30),

$$E(Y|X = \mathbf{x}, F = j) = \eta_{0j} + \mathbf{z}_j'\boldsymbol{\beta}^*$$

We recognize this as a mean function for *parallel regressions* with common slopes $\boldsymbol{\beta}^*$ and a separate intercept for each level of F. This mean function can be fit using standard OLS linear regression software. Save the estimate $\hat{\boldsymbol{\beta}}^*$ of $\boldsymbol{\beta}^*$.

2. Let $v = \mathbf{x}'\hat{\boldsymbol{\beta}}^*$, where $\hat{\boldsymbol{\beta}}^*$ was computed in step 1. Substitute v for $\mathbf{x}'\boldsymbol{\beta}^*$ in (6.30) to get

$$E(Y|X = \mathbf{x}, F = j) = \eta_{0j} + \eta_{1j}v$$

which we recognize as a mean function with a separate intercept and slope for each level of F. This mean function can also be fit using OLS linear regression software. Save the estimates of η_{1j} and use them in the next iteration of step 1.

Repeat this algorithm until the residual sum of squares obtained at the two steps is essentially the same. The estimates obtained at the last step will be the OLS estimates for the original mean function, and the residual sum of squares will be the residual sum of squares that would be obtained by fitting using nonlinear least squares. Estimated standard errors of the coefficients will be too small, so t-tests should not be used, but F-tests can be used to compare models.

    Write a computer program that implements this algorithm.

6.22. Using the computer program written in the last problem or some other computational tool, verify the results obtained in the text for the Australian Athletes data. Also, obtain tests for the general POD mean function versus the POD mean function with parallel mean functions.

6.23. The Minnesota Twins professional baseball team plays its games in the Metrodome, an indoor stadium with a fabric roof. In addition to the large air fans required to keep to roof from collapsing, the baseball field is surrounded by ventilation fans that blow heated or cooled air into the stadium. Air is normally blown into the center of the field equally from all directions.

According to a retired supervisor in the Metrodome, in the late innings of some games the fans would be modified so that the ventilation air would blow out from home plate toward the outfield. The idea is that the air flow might increase the length of a fly ball. For example, if this were done in the middle of the eighth inning, then the air-flow advantage would be in favor of the home team for six outs, three in each of the eighth and ninth innings, and in favor of the visitor for three outs in the ninth inning, resulting in a slight advantage for the home team.

To see if manipulating the fans could possibly make any difference, a group of students at the University of Minnesota and their professor built a "cannon" that used compressed air to shoot baseballs. They then did the following experiment in the Metrodome in March 2003:

1. A fixed angle of 50 degrees and velocity of 150 feet per second was selected. In the actual experiment, neither the velocity nor the angle could be controlled exactly, so the actual angle and velocity varied from shot to shot.

2. The ventilation fans were set so that to the extent possible all the air was blowing in from the outfield towards home plate, providing a headwind. After waiting about 20 minutes for the air flows to stabilize, 20 balls were shot into the outfield, and their distances were recorded. Additional variables recorded on each shot include the weight (in grams) and diameter (in cm) of the ball used on that shot, and the actual velocity and angle.

3. The ventilation fans were then reversed, so as much as possible air was blowing out toward the outfield, giving a tailwind. After waiting 20 minutes for air currents to stabilize, 15 balls were shot into the outfield, again measuring the ball weight and diameter, and the actual velocity and angle on each shot.

The data from this experiment are available in the file domedata.txt, courtesy of Ivan Marusic. The variable names are *Cond*, the condition, head or tail wind; *Velocity*, the actual velocity in feet per second; *Angle*, the actual angle; *BallWt*, the weight of the ball in grams used on that particular test; *BallDia*, the diameter in inches of the ball used on that test; *Dist*, distance in feet of the flight of the ball.

6.23.1. Summarize any evidence that manipulating the fans can change the distance that a baseball travels. Be sure to explain how you reached your conclusions, and provide appropriate summary statistics that might be useful for a newspaper reporter (a report of this experiment is given in the Minneapolis *StarTribune* for July 27, 2003).

6.23.2. In light of the discussion in Section 6.5, one could argue that this experiment by itself cannot provide adequate information to decide if the fans can affect length of a fly ball. The treatment is *manipulating the fans*; each condition was set up only once and then repeatedly observed. Unfortunately, resetting the fans after each shot is not practical because of the need to wait at least 20 minutes for the air flows to stabilize.

A second experiment was carried out in May 2003, using a similar experimental protocol. As before, the fans were first set to provide a headwind, and then, after several trials, the fans were switched to a tailwind. Unlike the first experiment, however, the nominal *Angle* and *Velocity* were varied according to a 3×2 factorial design. The data file domedata1.txt contains the results from both the first experiment and the second experiment, with an additional column called *Date* indicating which sample is which. Analyze these data, and write a brief report of your findings.

CHAPTER 7

Transformations

There are exceptional problems for which we know that the mean function $E(Y|X)$ is a linear regression mean function. For example, if (Y, X) has a joint normal distribution, then as in Section 4.3, the conditional distribution of $Y|X$ has a linear mean function. Sometimes, the mean function may be determined by a theory, apart from parameter values, as in the strong interaction data in Section 5.1.1. Often, there is no theory to tell us the correct form for the mean function, and any parametric form we use is little more than an *approximation* that we hope is adequate for the problem at hand. Replacing either the predictors, the response, or both by nonlinear transformations of them is an important tool that the analyst can use to extend the number of problems for which linear regression methodology is appropriate. This brings up two important questions: How do we choose transformations? How do we decide if an approximate model is adequate for the data at hand? We address the first of these questions in this chapter, and the second in Chapters 8 and 9.

7.1 TRANSFORMATIONS AND SCATTERPLOTS

The most frequent purpose of transformations is to achieve a mean function that is linear in the transformed scale. In problems with only one predictor and one response, the mean function can be visualized in a scatterplot, and we can attempt to select a transformation so the resulting scatterplot has an approximate straight-line mean function. With many predictors, selection of transformations can be harder, as the criterion to use for selecting transformations is less clear, so we consider the one predictor case first. We seek a transformation so if X is the *transformed* predictor and Y is the *transformed* response, then the mean function in the transformed scale is

$$E(Y|X = x) \approx \beta_0 + \beta_1 x$$

where we have used "$\approx$" rather than "$=$" to recognize that this relationship may be an approximation and not exactly true.

Applied Linear Regression, Third Edition, by Sanford Weisberg
ISBN 0-471-66379-4 Copyright © 2005 John Wiley & Sons, Inc.

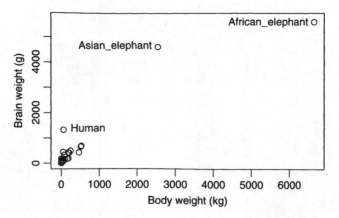

FIG. 7.1   Plot of *BrainWt* versus *BodyWt* for 62 mammal species.

Figure 7.1 contains a plot of body weight *BodyWt* in kilograms and brain weight *BrainWt* in grams for 62 species of mammals (Allison and Cicchetti, 1976), using the data in the file `brains.txt`. Apart from the three separated points for two species of elephants and for humans, the uneven distribution of points hides any useful visual information about the mean of *BrainWt*, given *BodyWt*. In any case, there is little or no evidence for a straight-line mean function here. Both variables range over several orders of magnitude from tiny species with body weights of just a few grams to huge animals of over 6600 kg. Transformations can help in this problem.

7.1.1   Power Transformations

A *transformation family* is a collection of transformations that are indexed by one or a few parameters that the analyst can select. The family that is used most often is called the *power family*, defined for a strictly positive variable U by

$$\psi(U, \lambda) = U^\lambda \qquad (7.1)$$

As the power parameter λ is varied, we get the members of this family, including the square root and cube root transformations, $\lambda = 1/2$ or $1/3$, and the inverse, $\lambda = -1$. *We will interpret the value of $\lambda = 0$ to be a log transformation.* The usual values of λ that are considered are in the range from -2 to 2, but values in the range from -1 to $+1$ are ordinarily selected. The value of $\lambda = +1$ corresponds to *no transformation.* The variable U must be strictly positive for these transformations to be used, but we will have more to say later about transforming variables that may be zero or negative. We have introduced this ψ-notation[1] because we will later consider other families of transformations, and having this notation will allow more clarity in the discussion.

---

[1] ψ is the Greek letter *psi*.

Some statistical packages include graphical tools that can help you select power transformations of both the predictor and the response. For example, a plot could include slidebars to select values of the transformation parameters applied to the horizontal and vertical variables. As different values of the parameters are selected in the slidebars, the graph is updated to reflect the transformation of the data corresponding to the currently selected value of the transformation parameter. If a graphical interface is not available in your package, you can draw several figures to help select a transformation. A mean smoother and the OLS line added to each of the plots may be helpful in looking at these plots.

Figure 7.2 shows plots of $\psi(BrainWt, \lambda)$ versus $\psi(BodyWt, \lambda)$ with *the same λ for both variables*, for $\lambda = -1, 0, 1/3, 1/2$. There is no necessity for the transformation to be the same for the two variables, but it is reasonable here because both variables are the same type of measurements, one being the weight of an object, and the other a weight of a component of the object. If we allowed each variable to have its own transformation parameter, the visual search for a transformation is harder because more possibilities need to be considered.

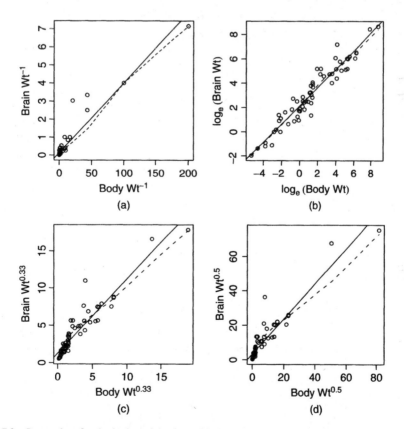

FIG. 7.2 Scatterplots for the brain weight data with four possible transformations. The solid line on each plot is the OLS line; the dashed line is a *loess* smooth.

From the four graphs in Figure 7.2, the clear choice is replacing the weights by their logarithms. In this scale, the mean function appears to be a straight line, with the smoother closely matching the OLS line shown on the graph in log scale but matching less well for the other transformations. As a bonus, the variance function in the log plot appears to be constant.

The use of logarithms for the brain weight data may not be particularly surprising, in light of the following two empirical rules that are often helpful in linear regression modeling:

The log rule If the values of a variable range over more than one order of magnitude and the variable is strictly positive, then replacing the variable by its logarithm is likely to be helpful.

The range rule If the range of a variable is considerably less than one order of magnitude, then any transformation of that variable is unlikely to be helpful.

The log rule is satisfied for both *BodyWt*, with range 0.005 kg to 6654 kg, and for *BrainWt*, with range 0.14 g to 5712 g, so log transformations would have been indicated as a starting point for examining these variables for transformations.

Simple linear regression seems to be appropriate with both variables in log scale. This corresponds to the *physical model*

$$BrainWt = \alpha \times BodyWt^{\beta_1} \times \delta \qquad (7.2)$$

where δ is a *multiplicative error*, meaning that the actual average brain weight for a particular species is obtained by taking the mean brain weight for species of a particular body weight and multiplying by δ. We would expect that δ would have mean 1 and a distribution concentrated on values close to 1. On taking logarithms and setting $\beta_1 = \log(\alpha)$ and $e = \log(\delta)$,

$$\log(BrainWt) = \beta_0 + \beta_1 \log(BodyWt) + e$$

which is the simple linear regression model. Scientists who study the relationships between attributes of individuals or species call (7.2) an *allometric* model (see, for example, Gould, 1966, 1973; Hahn, 1979), and the value of β_1 plays an important role in allometric studies. We emphasize, however, that not all useful transformations will correspond to interpretable physical models.

7.1.2 Transforming Only the Predictor Variable

In the brain weight example, transformations of both the response and the predictor are required to get a linear mean function. In other problems, transformation of only one variable may be desirable. If we want to use a family of power transformations, it is convenient to introduce the family of *scaled power transformations*, defined for strictly positive X by

$$\psi_S(X, \lambda) = \begin{cases} (X^\lambda - 1)/\lambda & \text{if } \lambda \neq 0 \\ \log(X) & \text{if } \lambda = 0 \end{cases} \qquad (7.3)$$

The scaled power transformations $\psi_S(X, \lambda)$ differ from the basic power transformations $\psi(X, \lambda)$ in several respects. First $\psi_S(X, \lambda)$ is continuous as a function of λ. Since $\lim_{\lambda \to 0} \psi_S(X, \lambda) = \log_e(X)$, the logarithmic transformation is a member of this family with $\lambda = 0$. Also, scaled power transformations preserve the direction of association, in the sense that if (X, Y) are positively related, then $(\psi_S(X, \lambda), Y)$ are positively related for all values of λ. With basic power transformations, the direction of association changes when $\lambda < 0$.

If we find an appropriate power to use for a scaled power transformation, we would in practice use the basic power transformation $\psi(X, \lambda)$ in regression modeling, since the two differ only by a scale, location, and possibly sign change. The scaled transformations are used to select a transformation only.

If transforming only the predictor and using a choice from the power family, we begin with the mean function

$$E(Y|X) = \beta_0 + \beta_1 \psi_S(X, \lambda) \tag{7.4}$$

If we know λ, we can fit (7.4) via OLS and get the residual sum of squares, $RSS(\lambda)$. The estimate $\hat{\lambda}$ of λ is simply the value of λ that minimizes $RSS(\lambda)$. As a practical matter, we do not need to know λ very precisely, and selecting λ to minimize $RSS(\lambda)$ from

$$\lambda \in \{-1, -1/2, 0, 1/3, 1/2, 1\} \tag{7.5}$$

is usually adequate.

As an example of transforming only the predictor, we consider the dependence of tree *Height* in decimeters on *Dbh*, the diameter of the tree in mm at 137 cm above the ground, for a sample of western cedar trees in 1991 in the Upper Flat Creek stand of the University of Idaho Experimental Forest (courtesy of Andrew Robinson). The data are in the file `ufcwc.txt`. Figure 7.3 is the scatterplot of

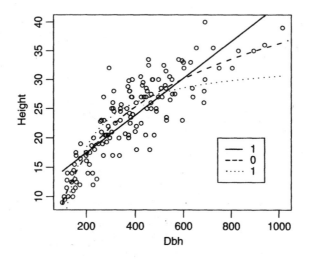

FIG. 7.3 *Height* versus *Dbh* for the red cedar data from Upper Flat Creek.

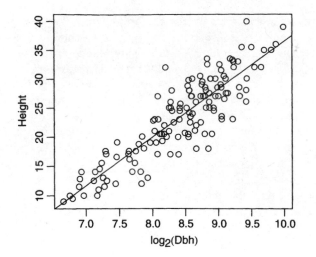

FIG. 7.4 The red cedar data from Upper Flat Creek transformed.

the data, and on this plot we have superimposed three curved lines. For each λ, we computed fitted values $\hat{y}(\lambda)$ from the OLS regression of *Height* on $\psi_S(Dbh, \lambda)$. The line for a particular value of λ is obtained by plotting the points $(Dbh, \hat{y}(\lambda))$ and joining them with a line. Only three values of λ are shown in the figure because it gets too crowded to see much with more lines, but among these three, the choice of $\lambda = 0$ seems to match the data most closely. The choice of $\lambda = 1$ does not match the data for large and small trees, while the inverse is too curved to match the data for larger trees. This suggests replacing *Dbh* with log(*Dbh*), as we have done in Figure 7.4.

As an alternative approach, the value of the transformation parameter can be estimated by fitting using *nonlinear least squares*. The mean function (7.4) is a nonlinear function of the parameters because β_1 multiplies the nonlinear function $\psi_S(X, \lambda)$ of the parameter λ. Using the methods described in Chapter 11, the estimate of λ turns out to be $\hat{\lambda} = 0.05$ with a standard error of 0.15, so $\lambda = 0$ is close enough to believe that this is a sensible transformation to use.

7.1.3 Transforming the Response Only

A transformation of the response only can be selected using an *inverse fitted value plot*, in which we put the fitted values from the regression of Y on X on the vertical axis and the response on the horizontal axis. In simple regression the fitted values are proportional to the predictor X, so an equivalent plot is of X on the vertical axis versus Y on the horizontal axis. The method outlined in Section 7.1.2 can then be applied to this inverse problem, as suggested by Cook and Weisberg (1994). Thus, to estimate a transformation $\psi_S(Y, \lambda_y)$, start with the mean function

$$E(\hat{y}|Y) = \alpha_0 + \alpha_1 \psi_S(Y, \lambda_y)$$

and estimate λ_y. An example of the use of an inverse response plot will be given in Section 7.3.

7.1.4 The Box and Cox Method

Box and Cox (1964) provided another general method for selecting transformations of the response that is applicable both in simple and multiple regression. As with the previous methods, we will select the transformation from a family indexed by a parameter λ. For the Box–Cox method, we need a slightly more complicated version of the power family that we will call the *modified power family*, defined by Box and Cox (1964) for strictly positive Y to be

$$\psi_M(Y, \lambda_y) = \psi_S(Y, \lambda_y) \times \text{gm}(Y)^{1-\lambda_y} \qquad (7.6)$$

$$= \begin{cases} \text{gm}(Y)^{1-\lambda_y} \times (Y^{\lambda_y} - 1)/\lambda_y & \text{if } \lambda_y \neq 0 \\ \text{gm}(Y) \times \log(Y) & \text{if } \lambda_y = 0 \end{cases}$$

where $\text{gm}(Y)$ is the *geometric mean* of the untransformed variable[2].

In the Box–Cox method, we assume that the mean function

$$E(\psi_M(Y, \lambda_y)|X = \mathbf{x}) = \boldsymbol{\beta}'\mathbf{x} \qquad (7.7)$$

holds for some λ_y. If λ_y were known, we could fit the mean function (7.7) using OLS because the transformed response $\psi_M(Y, \lambda_y)$ would then be completely specified. Write the residual sum of squares from this regression as $RSS(\lambda_y)$. Multiplication of the scaled power transformation by $\text{gm}(Y)^{1-\lambda}$ guarantees that the units of $\psi_M(Y, \lambda_y)$ are the same for all values of λ_y, and so all the $RSS(\lambda_y)$ are in the same units. We estimate λ_y to be the value of the transformation parameter that minimizes $RSS(\lambda_y)$. From a practical point of view, we can again select λ_y from among the choices in (7.5).

The Box–Cox method is not transforming for linearity, but rather it is transforming for *normality*: λ is chosen to make the residuals from the regression of $\psi(Y, \lambda_y)$ on X as close to normally distributed as possible. Hernandez and Johnson (1980) point out that "as close to normal as possible" need not be very close to normal, and so graphical checks are desirable after selecting a transformation. The Box and Cox method will also produce a confidence interval for the transformation parameter; see Appendix A.11.1 for details.

7.2 TRANSFORMATIONS AND SCATTERPLOT MATRICES

The data described in Table 7.1 and given in the data file highway.txt are taken from an unpublished master's paper in civil engineering by Carl Hoffstedt.

---

[2]If the values of Y are $y_1, \ldots, y_n$, the geometric mean of Y is $\text{gm}(Y) = \exp(\sum \log(y_i)/n)$, using natural logarithms.

TABLE 7.1   The Highway Accident Data[a]

Variable	Description
Rate	1973 accident rate per million vehicle miles
Len	Length of the segment in miles
ADT	Estimated average daily traffic count in thousands
Trucks	Truck volume as a percent of the total volume
Slim	1973 speed limit
Shld	Shoulder width in feet of outer shoulder on the roadway
Sigs	Number of signalized interchanges per mile in the segment

[a]Additional variables appear in the file **highway.txt**, and will be described in Table 10.5.

They relate the automobile accident rate in accidents per million vehicle miles to several potential terms. The data include 39 sections of large highways in the state of Minnesota in 1973. The goal of this analysis was to understand the impact of the design variables, *Acpts*, *Slim*, *Sigs*, and *Shld* that are under the control of the highway department, on accidents. The other variables are thought to be important determinants of accidents but are more or less beyond the control of the highway department and are included to reduce variability due to these uncontrollable factors. We have no particular reason to believe that *Rate* will be a linear function of the predictors, or any theoretical reason to prefer any particular form for the mean function.

An important first step in this analysis is to examine the scatterplot matrix of all the predictors and the response, as given in Figure 7.5. Here are some observations about this scatterplot matrix that might help in selecting transformations:

1. The variable *Sigs*, the number of traffic lights per mile, is zero for freeway-type road segments but can be well over 2 for other segments. Transformations may help with this variable, but since it has non positive values, we cannot use the power transformations directly. Since *Sigs* is computed as the number of signals divided by *Len*, we will replace *Sigs* by a related variable *Sigs1* defined by

$$Sigs1 = \frac{Sigs \times Len + 1}{Len}$$

This variable is always positive and can be transformed using the power family.

2. *ADT* and *Len* have a large range, and logarithms are likely to be appropriate for them.

3. *Slim* varies only from 40 mph to 70 mph, with most values in the range 50 to 60. Transformations are unlikely to be much use here.

4. Each of the predictors seems to be at least modestly associated with *Rate*, as the mean function for each of the plots in the top row of Figure 7.5 is not flat.

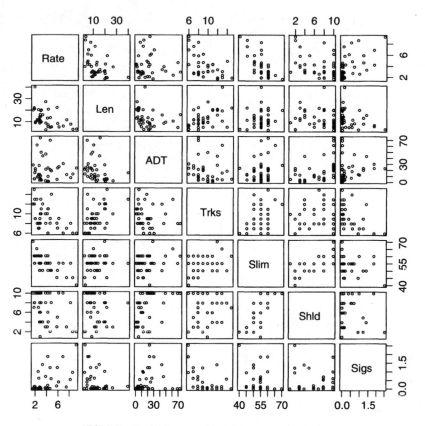

FIG. 7.5   The highway accident data, no transformations.

5. Many of the predictors are also related to each other. In some cases, the mean functions for the plots of predictor versus predictor appear to be linear; in other cases,. they are not linear.

Given these preliminary views of the scatterplot matrix, we now have the daunting task of finding good transformations to use. This raises immediate questions: What are the goals in selecting transformations? How can we decide if we have made a good choice?

The overall goal of transforming in linear regression is to find transformations in which multiple linear regression matches the data to a good approximation. The connection between this goal and choosing transformations that make the 2D plots of predictors have linear mean functions is not entirely obvious. Important work by Brillinger (1983) and Li and Duan (1989) provides a theoretical connection. Suppose we have a response variable Y and a set of predictors X, and suppose it were true that

$$E(Y|X = \mathbf{x}) = g(\boldsymbol{\beta}'\mathbf{x}) \qquad (7.8)$$

for some *completely unknown and unspecified function g*. According to this, the mean of Y depends on X only through a linear combination of the terms in X, and if we could draw a graph of Y versus $\beta'x$, this graph would have g as its mean function. We could then either estimate g, or we could transform Y to make the mean function linear. All this depends on estimating β without specifying anything about g. Are there conditions under which the OLS regression of Y on X can help us learn about β?

7.2.1 The 1D Estimation Result and Linearly Related Predictors

Suppose that $A = a'X$ and $B = b'X$ were any two linear combinations of the terms in X, such that

$$E(A|B) = \gamma_0 + \gamma_1 B \tag{7.9}$$

so the graph of A versus B has a straight-line mean function. We will say that X is a set of *linear predictors* if (7.9) holds for all linear combinations A and B. The condition that all the graphs in a scatterplot matrix of X have straight-line mean functions is weaker than (7.9), but it is a reasonable condition that we can check in practice. Requiring that X has a multivariate normal distribution is much stronger than (7.9). Hall and Li (1993) show that (7.9) holds approximately as the number of predictors grows large, so in very large problems, transformation becomes less important because (7.9) will hold approximately without any transformations.

Given that (7.9) holds at least to a reasonable approximation, and assuming that $E(Y|X = x) = g(\beta'x)$, then the OLS estimate $\hat{\beta}$ is a consistent estimate of $c\beta$ for some constant c that is usually nonzero (Li and Duan, 1989; see also Cook, 1998). Given this theorem, a useful general procedure for applying multiple linear regression analysis is:

1. Transform predictors to get terms for which (7.9) holds, at least approximately. The terms in X may include dummy variables that represent factors, which should not be transformed, as well as transformations of continuous predictors.

2. We can estimate g from the 2D scatterplot of Y versus $\hat{\beta}'x$, where $\hat{\beta}$ is the OLS estimator from the regression of Y on X. Almost equivalently, we can estimate a transformation of Y either from the inverse plot of $\hat{\beta}'x$ versus Y or from using the Box–Cox method.

This is a general and powerful approach to building regression models that match data well, based on the assumption that (7.8) is appropriate for the data. We have already seen mean functions in Chapter 6 for which (7.8) does not hold because of the inclusion of interaction terms, and so transformations chosen using the methods discussed here may not provide a comprehensive mean function when interactions are present.

The Li–Duan theorem is actually much more general and has been extended to problems with interactions present and to many other estimation methods beyond

OLS. See Cook and Weisberg (1999a, Chapters 18–20) and, at a higher mathematical level, Cook (1998).

7.2.2 Automatic Choice of Transformation of Predictors

Using the results of Section 7.2.1, we seek to transform the predictors so that all plots of one predictor versus another have a linear mean function, or at least have mean functions that are not too curved. Without interactive graphical tools, or some automatic method for selecting transformations, this can be a discouraging task, as the analyst may need to draw many scatterplot matrices to get a useful set of transformations.

Velilla (1993) proposed a multivariate extension of the Box and Cox method to select transformations to linearity, and this method can often suggest a very good starting point for selecting transformations of predictors. Starting with k untransformed strictly positive predictors $X = (X_1, \ldots, X_k)$, we will apply a modified power transformation to each X_j, and so there will be k transformation parameters collected into $\lambda = (\lambda_1, \lambda_2, \ldots, \lambda_k)'$. We will write $\psi_M(X, \lambda)$ to be the set of variables

$$\psi_M(X, \lambda) = (\psi_M(X_1, \lambda_1), \ldots, \psi_M(X_k, \lambda_k))$$

Let $V(\lambda)$ be the *sample covariance matrix of the transformed data* $\psi_M(X, \lambda)$. The value $\hat{\lambda}$ is selected as the value of λ that minimizes the logarithm of the determinant of $V(\lambda)$. This minimization can be carried using a general function minimizer included in high-level languages such as R, S-plus, Maple, Mathematica, or even Excel. The minimizers generally require only specification of the function to be minimized and a set of starting values for the algorithm. The starting values can be taken to be $\lambda = 0$, $\lambda = 1$, or some other appropriate vector of zeros and ones.

Returning to the highway data, we eliminate *Slim* as a variable to be transformed because its range is too narrow. For the remaining terms, we get the summary of transformations using the multivariate Box–Cox method in Table 7.2. The table

TABLE 7.2 Power Transformations to Normality for the Highway Data

```
Box Cox Transformations to Multivariate normality

        Est.Power Std.Err. Wald(Power=0) Wald(Power=1)
Len      0.1429    0.2124       0.6728        -4.0349
ADT      0.0501    0.1204       0.4162        -7.8904
Trks    -0.7019    0.6132      -1.1447        -2.7754
Shld     1.3455    0.3630       3.7065         0.9518
Sigs1   -0.2440    0.1488      -1.6402        -8.3621


L.R. test, all powers = 0:      23.373   df = 5   p = 3e-04
L.R. test, all powers = 1:     133.179   df = 5   p = 0
L.R. test, of (0,0,0,1,0):       6.143   df = 5   p = 0.29
```

gives the value of $\hat{\lambda}$ in the column marked "Est. power." The standard errors are computed as outlined in Appendix A.11.2. For our purposes, the standard errors can be treated like standard errors of regression coefficients. The next two columns are like t-tests of the transformation parameter equal to zero or to one. These tests should be compared with a normal distribution, so values larger in absolute value than 1.96 correspond to p-values less than 0.05. The power parameters for *Len*, *ADT*, *Trks* and *Sigs1* do not appear to be different from zero, and *Shld* does not appear to be different from one. At the foot of the table are three *likelihood ratio tests*. The first of these tests is that all powers are zero; this is firmly rejected as the approximate $\chi^2(5)$ is very large. Similarly, the test for no transformation ($\lambda = 1$) is firmly rejected. The test that the first three variables should be in log scale, the next untransformed, and the last in log scale, has a p-value 0.29 and suggests using these simple transformations in further analysis with these data. The predictors in transformed scale, along with the response, are shown in Figure 7.6. All these 2D plots have a linear mean function, or at least are not strongly nonlinear. They provide a good place to start regression modeling.

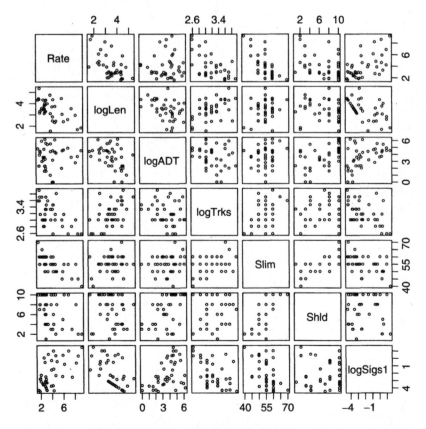

FIG. 7.6 Transformed predictors for the highway data.

7.3 TRANSFORMING THE RESPONSE

Once the terms are transformed, we can turn our attention to transforming the response. Figure 7.7 is the inverse fitted value plot for the highway data using the transformed terms determined in the last section. This plot has the response *Rate* on the horizontal axis and the fitted values from the regression of *Rate* on the transformed predictors on the vertical axis. Cook and Weisberg (1994) have shown that if the predictors are approximately linearly related, then we can use the method of Section 7.1.2 to select a transformation for *Rate*. Among the three curves shown on this plot, the logarithmic seems to be the most appropriate.

The Box–Cox method provides an alternative procedure for finding a transformation of the response. It is often summarized by a graph with λ_y on the horizontal axis and either $RSS(\lambda_y)$ or better yet $-(n/2)\log(RSS(\lambda_y)/n)$ on the vertical axis. With this latter choice, the estimate $\hat{\lambda}_y$ is the point that maximizes the curve, and a confidence interval for the estimate is given by the set of all λ_y with $\log(L(\hat{\lambda}_y)) - \log(L(\lambda_y) < 1.92$; see Appendix A.11.1. This graph for the highway data is shown in Figure 7.8, with $\hat{\lambda} \approx -0.2$ and the confidence interval of about -0.8 to $+0.3$. The log transformation is in the confidence interval, agreeing with the inverse fitted value plot.

In the highway data, the two transformation methods for the response seem to agree, but there is no theoretical reason why they need to give the same transformation. The following path is recommended for selecting a response transformation:

1. With approximately linear predictors, draw the inverse response plot of $\hat{y}$ versus the response. If this plot shows a clear nonlinear trend, then the response should be transformed to match the nonlinear trend. There is no reason why only power transformations should be considered. For example,

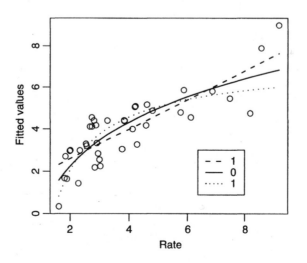

FIG. 7.7 Inverse fitted value plot for the highway data.

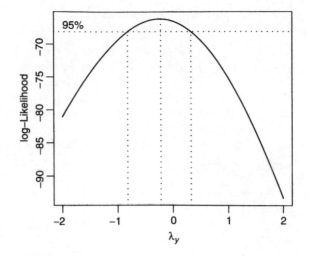

FIG. 7.8   Box–Cox summary graph for the highway data.

the transformation could be selected using a smoother. If there is no clear nonlinear trend, transformation of the response is unlikely to be helpful.

2. The Box–Cox procedure can be used to select a transformation to normality. It requires the use of a transformation family.

For the highway data, we now have a reasonable starting point for regression, with several of the predictors and the response all transformed to log scale. We will continue with this example in later chapters.

7.4   TRANSFORMATIONS OF NONPOSITIVE VARIABLES

Several transformation families for a variable U that includes negative values have been suggested. The central idea is to use the methods discussed in this chapter for selecting a transformation from a family but to use a family that permits U to be non positive. One possibility is to consider transformations of the form $(U + \gamma)^\lambda$, where γ is sufficiently large to ensure that $U + \gamma$ is strictly positive. We used a variant of this method with the variable *Sigs* in the highway data. In principle, (γ, λ) could be estimated simultaneously, although in practice estimates of γ are highly variable and unreliable. Alternatively, Yeo and Johnson (2000) proposed a family of transformations that can be used without restrictions on U that have many of the good properties of the Box–Cox power family. These transformations are defined by

$$\psi_{YJ}(U, \lambda) = \begin{cases} \psi_M(U + 1, \lambda) & \text{if } U \geq 0 \\ \psi_M(-U + 1, 2 - \lambda) & \text{if } U < 0 \end{cases} \tag{7.10}$$

If U is strictly positive, then the Yeo–Johnson transformation is the same as the Box–Cox power transformation of $(U + 1)$. If U is strictly negative, then the Yeo–Johnson transformation is the Box–Cox power transformation of $(-U + 1)$,

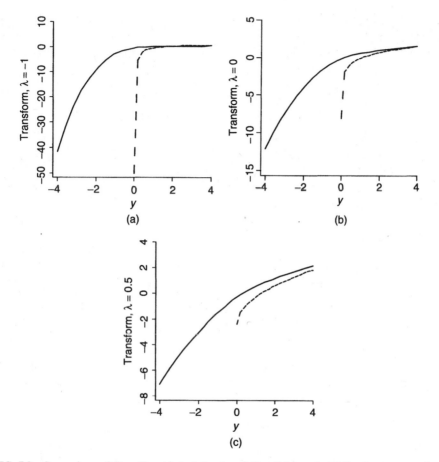

FIG. 7.9 Comparison of Box–Cox (dashed lines) and Yeo–Johnson (solid lines) power transformations for $\lambda = -1, 0, 0.5$. The Box–Cox transformations and Yeo–Johnson transformations behave differently for values of y close to zero.

but with power $2 - \lambda$. With both negative and positive values, the transformation is a mixture of these two, so different powers are used for positive and negative values. In this latter case, interpretation of the transformation parameter is difficult, as it has a different meaning for $U \geq 0$ and for $U < 0$. Figure 7.9 shows the Box–Cox transformation and Yeo–Johnson transformation for the values of $\lambda = -1, 0, .5$. For positive values, the two transformations differ in their behavior with values close to zero, with the Box–Cox transformations providing a much larger change for small values than do the Yeo–Johnson transformations.

PROBLEMS

7.1. The data in the file `baeskel.txt` were collected in a study of the effect of dissolved sulfur on the surface tension of liquid copper (Baes and Kellogg,

1953). The predictor *Sulfur* is the weight percent sulfur, and the response is *Tension*, the decrease in surface tension in dynes per cm. Two replicate observations were taken at each value of *Sulfur*. These data were previously discussed by Sclove (1968).

7.1.1. Draw the plot of *Tension* versus *Sulfur* to verify that a transformation is required to achieve a straight-line mean function.

7.1.2. Set $\lambda = -1$, and fit the mean function

$$\text{E}(Tension|Sulfur) = \beta_0 + \beta_1 Sulfur^\lambda$$

using OLS; that is, fit the OLS regression with *Tension* as the response and $1/Sulfur$ as the predictor. Let *new* be a vector of 100 equally spaced values between the minimum value of *Sulfur* and its maximum value. Compute the fitted values from the regression you just fit, given by $Fit.new = \hat{\beta}_0 + \hat{\beta}_1 new^\lambda$. Then, add to the graph you drew in Problem 7.1.1 the line joining the points (*new*, *Fit.new*). Repeat for $\lambda = 0, 1$. Which of these three choices of λ gives fitted values that match the data most closely?

7.1.3. Replace *Sulfur* by its logarithm, and consider transforming the response *Tension*. To do this, draw the inverse fitted value plot with the fitted values from the regression of *Tension* on log log(*Sulphur*) on the vertical axis and *Tension* on the horizontal axis. Repeat the methodology of Problem 7.1.2 to decide if further transformation of the response will be helpful.

7.2. The (hypothetical) data in the file `stopping.txt` give stopping times for $n = 62$ trials of various automobiles traveling at *Speed* miles per hour and the resulting stopping *Distance* in feet (Ezekiel and Fox, 1959).

7.2.1. Draw the scatterplot of *Distance* versus *Speed*. Add the simple regression mean function to your plot. What problems are apparent? Compute a test for lack of fit, and summarize results.

7.2.2. Find an appropriate transformation for *Distance* that can linearize this regression.

7.2.3. Hald (1960) has suggested on the basis of a theoretical argument that the mean function $\text{E}(Distance|Speed) = \beta_0 + \beta_1 Speed + \beta_2 Speed^2$, with $\text{Var}(Distance|Speed) = \sigma^2 Speed^2$ is appropriate for data of this type. Compare the fit of this model to the model found in Problem 7.2.2. For *Speed* in the range 0 to 40 mph, draw the curves that give the predicted *Distance* from each model, and qualitatively compare them.

7.3. This problem uses the data discussed in Problem 1.5. A major source of water in Southern California is the Owens Valley. This water supply is in turn replenished by spring runoff from the Sierra Nevada mountains. If runoff could be predicted, engineers, planners, and policy makers could do their jobs more efficiently. The data in the file water.txt contains 43 years' of precipitation measurements taken at six sites in the mountains, in inches of water, and stream runoff volume at a site near Bishop, California. The three sites with name starting with "O" are fairly close to each other, and the three sites starting with "A" are also fairly close to each other.

7.3.1. Load the data file, and construct the scatterplot matrix of the six snowfall variables, which are the predictors in this problem. Using the methodology for automatic choice of transformations outlined in Section 7.2.2, find transformations to make the predictors as close to linearly related as possible. Obtain a test of the hypothesis that all $\lambda_j = 0$ against a general alternative, and summarize your results. Do the transformations you found appear to achieve linearity? How do you know?

7.3.2. Given log transformations of the predictors, show that a log transformation of the response is reasonable.

7.3.3. Consider the multiple linear regression model with mean function given by

$$E(\log(y)|\mathbf{x}) = \beta_0 + \beta_1 \log(APMAM) + \beta_2 \log(APSAB)$$
$$+ \beta_3 \log(APSLAKE) + \beta_4 \log(OPBPC)$$
$$+ \beta_5 \log(OPRC) + \beta_6 \log(OPSLAKE)$$

with constant variance function. Estimate the regression coefficients using OLS. You will find that two of the estimates are negative; Which are they? Does a negative coefficient make any sense? Why are the coefficients negative?

7.3.4. In the OLS fit, the regression coefficient estimates for the three predictors beginning with "O" are approximately equal. Are there conditions under which one might expect these coefficients to be equal? What are they? Test the hypothesis that they are equal against the alternative that they are not all equal.

7.3.5. Write one or two paragraphs that summarize the use of the snowfall variables to predict runoff. The summary should discuss the important predictors, give useful graphical summaries, and give an estimate of variability. Be creative.

7.4. The data in the file salarygov.txt give the maximum monthly salary for 495 non-unionized job classes in a midwestern governmental unit in 1986. The variables are described in Table 7.3.

7.4.1. The data as given has as its unit of analysis the *job class*. In a study of the dependence of maximum salary on skill, one might prefer to have as unit of analysis the *employee*, not the job class. Discuss how this preference would change the analysis.

TABLE 7.3 **The Governmental Salary Data**

Variable	Description
MaxSalary	Maximum salary in dollars for employees in this job class, the response
NE	Total number of employees currently employed in this job class
NW	Number of women employees in the job class
Score	Score for job class based on difficulty, skill level, training requirements and level of responsibility as determined by a consultant to the governmental unit. This value for these data is in the range between 82 to 1017.
JobClass	Name of the job class; a few names were illegible or partly illegible

7.4.2. Examine the scatterplot of *MaxSalary* versus *Score*. Find transformation(s) that would make the mean function for the resulting scatterplot approximately linear. Does the transformation you choose also appear to achieve constant variance?

7.4.3. According to Minnesota statutes, and probably laws in other states as well, a job class is considered to be female dominated if 70% of the employees or more in the job class are female. These data were collected to examine whether female-dominated positions are compensated at a lower level, adjusting for *Score*, than are other positions. Create a factor with two levels that divides the job classes into female dominated or not, fit appropriate models, and summarize your results. Be mindful of the need to transform variables and the possibility of weighting.

7.4.4. An alternative to using a factor for female-dominated jobs is to use a term *NW/NE*, the fraction of women in the job class. Repeat the last problem, but encoding the information about sex using this variable in place of the factor.

7.5. World cities The Union Bank of Switzerland publishes a report entitled *Prices and Earnings Around the Globe* on their internet web site, www.ubs.com. The data in the file `BigMac2003.txt` and described in Table 7.4 are taken from their 2003 version for 70 world cities.

7.5.1. Draw the scatterplot with *BigMac* on the vertical axis and *FoodIndex* on the horizontal axis. Provide a qualitative description of this graph. Use an inverse fitted value plot and the Box–Cox method to find a transformation of *BigMac* so that the resulting scatterplot has a linear mean function. Two of the cities, with very large values for *BigMac*, are very influential for selecting a transformation. You should do this exercise with all the cities and with those two cities removed.

7.5.2. Draw the scatterplot matrix of the three variables (*BigMac*, *Rice*, *Bread*), and use the multivariate Box–Cox procedure to decide on normalizing transformations. Test the null hypothesis that $\lambda = (1, 1, 1)'$ against a

TABLE 7.4   Global Price Comparison Data

Variable	Description
BigMac	Minutes of labor to buy a Big Mac hamburger based on a typical wage averaged over 13 occupations
Bread	Minutes of labor to buy 1 kg bread
Rice	Minutes of labor to buy 1 kg of rice
Bus	Lowest cost of 10 km public transit
FoodIndex	Food price index, Zurich=100
TeachGI	Primary teacher's gross annual salary, thousands of US dollars
TeachNI	Primary teacher's net annual salary, thousands of US dollars
TaxRate	$100 \times (TeachGI - TeachNI)/TeackGI$. In some places, this is negative, suggesting a government subsidy rather than tax
TH	Teacher's hours per week of work
Apt	Monthly rent in US dollars of a typical three-room apartment
City	City name

Source: Most of the data are from the Union Bank of Switzerland publication *Prices and Earnings Around the Globe*, 2003 edition, from www.ubs.com.

general alternative. Does deleting Karachi and Nairobi change your conclusions?

7.5.3. Set up the regression using the four terms, $\log(Bread)$, $\log(Bus)$, $\log(TeachGI)$, and $Apt^{0.33}$, and with response *BigMac*. Draw the inverse fitted value plot of $\hat{y}$ versus *BigMac*. Estimate the best power transformation. Check on the adequacy of your estimate by refitting the regression model with the transformed response and drawing the inverse fitted value plot again. If transformation was successful, this second inverse fitted value plot should have a linear mean function.

7.6. The data in the file `wool.txt` were introduced in Section 6.3. For this problem, we will start with *Cycles*, rather than its logarithm, as the response.

7.6.1. Draw the scatterplot matrix for these data and summarize the information in this plot.

7.6.2. View all three predictors as factors with three levels, and *without transforming* Cycles, fit the second-order mean function with terms for all main effects and all two-factor interactions. Summarize results.

7.6.3. Fit the first-order mean function consisting only of the main effects. From Problem 7.6.2, this mean function is not adequate for these data based on using *Cycles* as the response. Use both the inverse fitted value plot and the Box–Cox method to select a transformation for *Cycles* based on the first-order mean function.

7.6.4. In the transformed scale, refit the second-order model, and show that none of the interactions are required in this scale. For this problem, the transformation leads to a much simpler model than is required for the response in the original scale. This is an example of *removable nonadditivity*.

7.7. Justify transforming *Miles* in the Fuel data.

7.8. The data file UN3.txt contains data described in Table 7.5. There are data for $n = 125$ localities, mostly UN member countries, for which values are observed for all the variables recorded.

Consider the regression problem with *ModernC* as the response variable and the other variables in the file as defining terms.

7.8.1. Select appropriate transformations of the predictors to be used as terms. (*Hint*: Since *Change* is negative for some localities, the Box–Cox family of transformations cannot be used directly.)

7.8.2. Given the transformed predictors as terms, select a transformation for the response.

7.8.3. Fit the regression using the transformations you have obtained, and summarize your results.

TABLE 7.5 Description of Variables in the Data File UN3.txt

Variable	Description
Locality	Country/locality name
ModernC	Percent of unmarried women using a modern method of contraception
Change	Annual population growth rate, percent
PPgdp	Per capita gross national product, US dollars
Frate	Percent of females over age 15 economically active
Pop	Total 2001 population, 1000s
Fertility	Expected number of live births per female, 2000
Purban	Percent of population that is urban, 2001

Source: The data were collected from http://unstats.un.org/unsd/demographic and refer to values collected between 2000 and 2003.

CHAPTER 8

Regression Diagnostics: Residuals

So far in this book, we have mostly used graphs to help us decide what to do before fitting a regression model. *Regression diagnostics* are used *after* fitting to check if a fitted mean function and assumptions are consistent with observed data. The basic statistics here are the residuals or possibly rescaled residuals. If the fitted model does not give a set of residuals that appear to be reasonable, then some aspect of the model, either the assumed mean function or assumptions concerning the variance function, may be called into doubt. A related issue is the importance of each case on estimation and other aspects of the analysis. In some data sets, the observed statistics may change in important ways if one case is deleted from the data. Such a case is called *influential*, and we shall learn to detect such cases. We will be led to study and use two relatively unfamiliar diagnostic statistics, called *distance measures* and *leverage values*. We concentrate on graphical diagnostics but include numerical quantities that can aid in interpretation of the graphs.

8.1 THE RESIDUALS

Using the matrix notation outlined in Chapter 3, we begin by deriving the properties of residuals. The basic multiple linear regression model is given by

$$\mathbf{Y} = \mathbf{X}\boldsymbol{\beta} + \mathbf{e} \quad \text{Var}(\mathbf{e}) = \hat{\sigma}^2\mathbf{I} \tag{8.1}$$

where $\mathbf{X}$ is a known matrix with n rows and p' columns, including a column of 1s for the intercept if the intercept is included in the mean function. We will further assume that we have selected a parameterization for the mean function so that $\mathbf{X}$ has full column rank, meaning that the inverse $(\mathbf{X}'\mathbf{X})^{-1}$ exists; as we have seen previously, this is not an important limitation on regression models because we can always delete terms from the mean function, or equivalently delete columns from $\mathbf{X}$, until we have full rank. The $p' \times 1$ vector $\boldsymbol{\beta}$ is the unknown parameter vector.

Applied Linear Regression, Third Edition, by Sanford Weisberg
ISBN 0-471-66379-4   Copyright © 2005 John Wiley & Sons, Inc.

The vector **e** consists of unobservable errors that we assume are equally variable and uncorrelated, unless stated otherwise.

In fitting model (8.1), we estimate β by $\hat{\beta} = (\mathbf{X'X})^{-1}\mathbf{X'Y}$, and the fitted values $\hat{\mathbf{Y}}$ corresponding to the observed values **Y** are then given by

$$\hat{\mathbf{Y}} = \mathbf{X}\hat{\beta}$$
$$= \mathbf{X}(\mathbf{X'X})^{-1}\mathbf{X'Y}$$
$$= \mathbf{HY} \tag{8.2}$$

where **H** is the $n \times n$ matrix defined by

$$\mathbf{H} = \mathbf{X}(\mathbf{X'X})^{-1}\mathbf{X'} \tag{8.3}$$

H is called the *hat matrix* because it transforms the vector of observed responses **Y** into the vector of fitted responses $\hat{\mathbf{Y}}$. The vector of residuals $\hat{\mathbf{e}}$ is defined by

$$\hat{\mathbf{e}} = \mathbf{Y} - \hat{\mathbf{Y}}$$
$$= \mathbf{Y} - \mathbf{X}\hat{\beta}$$
$$= \mathbf{Y} - \mathbf{X}(\mathbf{X'X})^{-1}\mathbf{X'Y}$$
$$= (\mathbf{I} - \mathbf{H})\mathbf{Y} \tag{8.4}$$

8.1.1   Difference Between $\hat{\mathbf{e}}$ and **e**

The errors **e** are unobservable random variables, assumed to have zero mean and uncorrelated elements, each with common variance σ^2. The residuals $\hat{\mathbf{e}}$ are computed quantities that can be graphed or otherwise studied. Their mean and variance, using (8.4) and Appendix A.7, are

$$\mathrm{E}(\hat{\mathbf{e}}) = \mathbf{0}$$
$$\mathrm{Var}(\hat{\mathbf{e}}) = \sigma^2(\mathbf{I} - \mathbf{H}) \tag{8.5}$$

Like the errors, each of the residuals has zero mean, but each residual may have a different variance. Unlike the errors, the residuals are correlated. From (8.4), the residuals are linear combinations of the errors. If the errors are normally distributed, so are the residuals. If the intercept is included in the model, then the sum of the residuals is 0, $\hat{\mathbf{e}}'\mathbf{1} = \sum \hat{e}_i = 0$. In scalar form, the variance of the ith residual is

$$\mathrm{Var}(\hat{e}_i) = \hat{\sigma}^2(1 - h_{ii}) \tag{8.6}$$

where h_{ii} is the ith diagonal element of **H**. Diagnostic procedures are based on the computed residuals, which we would like to assume behave as the unobservable errors would. The usefulness of this assumption depends on the hat matrix, since it is **H** that relates **e** to $\hat{\mathbf{e}}$ and also gives the variances and covariances of the residuals.

8.1.2   The Hat Matrix

H is $n \times n$ and symmetric with many special properties that are easy to verify directly from (8.3). Multiplying X on the left by H leaves X unchanged, $HX = X$. Similarly, $(I - H)X = 0$. The property $HH = H^2 = H$ also shows that $H(I - H) = 0$, so the covariance between the fitted values HY and residuals $(I - H)Y$ is

$$Cov(\hat{e}, \hat{Y}) = Cov(HY, (I - H)Y)$$

$$= \sigma^2 H(I - H) = 0$$

Another name for H is the *orthogonal projection* on the column space of X. The elements of H, the h_{ij}, are given by

$$h_{ij} = x_i'(X'X)^{-1}x_j = x_j'(X'X)^{-1}x_i = h_{ji} \qquad (8.7)$$

Many helpful relationships can be found between the h_{ij}. For example,

$$\sum_{i=1}^{n} h_{ii} = p' \qquad (8.8)$$

and, if the mean function includes an intercept,

$$\sum_{i=1}^{n} h_{ij} = \sum_{j=1}^{n} h_{ij} = 1 \qquad (8.9)$$

Each diagonal element h_{ii} is bounded below by $1/n$ and above by $1/r$, if r is the number of rows of X that are identical to x_i.

As can be seen from (8.6), cases with large values of h_{ii} will have small values for $Var(\hat{e}_i)$; as h_{ii} gets closer to 1, this variance will approach 0. For such a case, no matter what value of y_i is observed for the ith case, we are nearly certain to get a residual near 0. Hoaglin and Welsch (1978) pointed this out using a scalar version of (8.2),

$$\hat{y}_i = \sum_{j=1}^{n} h_{ij} y_j = h_{ii} y_i + \sum_{j \neq i}^{n} h_{ij} y_j \qquad (8.10)$$

In combination with (8.9), equation (8.10) shows that as h_{ii} approaches 1, $\hat{y}_i$ gets closer to y_i. For this reason, they called h_{ii} the *leverage* of the ith case.

Cases with large values of h_{ii} will have unusual values for x_i. Assuming that the intercept is in the mean function, and using the notation of the deviations from the average cross-products matrix discussed in Chapter 3, h_{ii} can be written as

$$h_{ii} = \tfrac{1}{n} + (x_i^* - \bar{x})'(\mathcal{X}'\mathcal{X})^{-1}(x_i^* - \bar{x}) \qquad (8.11)$$

The second term on the right-hand side of (8.11) is the equation of an ellipsoid centered at $\bar{x}$, and $x_i' = (1, x_i^{*'})$.

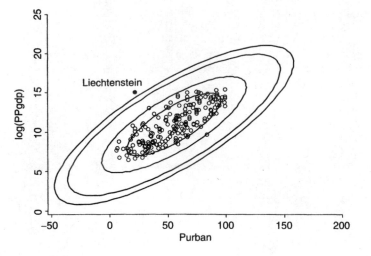

FIG. 8.1 Contours of constant leverage in two dimensions.

For example, consider again the United Nations data, Section 3.1. The plot of log(*PPgdp*) versus *Purban* is given in the scatterplot in Figure 8.1. The ellipses drawn on graph correspond to elliptical contours of constant h_{ii} for $h_{ii} = 0.02$, 0.04, 0.06, 0.08, and 0.10. Any point that falls exactly on the outer contour would have $h_{ii} = 0.10$, while points on the innermost contour have $h_{ii} = 0.02$. Points near the long or major axis of the ellipsoid need to be much farther away from $\bar{\mathbf{x}}$, in the usual Euclidean distance sense, than do points closer to the minor axis, to have the same values for h_{ii}[1].

In the example, the localities with the highest level of urbanization, which are Bermuda, Hong Kong, Singapore, and Guadalupe, all with 100% urbanization, do not have particularly high leverage, as all the points for these places are between the contour for $h_{ii} = 0.04$ and 0.06. None of the h_{ii} is very large, with the largest value for the marked point for Liechtenstein, which has relatively high income for relatively low urbanization. In other problems, high-leverage points with values close to 1 can occur, and identifying these cases is very useful in understanding a regression problem.

8.1.3 Residuals and the Hat Matrix with Weights

When $\text{Var}(\mathbf{e}) = \sigma^2 \mathbf{W}^{-1}$ with $\mathbf{W}$ a known diagonal matrix of positive weights as in Section 5.1, all the results so far in this section require some modification. A useful version of the hat matrix is given by (see Problem 8.9)

$$\mathbf{H} = \mathbf{W}^{1/2}\mathbf{X}(\mathbf{X}'\mathbf{W}\mathbf{X})^{-1}\mathbf{X}'\mathbf{W}^{1/2} \tag{8.12}$$

[1]The term *Purban* is a percentage between 0 and 100. Contours of constant leverage corresponding to *Purban* > 100 are shown to give the shape of the contours, even though in this particular problem points could not occur in this region.

and the leverages are the diagonal elements of this matrix. The fitted values are given as usual by $\hat{\mathbf{Y}} = \mathbf{X}\hat{\boldsymbol{\beta}}$, where now $\hat{\boldsymbol{\beta}}$ is the WLS estimator.

The definition of the residuals is a little trickier. The "obvious" definition of a residual is, in scalar version, $y_i - \hat{\boldsymbol{\beta}}' \mathbf{x}_i$, but this choice has important deficiencies. First, the sum of squares of these residuals will *not* equal the residual sum of squares because the weights are ignored. Second, the variance of the ith residual will depend on the weight of case i.

Both of these problems can be solved by defining residuals for weighted least squares for $i = 1, \ldots, n$ by

$$\hat{e}_i = \sqrt{w_i}(y_i - \hat{\boldsymbol{\beta}}' \mathbf{x}_i) \tag{8.13}$$

The sum of squares of these residuals is the residual sum of squares, and the variance of the residuals does not depend on the weight. When all the weights are equal to 1, (8.13) reduces to (8.4). In drawing graphs and other diagnostic procedures discussed in this book, (8.13) should be used to define residuals. Some computer packages use the unweighted residuals rather than (8.13) by default. There is no consistent name for these residuals. For example, in R and S-plus, the residuals defined by (8.13) are called *Pearson residuals* in some functions, and *weighted residuals* elsewhere. *In this book, the symbols $\hat{e}_i$ or $\hat{\mathbf{e}}$ always refer to the residuals defined by (8.13).*

8.1.4 The Residuals When the Model Is Correct

Suppose that U is equal to one of the terms in the mean function, or some linear combination of the terms. Residuals are generally used in scatterplots of the residuals $\hat{\mathbf{e}}$ against U. The key features of these residual plots when the correct model is fit are as follows:

1. The mean function is $E(\hat{\mathbf{e}}|U) = 0$. This means that the scatterplot of residuals on the horizontal axis versus *any linear combination of the terms* should have a constant mean function equal to 0.

2. Since $\text{Var}(\hat{\mathbf{e}}|U) = \sigma^2(1 - h_{ii})$ even if the fitted model is correct, the variance function is not quite constant. The variability will be smaller for high-leverage cases with h_{ii} close to 1.

3. The residuals are correlated, but this correlation is generally unimportant and not visible in residual plots.

When the model is correct, residual plots should look like null plots.

8.1.5 The Residuals When the Model Is Not Correct

If the fitted model is based on incorrect assumptions, there will be a plot of residuals versus some term or combination of terms that is not a null plot. Figure 8.2 shows several generic residual plots for a simple linear regression problem. The first plot is

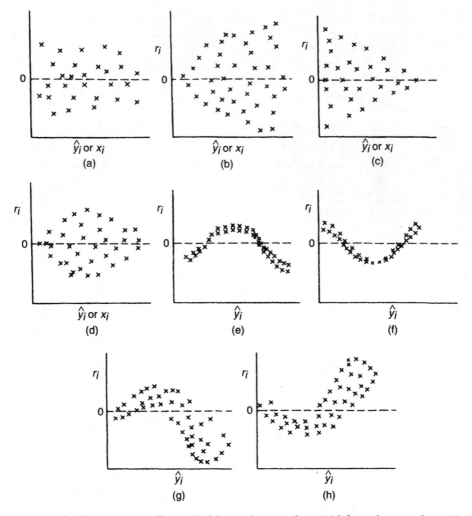

FIG. 8.2 Residual plots: (a) null plot; (b) right-opening megaphone; (c) left-opening megaphone; (d) double outward box; (e)–(f) nonlinearity; (g)–(h) combinations of nonlinearity and nonconstant variance function.

a null plot that indicates no problems with the fitted model. From Figures 8.2b–d, in simple regression, we would infer nonconstant variance as a function of the quantity plotted on the horizontal axis. The curvature apparent in Figures 8.2e–h suggests an incorrectly specified mean function. Figures 8.2g–h suggest both curvature and nonconstant variance.

In models with many terms, we cannot necessarily associate shapes in a residual plot with a particular problem with the assumptions. For example, Figure 8.3 shows a residual plot for the fit of the mean function $E(Y|X = x) = \beta_0 + \beta_1 x_1 + \beta_2 x_2$ for the artificial data given in the file `caution.txt` from Cook and Weisberg

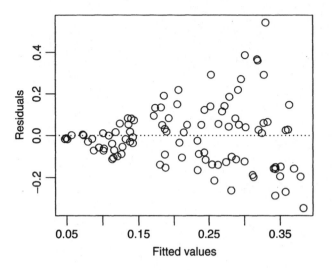

FIG. 8.3 Residual plot for the caution data.

(1999b). The right-opening megaphone is clear in this graph, suggesting nonconstant variance. But these data were actually generated using a mean function

$$E(Y|X = \mathbf{x}) = \frac{|x_1|}{2 + (1.5 + x_2)^2} \tag{8.14}$$

with constant variance, with scatterplot matrix given in Figure 8.4. The real problem is that the mean function is wrong, even though from the residual plot, nonconstant variance appears to be the problem. A nonnull residual plot in multiple regression be indicates that something is wrong but does not necessarily tell what is wrong.

Residual plots in multiple regression *can* be interpreted just as residual plots in simple regression if two conditions are satisfied. First, the predictors should be approximately linearly related, Section 7.2.1. The second condition is on the mean function: we must be able to write the mean function in the form $E(Y|X = \mathbf{x}) = g(\boldsymbol{\beta}'\mathbf{x})$ for some unspecified function g. If either of these conditions fails, then residual plots cannot be interpreted as in simple regression (Cook and Weisberg, 1999b). In the caution data, the first condition is satisfied, as can be verified by looking at the plot of X_1 versus X_2 in Figure 8.4, but the second condition fails because (8.14) cannot be written as a function of a single linear combination of the terms.

8.1.6 Fuel Consumption Data

According to theory, if the mean function and other assumptions are correct, then *all possible residual plots* of residuals versus any function of the terms should resemble a null plot, so many plots of residuals should be examined. Usual choices

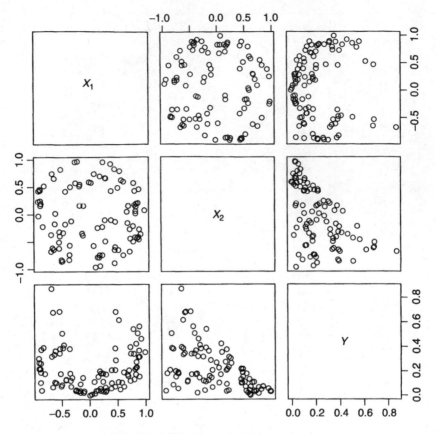

FIG. 8.4   Scatterplot matrix for the caution data.

include plots versus each of the terms and versus fitted values, as shown in Figure 8.5 for the fuel consumption data. None of the plots versus individual terms in Figure 8.5a–d suggest any particular problems, apart from the relatively large positive residual for Wyoming and large negative residual for Alaska. In some of the graphs, the point for the District of Columbia is separated from the others. Wyoming is large but sparsely populated with a well-developed road system. Driving long distances for the necessities of life like going to see a doctor will be common in this state. While Alaska is also very large and sparsely populated, most people live in relatively small areas around cities. Much of Alaska is not accessible by road. These conditions should result in lower use of motor fuel than might otherwise be expected. The District of Columbia is a very compact urban area with good rapid transit, so use of cars will generally be less. It has a small residual but unusual values for the terms in the mean function, so it is separated horizontally from most of the rest of the data. The District of Columbia has high leverage ($h_{9,9} = 0.415$), while the other two are candidates for outliers. We will return to these issues in the next chapter.

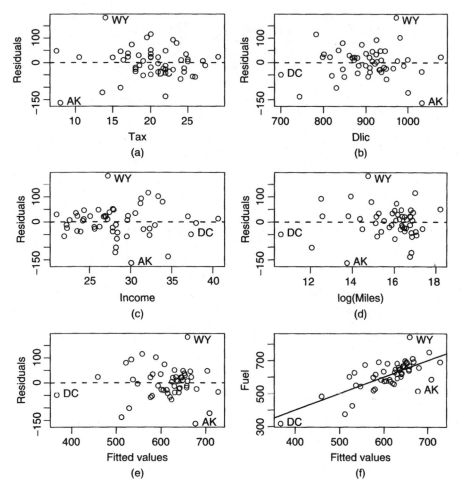

FIG. 8.5 Residual plots for the fuel consumption data.

Figure 8.5e is a plot of residuals versus the fitted values, which are just a linear combination of the terms. Some computer packages will produce this graph as the *only* plot of residuals, and if only one plot were possible, this would be the plot to draw, as it contains some information from all the terms in the mean function. There is a hint of curvature in this plot, possibly suggesting that the mean function is not adequate for the data. We will look at this more carefully in the next section.

Figure 8.5f is different from the others because it is not a plot of residuals but rather a plot of the response versus the fitted values. This is really just a rescaling of Figure 8.5e. If the mean function and the assumptions appear to be sensible, then this figure is a summary graph for the regression problem. The mean function for the graph should be the straight line shown in the figure, which is just the fitted OLS simple regression fit of the response on the fitted values.

TABLE 8.1   Significance Levels for the Lack-of-Fit Tests for the Residual Plots in Figure 8.5

Term	Test Stat.	$Pr(> |t|)$
Tax	−1.08	0.29
Dlic	−1.92	0.06
Income	−0.09	0.93
log(Miles)	−1.35	0.18
Fitted values	−1.45	0.15

8.2   TESTING FOR CURVATURE

Tests can be computed to help decide if residual plots such as those in Figure 8.5 are null plots or not. One helpful test looks for curvature in this plot. Suppose we have a plot of residuals $\hat{e}$ versus a quantity U on the horizontal axis, where U could be a term in the mean function or a combination of terms[2]. A simple test for curvature is to refit the original mean function with an additional term for U^2 added. The test for curvature is then based on the t-statistic for testing the coefficient for U^2 to be 0. If U does not depend on estimated coefficients, then a usual t-test of this hypothesis can be used. If U is equal to the fitted values so that it depends the estimated coefficients, then the test statistic should be compared with the standard normal distribution to get significance levels. This latter case is called *Tukey's test for nonadditivity* (Tukey, 1949).

Table 8.1 gives the lack-of-fit tests for the residual plots in Figure 8.5. None of the tests have small significance levels, providing no evidence against the mean function.

As a second example, consider again the United Nations data from Section 3.1 with response log(*Fertility*) and two predictors log(*PPgdp*) and *Purban*. The apparent linearity in all the frames of the scatterplot matrix in Figure 8.6 suggests that the mean function

$$E(\log(Fertility)|\log(PPgdp), Purban) = \beta_0 + \beta_1\log(PPgdp) + \beta_2 Purban \quad (8.15)$$

should be appropriate for these data. Plots of residuals versus the two terms and versus fitted values are shown in Figure 8.7. Without reference to the curved lines shown on the plot, the visual appearance of these plots is satisfactory, with no obvious curvature or nonconstant variance. However, the curvature tests tell a different story. In each of the graphs, the value of the test statistics shown in Table 8.2 has a p-value of 0 to 2 decimal places, suggesting that the mean function (8.15) is not adequate for these data. We will return to this example in Section 8.4.

---

[2]If U is a polynomial term, for example, $U = X_1^2$, where X_1 is another term in the mean function, this procedure is not recommended.

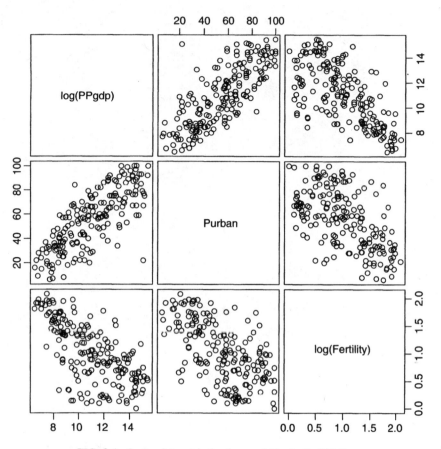

FIG. 8.6　Scatterplot matrix for three variables in the UN data.

TABLE 8.2　Lack-of-Fit Tests for the UN Data

	Test Stat.	Pr($>$ |t|)
log($PPgdp$)	3.22	0.00
Purban	3.37	0.00
Tukey test	3.65	0.00

8.3　NONCONSTANT VARIANCE

A nonconstant variance function in a residual plot may indicate that a constant variance assumption is false. There are at least four basic remedies for nonconstant variance. The first is to use a *variance stabilizing transformation* since replacing Y by Y_T may induce constant variance in the transformed scale. A second option is to find empirical weights that could be used in weighted least squares. Weights

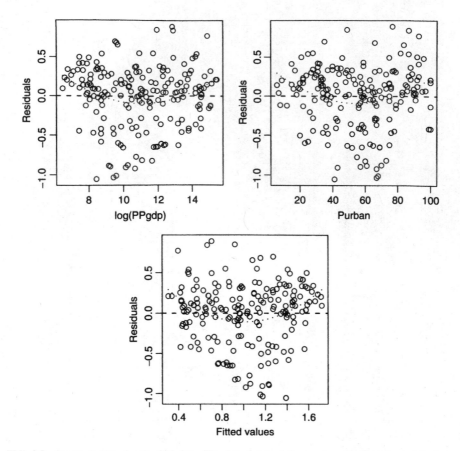

FIG. 8.7 Residual plots for the UN data. The dotted curved lines are quadratic polynomials fit to the residual plot and do not correspond exactly to the lack-of-fit tests that add a quadratic term to the original mean function.

that are simple functions of single predictors, such as $\text{Var}(Y|X) = \sigma^2 X_1$, with $X_1 > 0$, can sometimes be justified theoretically. If replication is available, then within group variances may be used to provide approximate weights. Although beyond the scope of this book, it is also possible to use *generalized least squares* and estimate weights and coefficients simultaneously (see, for example, Pinheiro and Bates, 2000, Section 5.1.2).

The third option is to do nothing. Estimates of parameters, given a misspecified variance function, remain unbiased, if somewhat inefficient. Tests and confidence intervals computed with the wrong variance function will be inaccurate, but the bootstrap may be used to get more accurate results.

The final option is to use regression models that account for the nonconstant variance that is a function of the mean. These are called *generalized linear models* and are discussed in the context of logistic regression in Chapter 12. In this section, we consider primarily the first two options.

8.3.1 Variance Stabilizing Transformations

Suppose that the response is strictly positive, and the variance function before transformation is

$$\text{Var}(Y|X = \mathbf{x}) = \sigma^2 g(\text{E}(Y|X = \mathbf{x})) \qquad (8.16)$$

where $g(\text{E}(Y|X = \mathbf{x}))$ is a function that is increasing with the value of its argument. For example, if the distribution of $Y|X$ has a Poisson distribution, then $g(\text{E}(Y|X = \mathbf{x})) = \text{E}(Y|X = \mathbf{x})$, since for Poisson variables, the mean and variance are equal.

For distributions in which the mean and variance are functionally related as in (8.16), Scheffé (1959, Section 10.7) provides a general theory for determining transformations that can stabilize variance, so that $\text{Var}(Y_T|X = \mathbf{x})$ will be approximately constant. Table 8.3 lists the common variance stabilizing transformations. Of course, transforming away nonconstant variance can introduce nonlinearity into the mean function, so this option may not always be reasonable.

The square root, $\log(Y)$, and $1/Y$ are appropriate when variance increases or decreases with the response, but each is more severe than the one before it. The square-root transformation is relatively mild and is most appropriate when the response follows a Poisson distribution, usually the first model considered for errors in counts. The logarithm is the most commonly used transformation; the base of the logarithms is irrelevant. It is appropriate when the error standard deviation is a percent of the response, such as $\pm 10\%$ of the response, not ± 10 units, so $\text{Var}(Y|X) \propto \sigma^2[\text{E}(Y|X)]^2$.

The reciprocal or inverse transformation is often applied when the response is a time until an event, such as time to complete a task, or until healing. This converts times per event to a rate per unit time; often the transformed measurements may be multiplied by a constant to avoid very small numbers. Rates can provide a natural measurement scale.

TABLE 8.3 Common Variance Stabilizing Transformations

Y_T	Comments
$\sqrt{Y}$	Used when $\text{Var}(Y|X) \propto \text{E}(Y|X)$, as for Poisson distributed data. $Y_T = \sqrt{Y} + \sqrt{Y+1}$ can be used if all the counts are small (Freeman and Tukey, 1950).
$\log(Y)$	Use if $\text{Var}(Y|X) \propto [\text{E}(Y|X)]^2$. In this case, the errors behave like a percentage of the response, $\pm 10\%$, rather than an absolute deviation, ± 10 units.
$1/Y$	The inverse transformation stabilizes variance when $\text{Var}(Y|X) \propto [\text{E}(Y|X)]^4$. It can be appropriate when responses are mostly close to 0, but occasional large values occur.
$\sin^{-1}(\sqrt{Y})$	The *arcsine square-root* transformation is used if Y is a proportion between 0 and 1, but it can be used more generally if y has a limited range by first transforming Y to the range $(0, 1)$, and then applying the transformation.

8.3.2 A Diagnostic for Nonconstant Variance

Cook and Weisberg (1983) provided a diagnostic test for nonconstant variance. Suppose now that $\mathrm{Var}(Y|X)$ depends on an unknown vector parameter λ and a known set of terms Z with observed values for the ith case z_i. For example, if $Z = Y$, then variance depends on the response. Similarly, Z may be the same as X, a subset of X, or indeed it could be completely different from X, perhaps indicating spatial location or time of observation. We assume that

$$\mathrm{Var}(Y|X, Z = \mathbf{z}) = \sigma^2 \exp(\lambda'\mathbf{z}) \tag{8.17}$$

This complicated form says that (1) $\mathrm{Var}(Y|Z = \mathbf{z}) > 0$ for all $\mathbf{z}$ because the exponential function is never negative; (2) variance depends on $\mathbf{z}$ and λ but only through the linear combination $\lambda'\mathbf{z}$; (3) $\mathrm{Var}(Y|Z = \mathbf{z})$ is monotonic, either increasing or decreasing, in each component of Z; and (4) if $\lambda = \mathbf{0}$, then $\mathrm{Var}(Y|Z = \mathbf{z}) = \sigma^2$. The results of Chen (1983) suggest that the tests described here are not very sensitive to the exact functional form used in (8.17), and so the use of the exponential function is relatively benign, and any form that depends on the linear combination $\lambda'\mathbf{z}$ would lead to very similar inference.

Assuming that errors are normally distributed, a *score test* of $\lambda = \mathbf{0}$ is particularly simple to compute using standard regression software. The test is carried out using the following steps:

1. Compute the OLS fit with the mean function

$$E(Y|X = \mathbf{x}) = \beta'\mathbf{x}$$

   as if

$$\mathrm{Var}(Y|X, Z = \mathbf{z}) = \sigma^2$$

   or equivalently, $\lambda = \mathbf{0}$. Save the residuals $\hat{e}_i$.

2. Compute scaled squared residuals $u_i = \hat{e}_i^2/\tilde{\sigma}^2 = n\hat{e}_i^2/[(n - p')\hat{\sigma}^2]$, where $\tilde{\sigma}^2 = \sum \hat{e}_j^2/n$ is the maximum likelihood estimate of σ^2, and differs from $\hat{\sigma}^2$ only by the divisor of n rather than $n - p'$. We combine the u_i into a variable U.

3. Compute the regression with the mean function $E(U|Z = \mathbf{z}) = \lambda_0 + \lambda'\mathbf{z}$. Obtain *SSreg* for this regression with df $= q$, the number of components in Z. If variance is thought to be a function of the responses, then in this regression, replace Z by the fitted values from the regression in step 1. The *SSreg* then will have 1 df.

4. Compute the score test, $S = SSreg/2$. The significance level for the test can be obtained by comparing S with its asymptotic distribution, which, under the hypothesis $\lambda = \mathbf{0}$, is $\chi^2(q)$. If $\lambda \neq \mathbf{0}$, then S will be too large, so large values of S provide evidence against the hypothesis of constant variance.

If we had started with a set of known weights, then the score test could be based on fitting the variance function

$$\text{Var}(Y|X, Z = \mathbf{z}) = \frac{\sigma^2}{w} \exp(\boldsymbol{\lambda}'\mathbf{z}) \tag{8.18}$$

The null hypothesis for the score test is then $\text{Var}(Y|X, Z = \mathbf{x}) = \sigma^2/w$ versus the alternative given by (8.18). The test is exactly the same as outlined above except that in step one, the WLS fit with weights w is used in place of the OLS fit, and in the remaining steps, the weighted or Pearson residuals given by (8.13) are used, not unweighted residuals.

Snow Geese

The relationship between *photo* = photo count, *obs1* = count by observer 1, and *obs2* = count by observer 2 of flocks of snow geese in the Hudson Bay area of Canada is discussed in Problem 5.5 of Chapter 5. The data are displayed in Figure 8.8. We see in the graph that (1) there is substantial disagreement between the observers; (2) the observers cannot predict the photo count very well, and (3) the variability appears to be larger for larger flocks.

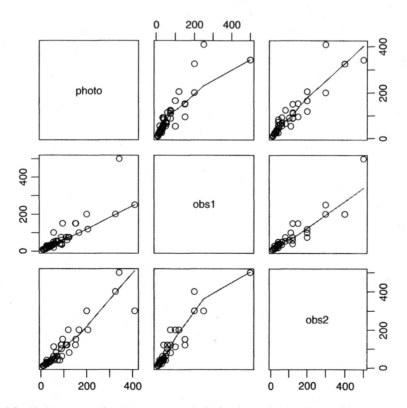

FIG. 8.8 The snow geese data. The line on each plot is a *loess* smooth with smoothing parameter 2/3.

Using the first observer only, we illustrate computation of the score test for constant variance. The first step is to fit the OLS regression of *photo* on *obs1*. The fitted OLS mean function is $\hat{E}(photo|obs1) = 26.55 + 0.88obs1$. From this, we can compute the residuals $\hat{e}_i$, $\tilde{\sigma} = \sum \hat{e}_i^2/n$, and then compute the $u_i = \hat{e}_i^2/\tilde{\sigma}$. We then compute the regression of U on *obs1*, under the hypothesis suggested by Figure 8.8 that variance increases with *obs1*. The analysis of variance table for this second regression is:

```
Response: U
           Df  Sum Sq Mean Sq F value    Pr(>F)
obs1        1 162.826 162.826  50.779 8.459e-09
Residuals  43 137.881   3.207
```

The score test for nonconstant variance is $S = (1/2)SSreg = (1/2)162.83 = 81.41$, which, when C compared with the chi-squared distribution with one df, gives an extremely small *p*-value. The hypothesis of constant residual variance is not tenable. The analyst must now cope with the almost certain nonconstant variance evident in the data. Two courses of action are outlined in Problems 5.5.3 and 5.5.4.

Sniffer Data

When gasoline is pumped into a tank, hydrocarbon vapors are forced out of the tank and into the atmosphere. To reduce this significant source of air pollution, devices are installed to capture the vapor. In testing these vapor recovery systems, a "sniffer" measures the amount recovered. To estimate the efficiency of the system, some method of estimating the total amount given off must be used. To this end, a laboratory experiment was conducted in which the amount of vapor given off was measured under controlled conditions. Four predictors are relevant for modeling:

$TankTemp$ = initial tank temperature (degrees F)

$GasTemp$ = temperature of the dispensed gasoline (degrees F)

$TankPres$ = initial vapor pressure in the tank (psi)

$GasPres$ = vapor pressure of the dispensed gasoline (psi)

The response is the hydrocarbons Y emitted in grams. The data, kindly provided by John Rice, are given in the data file `sniffer.txt`, and are shown in Figure 8.9. The clustering of points in many of the frames of this scatterplot is indicative of the attempt of the experimenters to set the predictors at a few nominal values, but the actual values of the predictors measured during the experiment was somewhat different from the nominal. We also see that (1) the predictors are generally linearly related, so transformations are unlikely to be desirable here, and (2) some of the predictors, notably the two pressure variables, are closely linearly related, suggesting, as we will see in Chapter 10, that using both in the mean function may not be desirable. For now, however, we will use all four terms and begin with the mean function including all four predictors as terms and fit via OLS as if the

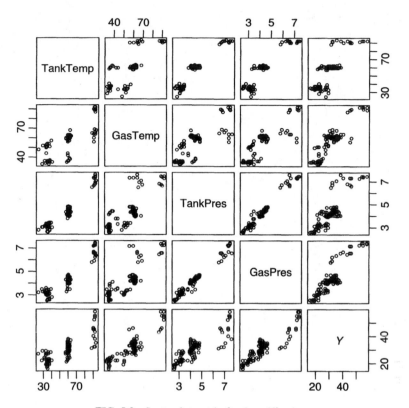

FIG. 8.9 Scatterplot matrix for the sniffer data.

variance function were constant. Several plots of the residuals for this regression are shown in Figure 8.10.

Figure 8.10a is the plot of residuals versus fitted values. While this plot is far from perfect, it does not suggest the need to worry much about the assumption of nonconstant variance. Figures 8.10b and c, which are plots of residuals against *TankTemp* and *GasPres*, respectively, give a somewhat different picture, as particularly in Figure 8.10c variance does appear to increase from left to right. Because none of the graphs in Figure 8.9 have clearly nonlinear mean functions, the inference that variance may not be constant can be tentatively adopted from the residual plots.

Table 8.4 gives the results of several nonconstant variance score tests, each computed using a different choice for Z. Each of these tests is just half the sum of squares for regression for U on the choice of Z shown. The plot shown in Figure 8.10d has the fitted values from the regression of U on all four predictors, and corresponds to the last line of Table 8.4.

From Table 8.4, we would diagnose nonconstant variance as a function of various choices of Z. We can compare *nested* choices for Z by taking the difference between the score tests and comparing the result with the χ^2 distribution with df

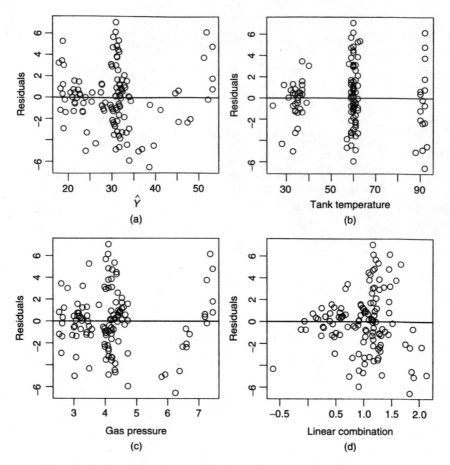

FIG. 8.10   Residuals plots for the sniffer data with variance assumed to be constant.

TABLE 8.4   Score Tests for Sniffer Data

Choice for Z	df	S	p-value
TankTemp	1	5.50	.019
GasPres	1	9.71	.002
TankTemp, GasPres	2	11.78	.003
TankTemp, GasTemp TankPres, GasPres	4	13.76	.008
Fitted values	1	4.80	.028

equal to the difference in their df (Hinkley, 1985). For example, to compare the 4 df choice of Z to $Z = $ (*TankTemp, GasPres*), we can compute $13.76 - 11.78 = 1.98$ with $4 - 2 = 2$ df, to get a p-value of about 0.37, and so the simpler Z with two terms is adequate. Comparing $Z = $ (*TankTemp, GasPres*) with $Z = GasPres$, the test statistic is $11.78 - 9.71 = 2.07$ with $2 - 1 = 1$ df, giving a p-value of about .15, so once again the simpler choice of Z seems adequate. Combining these tests, we would be led to assessing that the variance is primarily a function of *GasPres*.

A reasonable approach to working with these data is to assume that

$$\mathrm{Var}(Y|X, Z) = \sigma^2 \times GasPres$$

and use $1/GasPres$ as weights in weighted least squares.

8.3.3 Additional Comments

Some computer packages will include functions for the score test for nonconstant variance. With other computer programs, it may be more convenient to compute the score test as follows: (1) Compute the residuals $\hat{e}_i$ from the regression of Y on X; let $\hat{\sigma}^2$ be the usual residual mean square from this regression. (2) Compute the regression of $\hat{e}_i^2$ on Z for the choice of Z of interest, and let $SSreg(Z)$ be the resulting sum of squares for regression. (3) Compute $S = (1/2)SSreg(Z)/[(n - p')\hat{\sigma}^2/n]^2$. Pinheiro and Bates (2000, Section 5.2.1) present methodology and software for estimating weights using models similar to those discussed here.

8.4 GRAPHS FOR MODEL ASSESSMENT

Residual plots are used to examine regression models to see if they fail to match observed data. If systematic failures are found, then models may need to reformulated to find a better fitting model.

A closely related problem is assessing how *well* a model matches the data. Let us first think about a regression with one predictor in which we have fitted a simple linear regression model, and the goal is to summarize how well a fitted model matches the observed data. The lack-of-fit tests developed in Section 5.3 and 6.1 approach the question of goodness of fit from a testing point of view. We now look at this issue from a graphical point of view using *marginal model plots*.

We illustrate the idea first with a problem with just one predictor. In Section 7.1.2, we discussed the regression of *Height* on *Dbh* for a sample of western red cedar trees from Upper Flat Creek, Idaho. The mean function

$$\mathrm{E}(Height|Dbh) = \beta_0 + \beta_1 Dbh \tag{8.19}$$

was shown to be a poor summary of these data, as can be seen in Figure 8.11. Two smooths are given on the plot. The OLS fit of (8.19), shown as a dashed line, estimates the mean function only if the simple linear regression model is correct. The *loess* fit, shown as a solid line, estimates the mean function regardless of the fit of the simple linear regression model. If we judge these two fits to be different, then we have visual evidence against the simple linear regression mean function.

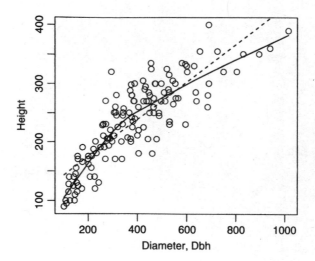

FIG. 8.11   Model checking plot for the simple linear regression for western red cedar trees at Upper Flat Creek. The dashed line is the OLS simple linear regression fit, and the solid line is a *loess* smooth.

The *loess* fit is clearly curved, so the mean function (8.19) is not a very good summary of this regression problem.

Although Figure 8.11 includes the data, the primary focus in this plot is comparing the two curves, using the data as background mostly to help choose the smoothing parameter for the *loess* smooth, to help visualize variation, and to locate any extreme or unusual points.

8.4.1   Checking Mean Functions

With more than one predictor, we will look at marginal models to get a sequence of two-dimensional plots to examine. Suppose that the model we have fitted has mean function $E(Y|X = x) = \beta'x$, although in what follows the exact form of the mean function is not important. We will draw a plot with the response Y on the vertical axis. On the horizontal axis, we will plot a quantity U that will consist of *any function of X we think is relevant*, such as fitted values, any of the individual terms in X, or even transformations of them. Fitting a smoother to the plot of Y versus U estimates $E(Y|U)$ without any assumptions. We want to compare this smooth to an estimate of $E(Y|U)$ based on the model.

Following Cook and Weisberg (1997), an estimate of $E(Y|U)$, given the model, can be based on application of equation (A.4) in Appendix A.2.4. Under the model, we have

$$E(Y|U = u) = E[E(Y|X = x)|U = u]$$

We need to substitute an estimate for $E(Y|X = x)$. On the basis of the model, this expectation is estimated at the observed values of the terms by the fitted values $\hat{Y}$.

We get

$$E(Y|\widehat{U} = \mathbf{u}) = E(\hat{\mathbf{Y}}|U = \mathbf{u}) \qquad (8.20)$$

The implication of this result is that *we can estimate* $E(Y|U = \mathbf{u})$ *by smoothing the scatterplot with U on the horizontal axis, and the fitted values* $\hat{\mathbf{Y}}$ *on the vertical axis.* If the model is correct, then the smooth of Y versus U and the smooth of $\hat{\mathbf{Y}}$ versus U should agree; if the model is not correct, these smooths may not agree.

As an example, we return to the United Nations data discussed in Section 8.2, starting with the mean function given by (8.15),

$$E(\log(Fertility)|\log(PPgdp), Purban) = \beta_0 + \beta_1 \log(PPgdp) + \beta_2 Purban \quad (8.21)$$

and suppose that $U = \log(PPgdp)$, one of the terms in the mean function. Figure 8.12 shows plots of $\log(Fertility)$ versus U and of $\hat{\mathbf{Y}}$ versus U. The smooth in Figure 8.12a estimates $E(\log(Fertility)|\log(PPgdp))$ whether the model is right or not, but the smooth in Figure 8.12b may not give a useful estimate of $E(\log(Fertility)|\log(PPgdp))$ if the linear regression model is wrong. Comparison of these two estimated mean functions provides a visual assessment of the adequacy of the mean function for the model. Superimposing the smooth in Figure 8.12b on Figure 8.12a gives the marginal model plot shown in Figure 8.13a. The two fitted curves do not match well, suggesting that the mean function (8.21) is inadequate.

Three additional marginal model plots are shown in Figure 8.13. The plots versus *Purban* and versus fitted values also exhibit a curved mean smooth based on the data compared to a straight smooth based on the fitted model, confirming the inadequacy of the mean function. The final plot in Figure 8.13d is a little different. The quantity on the horizontal axis is a randomly selected linear combination of *Purban* and

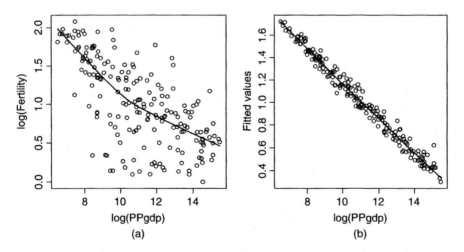

FIG. 8.12 Plots for $\log(Fertility)$ versus $\log(PPgdp)$ and $\hat{y}$ versus $\log(PPgdp)$. In both plots, the curves are *loess* smooths with smoothing parameters equal to 2/3. If the model has the correct mean function, then these two smooths estimate the same quantity.

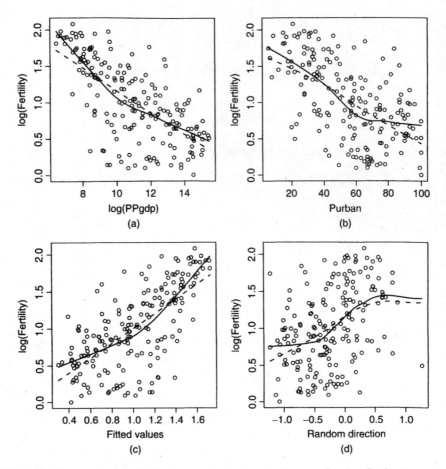

FIG. 8.13 Four marginal model plots, versus the two terms in the mean function, fitted values, and a random linear combination of the terms in the mean function.

log(*PPgdp*). In this direction, *both* smooths are curved, and they agree fairly well, except possibly at the left edge of the graph. If a fitted model is appropriate for data, then the two smooths in the marginal model plots will agree for any choice of U, including randomly selected ones. If the model is wrong, then for some choices of U, the two smooths will disagree.

Since the mean function (8.21) is inadequate, we need to consider further modification to get a mean function that matches the data well. One approach is to expand (8.15) by including both quadratic terms and an interaction between log(*PPgdp*) and *Purban*. Using the methods described elsewhere in this book, we conclude that the mean function

$$E(\log(Fertility)|\log(PPgdp), Purban) = \beta_0 + \beta_1\log(PPgdp) + \beta_2 Purban$$
$$+ \beta_{22} Purban^2 \qquad (8.22)$$

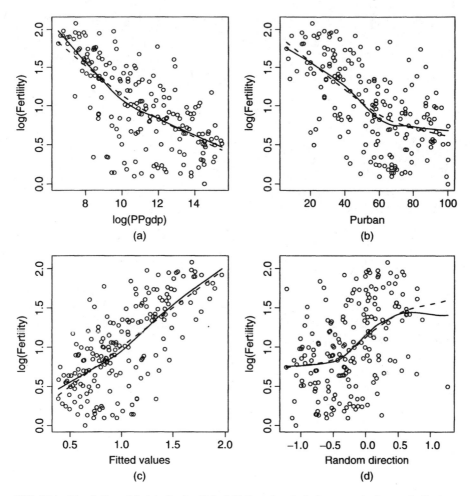

FIG. 8.14   Marginal model plots for the United Nations data, including a quadratic term in *Purban*.

matches the data well, as confirmed by the marginal model plots in Figure 8.14. Evidently, adding the quadratic in *Purban* allows the effect of increasing *Purban* on log(*Fertility*) to be smaller when *Purban* is large than when *Purban* is small.

8.4.2   Checking Variance Functions

Model checking plots can also be used to check for model inadequacy in the variance function, which for the multiple linear regression problem means checking the constant variance assumption. We call the square root of the variance function the standard deviation function. The plot of Y versus U can be used to get the estimate $SD_{data}(Y|U)$ of the standard deviation function, as discussed in Appendix A.5. This estimate of the square root of the variance function does not depend on a model.

We need a model-based estimate of the standard deviation function. Applying (A.5) and again substituting $\hat{\mathbf{Y}} \approx E(Y|U = \mathbf{u})$,

$$\text{Var}(Y|U) = E[\text{Var}(Y|X)|U] + \text{Var}[E(Y|X)|U] \qquad (8.23)$$

$$\approx E[\sigma^2|U] + \text{Var}[\hat{\mathbf{Y}}|U]$$

$$= \sigma^2 + \text{Var}[\hat{\mathbf{Y}}|U] \qquad (8.24)$$

Equation (8.23) is the general result that holds for any model. Equation (8.24) holds for the linear regression model in which the variance function $\text{Var}(Y|X) = \sigma^2$ is constant. According to this result, we can estimate $\text{Var}(Y|U)$ under the model by getting a variance smooth of $\hat{\mathbf{Y}}$ versus U, and then adding to this an estimate of σ^2, for which we use $\hat{\sigma}^2$ from the OLS fit of the model. We will call the square root of this estimated variance function $\text{SD}_{model}(Y|U)$. If the model is appropriate for the data, then apart from sampling error, $\text{SD}_{data}(Y|U) = \text{SD}_{model}(Y|U)$, but if the model is wrong, these two functions need not be equal.

For visual display, we show the mean function estimated from the plot $\pm\text{SD}_{data}$ $(Y|U)$ using solid lines and the mean function estimated from the model $\pm\text{SD}_{model}$ $(Y|U)$ using dashed lines; colored lines would be helpful here. The same smoothing parameter should be used for all the smooths, so any bias due to smoothing will tend to cancel. These smooths for the United Nations example are shown in Figure 8.15, first for the fit of (8.21) and then for the fit of (8.22). For both, the horizontal axis is fitted values, but almost anything could be put on this axis. Apart from the edges of the plot where the smooths are less accurate, these plots do not suggest any problem with nonconstant variance, as the estimated variances functions using the two methods are similar, particularly for the mean function (8.22) that matches the data.

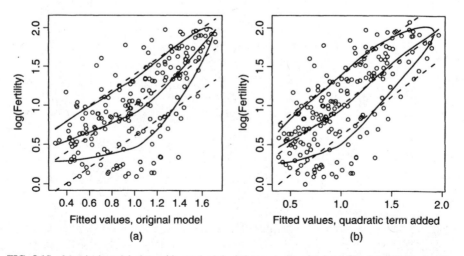

FIG. 8.15 Marginal model plots with standard deviation smooths added. (a) The fit of (8.21). (b) The fit of (8.22).

The marginal model plots described here can be applied in virtually any regression problem, not just for the linear regression. For example, Pan, Connett, Porzio, and Weisberg (2001) discuss application to longitudinal data, and Porzio (2002) discusses calibrating marginal model plots for binary regression.

PROBLEMS

8.1. Working with the hat matrix

    8.1.1. Prove the results given by (8.7).

    8.1.2. Prove that $1/n \leq h_{ii} \leq 1/r$, where h_{ii} is a diagonal entry in $\mathbf{H}$, and r is the number of rows in $\mathbf{X}$ that are exactly the same as $\mathbf{x}_i$.

8.2. If the linear trend were removed from Figure 8.5f, what would the resulting graph look like?

8.3. This example compares in-field ultrasonic measurements of the depths of defects in the Alaska oil pipeline to measurements of the same defects in a laboratory. The lab measurements were done in six different batches. The goal is to decide if the field measurement can be used to predict the more accurate lab measurement. In this analysis, the field measurement is the response variable and the laboratory measurement is the predictor variable. The data, in the file pipeline.txt, were given at www.itl.nist.gov/div898/handbook/pmd/section6/pmd621.htm. The three variables are called *Field*, the in-field measurement, *Lab*, the more accurate in-lab measurement, and *Batch*, the batch number.

    8.3.1. Draw the scatterplot of *Lab* versus *Field*, and comment on the applicability of the simple linear regression model.

    8.3.2. Fit the simple regression model, and get the residual plot. Compute the score test for nonconstant variance, and summarize your results.

    8.3.3. Fit the simple regression mean function again, but this time assume that $\text{Var}(Lab|Field) = \sigma^2/Field$. Get the score test for the fit of this variance function. Also test for nonconstant variance as a function of batch; since the batches are arbitrarily numbered, be sure to treat *Batch* as a factor. (*Hint:* Both these tests are extensions of the methodology outlined in the text. The only change required is to be sure that the residuals defined by (8.13) are used when computing the statistic.)

    8.3.4. Repeat Problem 8.3.3, but with $\text{Var}(Lab|Field) = \sigma^2/Field^2$.

8.4. Refer to Problem 7.2. Fit Hald's model, given in Problem 7.2.3, but with constant variance, $\text{Var}(Distance|Speed) = \sigma^2$. Compute the score test for nonconstant variance for the alternatives that (a) variance depends on the mean; (b) variance depends on *Speed*; and (c) variance depends on *Speed* and $Speed^2$. Is adding $Speed^2$ helpful?

8.5. Consider the simple regression model, $E(Y|X = x) = \beta_0 + \beta_1 x$ with Var $(Y|X = x) = \sigma^2$.

8.5.1. Find a formula for the h_{ij} and for the leverages h_{ii}.

8.5.2. In a 2D plot of the response versus the predictor in a simple regression problem, explain how high-leverage points can be identified.

8.5.3. Make up a predictor X so that the value of the leverage in simple regression for one of the cases is equal to 1.

8.6. Using the QR factorization defined in Appendix A.12, show that $\mathbf{H} = \mathbf{Q}\mathbf{Q}'$. Hence, if q_i is the ith row of $\mathbf{Q}$,

$$h_{ii} = q_i'q_i \qquad\qquad h_{ij} = q_i'q_j$$

Thus, if the QR factorization of $\mathbf{X}$ is computed, the h_{ii} and the h_{ij} are easily obtained.

8.7. Let $\mathbf{U}$ be an $n \times 1$ vector with 1 as its first element and 0s elsewhere. Consider computing the regression of $\mathbf{U}$ on an $n \times p'$ full rank matrix $\mathbf{X}$. As usual, let $\mathbf{H} = \mathbf{X}(\mathbf{X}'\mathbf{X})^{-1}\mathbf{X}'$ be the Hat matrix with elements h_{ij}.

8.7.1. Show that the elements of the vector of fitted values from the regression of $\mathbf{U}$ on $\mathbf{X}$ are the h_{1j}, $j = 1, 2, \ldots, n$.

8.7.2. Show that the vector of residuals have $1 - h_{11}$ as the first element, and the other elements are $-h_{1j}$, $j > 1$.

8.8. Two $n \times n$ matrices $\mathbf{A}$ and $\mathbf{B}$ are *orthogonal* if $\mathbf{AB} = \mathbf{BA} = 0$. Show that $\mathbf{I} - \mathbf{H}$ and $\mathbf{H}$ are orthogonal. Use this result to show that as long as the intercept is in the mean function, the slope of the regression of $\hat{\mathbf{e}}$ on $\hat{\mathbf{Y}}$ is 0. What is the slope of the regression of $\hat{\mathbf{e}}$ on $\mathbf{Y}$?

8.9. Suppose that $\mathbf{W}$ is a known positive diagonal matrix of positive weights, and we have a weighted least squares problem,

$$\mathbf{Y} = \mathbf{X}\beta + \mathbf{e} \quad \text{Var}(\mathbf{e}) = \hat{\sigma}^2\mathbf{W}^{-1} \tag{8.25}$$

Using the transformations as in Section 5.1, show that the hat matrix is given by (8.12).

8.10. Draw residuals plots for the mean function described in Problem 7.3.3 for the California water data, and comment on your results. Test for curvature as a function of fitted values. Also, get marginal model plots for this model.

8.11. Refer to the transactions data discussed in Section 4.6.1. Fit the mean function (4.16) with constant variance, and use marginal model plots to examine the fit. Be sure to consider both the mean function and the variance function. Comment on the results.

TABLE 8.5   Crustacean Zooplankton Species Data

Variable	Description
Species	Number of zooplankton species
MaxDepth	Maximum lake depth, m
MeanDepth	Mean lake depth, m
Cond	Specific conductance, micro Siemans
Elev	Elevation, m
Lat	N latitude, degrees
Long	W longitude, degrees
Dist	distance to nearest lake, km
NLakes	number of lakes within 20 km
Photo	Rate of photosynthesis, mostly by the ^{14}C method
Area	Surface area of the lake, in hectares
Lake	Name of lake

Source: From Dodson (1992).

8.12. The number of crustacean zooplankton species present in a lake can be different, even for two nearby lakes. The data in the file `lakes.txt`, provided by S. Dodson and discussed in part in Dodson (1992), give the number of known crustacean zooplankton species for 69 world lakes. Also included are a number of characteristics of each lake. There are some missing values, indicated with a "?" in the data file. The goal of the analysis is to understand how the number of species present depends on the other measured variables that are characteristics of the lake. The variables are described in Table 8.5.

Decide on appropriate transformations of the data to be used in this problem. Then, fit appropriate linear regression models, and summarize your results.

CHAPTER 9

Outliers and Influence

9.1 OUTLIERS

In some problems, the observed response for a few of the cases may not seem to correspond to the model fitted to the bulk of the data. In a simple regression problem such as displayed in Figure 1.9c, page 13, this may be obvious from a plot of the response versus the predictor, where most of the cases lie near a fitted line but a few do not. Cases that do not follow the same model as the rest of the data are called *outliers*, and identifying these cases can be useful.

We use the *mean shift outlier model* to define outliers. Suppose that the ith case is a candidate for an outlier. We assume that the mean function for all other cases is $E(Y|X = \mathbf{x}_j) = \mathbf{x}'_j\boldsymbol{\beta}$, but for case i the mean function is $E(Y|X = \mathbf{x}_i) = \mathbf{x}'_i\boldsymbol{\beta} + \delta$. The expected response for the ith case is shifted by an amount δ, and a test of $\delta = 0$ is a test for a single outlier in the ith case. In this development, we assume $\mathrm{Var}(Y|X) = \sigma^2$.

Cases with large residuals are candidates for outliers. Not all large residual cases are outliers, since large errors e_i will occur with the frequency prescribed by the generating probability distribution. Whatever testing procedure we develop must offer protection against declaring too many cases to be outliers. This leads to the use of simultaneous testing procedures. Also, not all outliers are bad. For example, a geologist searching for oil deposits may be looking for outliers, if the oil is in the places where a fitted model does not match the data. Outlier identification is done relative to a specified model. If the form of the model is modified, the status of individual cases as outliers may change. Finally, some outliers will have greater effect on the regression estimates than will others, a point that is pursued shortly.

9.1.1 An Outlier Test

Suppose that the ith case is suspected to be an outlier. First, define a new term, say U, with the jth element $u_j = 0$ for $j \neq i$, and the ith element $u_i = 1$. Thus, U is

Applied Linear Regression, Third Edition, by Sanford Weisberg
ISBN 0-471-66379-4   Copyright © 2005 John Wiley & Sons, Inc.

a dummy variable that is zero for all cases but the ith. Then, simply compute the regression of the response on both the terms in X and U. The estimated coefficient for U is the estimate of the mean shift δ. The t-statistic for testing $\delta = 0$ against a two-sided alternative is the appropriate test statistic. Normally distributed errors are required for this test, and then the test will be distributed as Student's t with $n - p' - 1$ df.

We will now consider an alternative approach that will lead to the same test, but from a different point of view. The equivalence of the two approaches is left as an exercise.

Again suppose that the ith case is suspected to be an outlier. We can proceed as follows:

1. Delete the ith case from the data, so $n - 1$ cases remain in the reduced data set.
2. Using the reduced data set, estimate β and σ^2. Call these estimates $\hat{\beta}_{(i)}$ and $\hat{\sigma}_{(i)}^2$ to remind us that case i was not used in estimation. The estimator $\hat{\sigma}_{(i)}^2$ has $n - p' - 1$ df.
3. For the deleted case, compute the fitted value $\hat{y}_{i(i)} = \mathbf{x}'_i \hat{\beta}_{(i)}$. Since the ith case was not used in estimation, y_i and $\hat{y}_{i(i)}$ are independent. The variance of $y_i - \hat{y}_{i(i)}$ is given by

$$\text{Var}(y_i - \hat{y}_{i(i)}) = \sigma^2 + \sigma^2 \mathbf{x}'_i (\mathbf{X}'_{(i)} \mathbf{X}_{(i)})^{-1} \mathbf{x}_i \qquad (9.1)$$

where $\mathbf{X}_{(i)}$ is the matrix $\mathbf{X}$ with the ith row deleted. This variance is estimated by replacing σ^2 with $\hat{\sigma}_{(i)}^2$ in (9.1).

4. Now $\text{E}(y_i - \hat{y}_{i(i)}) = \delta$, which is zero under the null hypothesis that case i is not an outlier but nonzero otherwise. Assuming normal errors, a Student's t-test of the hypothesis $\delta = 0$ is given by

$$t_i = \frac{y_i - \hat{y}_{i(i)}}{\hat{\sigma}_{(i)} \sqrt{1 + \mathbf{x}'_i (\mathbf{X}'_{(i)} \mathbf{X}_{(i)})^{-1} \mathbf{x}_i}} \qquad (9.2)$$

This test has $n - p' - 1$ df, and is identical to the t-test suggested in the first paragraph of this section.

There is a simple computational formula for t_i in (9.2). We first define an intermediate quantity often called a *standardized residual*, by

$$r_i = \frac{\hat{e}_i}{\hat{\sigma} \sqrt{1 - h_{ii}}} \qquad (9.3)$$

where the h_{ii} is the leverage for the ith case, defined at (8.7). Like the residuals $\hat{e}_i$, the r_i have mean zero, but unlike the $\hat{e}_i$, the variances of the r_i are all equal to one. Because the h_{ii} need not all be equal, the r_i are not just a rescaling of the $\hat{e}_i$. With the aid of Appendix A.12, one can show that t_i can be computed as

$$
t_i = r_i \left(\frac{n - p' - 1}{n - p' - r_i^2} \right)^{1/2} = \frac{\hat{e}_i}{\hat{\sigma}_{(i)} \sqrt{1 - h_{ii}}}
\tag{9.4}
$$

A statistic divided by its estimated standard deviation is usually called a *studentized statistic*, in honor of W. S. Gosset, who first wrote about the t-distribution using the pseudonym Student[1]. The residual t_i is called a *studentized residual*. We see that r_i and t_i carry the same information since one can be obtained from the other via a simple formula. Also, this result shows that t_i can be computed from the residuals, the leverages and $\hat{\sigma}^2$, so we don't need to delete the ith case, or to add a variable U, to get the outlier test.

9.1.2 Weighted Least Squares

If we initially assumed that $\mathrm{Var}(Y|X) = \sigma^2/w$ for known positive weights w, then in equation (9.3), we compute the residuals $\hat{e}_i$ using the correct weighted formula (8.13) and leverages are the diagonal elements of (8.12). Otherwise, no changes are required.

9.1.3 Significance Levels for the Outlier Test

If the analyst suspects in advance that the ith case is an outlier, then t_i should be compared with the central t-distribution with the appropriate number of df. The analyst rarely has a prior choice for the outlier. Testing the case with the largest value of $|t_i|$ to be an outlier is like performing n significance tests, one for each of n cases. If, for example, $n = 65$, $p' = 4$, the probability that a t statistic with 60 df exceeds 2.000 in absolute value is 0.05; however, the probability that the largest of 65 independent t-tests exceeds 2.000 is 0.964, suggesting quite clearly the need for a different critical value for a test based on the maximum of many tests. Since tests based on the t_i are correlated, this computation is only a guide. Excellent discussions of this and other multiple-test problems are presented by Miller (1981).

The technique we use to find critical values is based on the *Bonferroni inequality*, which states that for n tests each of size a, the probability of falsely labeling at least one case as an outlier is no greater than na. This procedure is *conservative* and provides an upper bound on the probability. For example, the Bonferroni inequality specifies only that the probability of the maximum of 65 tests exceeding 2.00 is no greater than 65(0.05), which is larger than 1. Choosing the critical value to be the $(\alpha/n) \times 100\%$ point of t will give a significance level of no more than $n(\alpha/n) = \alpha$. We would choose a level of $.05/65 = .00077$ for each test to give an overall level of no more than $65(.00077) = .05$.

---

[1] See www-gap.dcs.st-and.ac.uk/~history/Mathematicians/Gosset.html for a biography of Student.

Standard functions for the t-distribution can be used to compute p-values for the outlier test: simply compute the p-value as usual and then multiply by the sample size. If this number is smaller then one, then this is the p-value adjusted for multiple testing. If this number exceeds one, then the p-value is one.

In Forbes' data, Example 1.1, case 12 was suspected to be an outlier because of its large residual. To perform the outlier test, we first need the standardized residual, which is computed using (9.3) from $\hat{e}_i = 1.36$, $\hat{\sigma} = 0.379$, and $h_{12,12} = 0.0639$,

$$r_{12} = \frac{1.3592}{0.379\sqrt{1 - .0639}} = 3.7078$$

and the outlier test is

$$t_i = 3.7078 \left(\frac{17 - 2 - 1}{17 - 2 - 3.7078^2} \right)^{1/2} = 12.40$$

The nominal two-sided p-value corresponding to this test statistic when compared with the $t(14)$ distribution is 6.13×10^{-9}. If the location of the outlier was not selected in advance, the Bonferroni-adjusted p-value is $17 \times 6.13 \times 10^{-9} = 1.04 \times 10^{-7}$. This very small value supports case 12 as an outlier.

The test locates an outlier, but it does not tell us what to do about it. If we believe that the case is an outlier because of a blunder, for example, an unusually large measurement error, or a recording error, then we might delete the outlier and analyze the remaining cases without the suspected case. Sometimes, we can try to figure out why a particular case is outlying, and finding the cause may be the most important part of the analysis. All this depends on the context of the problem you are studying.

9.1.4   Additional Comments

There is a vast literature on methods for handling outliers, including Barnett and Lewis (2004), Hawkins (1980), and Beckman and Cook (1983). If a set of data has more than one outlier, a sequential approach can be recommended, but the cases may mask each other, making finding groups of outliers difficult. Cook and Weisberg (1982, p. 28) provide the generalization of the mean shift model given here to multiple cases. Hawkins, Bradu, and Kass (1984) provide a promising method for searching all subsets of cases for outlying subsets. Bonferroni bounds for outlier tests are discussed by Cook and Prescott (1981). They find that for one-case-at-a-time methods the bound is very accurate, but it is much less accurate for multiple-case methods.

The testing procedure helps the analyst in finding outliers, to make them available for further study. Alternatively, we could design robust statistical methods that can tolerate or accommodate some proportion of bad or outlying data; see, for example, Staudte and Sheather (1990).

9.2 INFLUENCE OF CASES

Single cases or small groups of cases can strongly influence the fit of a regression model. In Anscombe's examples in Figure 1.9d, page 13, the fitted model depends entirely on the one point with $x = 19$. If that case were deleted, we could not estimate the slope. If it were perturbed, moved around a little, the fitted line would follow the point. In contrast, if any of the other cases were deleted or moved around, the change in the fitted mean function would be quite small.

The general idea of *influence analysis* is to study changes in a specific part of the analysis when the data are slightly perturbed. Whereas statistics such as residuals are used to find problems with a model, influence analysis is done as if the model were correct, and we study the robustness of the conclusions, given a particular model, to the perturbations. The most useful and important method of perturbing the data is deleting the cases from the data one at a time. We then study the effects or influence of each individual case by comparing the full data analysis to the analysis obtained with a case removed. Cases whose removal causes major changes in the analysis are called influential.

Using the notation from the last section, a subscript (i) means "with the ith case deleted," so, for example, $\boldsymbol{\beta}_{(i)}$ is the estimate of $\boldsymbol{\beta}$ computed without case i, $\mathbf{X}_{(i)}$ is the $(n - 1) \times p'$ matrix obtained from $\mathbf{X}$ by deleting the ith row, and so on. In particular, then,

$$\hat{\boldsymbol{\beta}}_{(i)} = (\mathbf{X}'_{(i)}\mathbf{X}_{(i)})^{-1}\mathbf{X}'_{(i)}\mathbf{Y}_{(i)} \tag{9.5}$$

Figure 9.1 is a scatterplot matrix of coefficient estimates for the three parameters in the UN data from Section 3.1 obtained by deleting cases one at a time. Every time a case is deleted, different coefficient estimates may be obtained. All 2D plot in Figure 9.1 are more or less elliptically shaped, which is a common characteristic of the deletion estimates. In the plot for the coefficients for log($PPgdp$) and *Purban*, the points for Armenia and Ukraine are in one corner and the point for Djibouti is in the opposite corner; deleting any one of these localities causes the largest change in the values of the estimated parameters, although all the changes are small.

While the plots in Figure 9.1 are informative about the effects of deleting cases one at a time, looking at these plots can be bewildering, particularly if the number of terms in the model is large. A single summary statistic that can summarize these pictures is desirable, and this is provided by Cook's distance.

9.2.1 Cook's Distance

We can summarize the influence on the estimate of $\boldsymbol{\beta}$ by comparing $\hat{\boldsymbol{\beta}}$ to $\hat{\boldsymbol{\beta}}_{(i)}$. Since each of these is a p' vector, the comparison requires a method of combining information from each of the p' components into a single number. Several ways of doing this have been proposed in the literature, but most of them will result in roughly the same information, at least for multiple linear regression. The method we use is due to Cook (1977). We define *Cook's distance* D_i to be

$$D_i = \frac{(\hat{\boldsymbol{\beta}}_{(i)} - \hat{\boldsymbol{\beta}})'(\mathbf{X}'\mathbf{X})(\hat{\boldsymbol{\beta}}_{(i)} - \hat{\boldsymbol{\beta}})}{p'\hat{\sigma}^2} \tag{9.6}$$

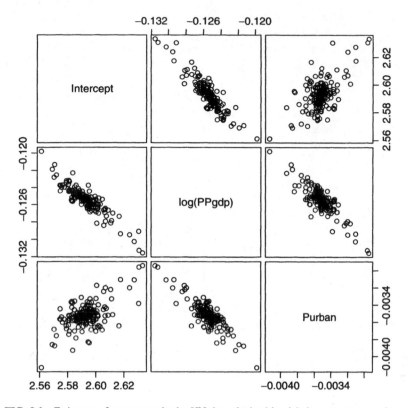

FIG. 9.1 Estimates of parameters in the UN data obtained by deleting one case at a time.

This statistic has several desirable properties. First, contours of constant D_i are ellipsoids, with the same shape as confidence ellipsoids. Second, the contours can be thought of as defining the distance from $\hat{\beta}_{(i)}$ to $\hat{\beta}$. Third, D_i does not depend on parameterization, so if the columns of $\mathbf{X}$ are modified by linear transformation, D_i is unchanged. Finally, if we define vectors of fitted values as $\hat{\mathbf{Y}} = \mathbf{X}\hat{\beta}$ and $\hat{\mathbf{Y}}_{(i)} = \mathbf{X}\hat{\beta}_{(i)}$, then (9.6) can be rewritten as

$$D_i = \frac{(\hat{\mathbf{Y}}_{(i)} - \hat{\mathbf{Y}})'(\hat{\mathbf{Y}}_{(i)} - \hat{\mathbf{Y}})}{p'\hat{\sigma}^2} \tag{9.7}$$

so D_i is the ordinary Euclidean distance between $\hat{\mathbf{Y}}$ and $\hat{\mathbf{Y}}_{(i)}$. Cases for which D_i is large have substantial influence on both the estimate of β and on fitted values, and deletion of them may result in important changes in conclusions.

9.2.2 Magnitude of D_i

Cases with large values of D_i are the ones whose deletion will result in substantial changes in the analysis. Typically, the case with the largest D_i, or in large data sets the cases with the largest few D_i, will be of interest. One method of calibrating

D_i is obtained by analogy to confidence regions. If D_i were exactly equal to the $\alpha \times 100\%$ point of the F distribution with p' and $n - p'$ df, then deletion of the ith case would move the estimate of $\hat{\boldsymbol{\beta}}$ to the edge of a $(1 - \alpha) \times 100\%$ confidence region based on the complete data. Since for most F distributions the 50% point is near one, a value of $D_i = 1$ will move the estimate to the edge of about a 50% confidence region, a potentially important change. If the largest D_i is substantially less than one, deletion of a case will not change the estimate of $\hat{\boldsymbol{\beta}}$ by much. To investigate the influence of a case more closely, the analyst should delete the large D_i case and recompute the analysis to see exactly what aspects of it have changed.

9.2.3 Computing D_i

From the derivation of Cook's distance, it is not clear that using these statistics is computationally convenient. However, the results sketched in Appendix A.12 can be used to write D_i using more familiar quantities. A simple form for D_i is

$$D_i = \frac{1}{p'} r_i^2 \frac{h_{ii}}{1 - h_{ii}} \tag{9.8}$$

D_i is a product of the square of the ith standardized residual r_i and a monotonic function of h_{ii}. If p' is fixed, the size of D_i will be determined by two different sources: the size of r_i, a random variable reflecting lack of fit of the model at the ith case, and h_{ii}, reflecting the location of $\mathbf{x}_i$ relative to $\bar{\mathbf{x}}$. A large value of D_i may be due to large r_i, large h_{ii}, or both.

Rat Data

An experiment was conducted to investigate the amount of a particular drug present in the liver of a rat. Nineteen rats were randomly selected, weighed, placed under light ether anesthesia and given an oral dose of the drug. Because large livers would absorb more of a given dose than smaller livers, the actual dose an animal received was approximately determined as 40 mg of the drug per kilogram of body weight. Liver weight is known to be strongly related to body weight. After a fixed length of time, each rat was sacrificed, the liver weighed, and the percent of the dose in the liver determined. The experimental hypothesis was that, for the method of determining the dose, there is no relationship between the percentage of the dose in the liver (Y) and the body weight *BodyWt*, liver weight *LiverWt*, and relative *Dose*. The data, provided by Dennis Cook and given in the file rat.txt, are shown in Figure 9.2. As had been expected, the marginal summary plots for Y versus each of the predictors suggests no relationship, and none of the simple regressions is significant, all having t-values less than one.

The fitted regression summary for the regression of Y on the three predictors is shown in Table 9.1. *BodyWt* and *Dose* have significant t-tests, with $p < 0.05$ in both cases, indicating that the two measurements combined are a useful indicator of Y; if *LiverWt* is dropped from the mean function, the same phenomenon appears. The analysis so far, based only on summary statistics, might lead to the conclusion

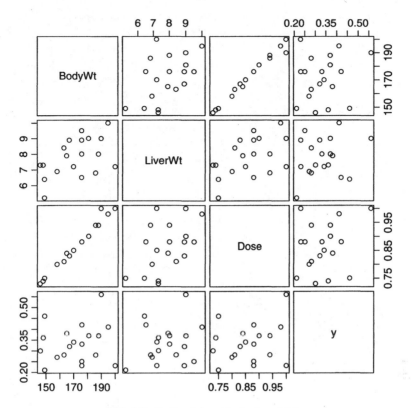

FIG. 9.2   Scatterplot matrix for the rat data.

TABLE 9.1   Regression Summary for the Rat Data

```
Coefficients:
             Estimate Std. Error t value Pr(>|t|)
(Intercept)  0.265922   0.194585   1.367   0.1919
BodyWt      -0.021246   0.007974  -2.664   0.0177
LiverWt      0.014298   0.017217   0.830   0.4193
Dose         4.178111   1.522625   2.744   0.0151

Residual standard error: 0.07729 on 15 degrees of freedom
Multiple R-Squared: 0.3639
F-statistic:  2.86 on 3 and 15 DF,  p-value: 0.07197
```

that while neither *BodyWt* or *Dose* are associated with the response when the other is ignored, in combination they are associated with the response. But, from Figure 9.2, *Dose* and *BodyWt* are almost perfectly linearly related, so they measure the same thing!

We turn to case analysis to attempt to resolve this paradox. Figure 9.3 displays diagnostic statistics for the mean function with all the terms included. The outlier

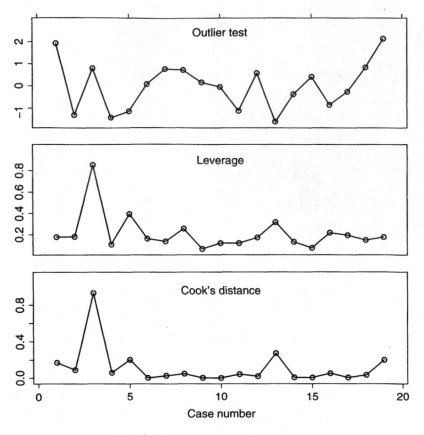

FIG. 9.3   Diagnostic statistics for the rat data.

statistics are not particularly large. However, Cook's distance immediately locates a possible cause: case three has $D_3 = .93$; no other case has D_i bigger than 0.27, suggesting that case number three alone may have large enough influence on the fit to induce the anomaly. The value of $h_{33} = 0.85$ indicates that the problem is an unusual set of predictors for case 3.

One suggestion at this point is to delete the third case and recompute the regression. These computations are given in Table 9.2. The paradox dissolves and the apparent relationship found in the first analysis can thus be ascribed to the third case alone.

Once again, the diagnostic analysis finds a problem, but does not tell us what to do next, and this will depend on the context of the problem. Rat number three, with weight 190 g, was reported to have received a full dose of 1.000, which was a larger dose than it should have received according to the rule for assigning doses; for example, rat eight with weight of 195 g got a lower dose of 0.98. A number of causes for the result found in the first analysis are possible: (1) the dose or weight recorded for case 3 was in error, so the case should probably be deleted

TABLE 9.2   Regression Summary for the Rat Data with Case 3 Deleted

```
Coefficients:
             Estimate Std. Error t value Pr(>|t|)
(Intercept)  0.311427   0.205094   1.518    0.151
BodyWt      -0.007783   0.018717  -0.416    0.684
LiverWt      0.008989   0.018659   0.482    0.637
Dose         1.484877   3.713064   0.400    0.695

Residual standard error: 0.07825 on 14 degrees of freedom
Multiple R-Squared: 0.02106
F-statistic: 0.1004 on 3 and 14 DF,  p-value: 0.9585
```

from the study, or (2) the regression fit in the second analysis is not appropriate except in the region defined by the 18 points excluding case 3. This has many implications concerning the experiment. It is possible that the combination of dose and rat weight chosen was fortuitous, and that the lack of relationship found would not persist for any other combinations of them, since inclusion of a data point apparently taken under different conditions leads to a different conclusion. This suggests the need for collection of additional data, with dose determined by some rule other than a constant proportion of weight.

9.2.4   Other Measures of Influence

The added-variable plots introduced in Section 3.1 provide a graphical diagnostic for influence. Cases corresponding to points at the left or right of an added-variable plot that do not match the general trend in the plot are likely to be influential for the variable that is to be added. For example, Figure 9.4 shows the added-variable plots for *BodyWt* and for *Dose* for the rat data. The point for case three is clearly

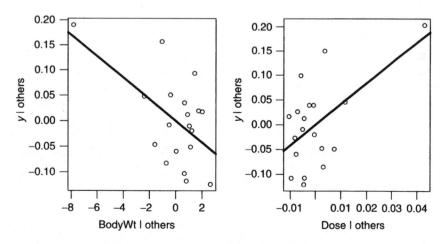

FIG. 9.4   Added-variable plots for *BodyWt* and *Dose*.

separated from the others, and is a likely influential point based on these graphs. The added-variable plot does not correspond exactly to Cook's distance, but to *local influence* defined by Cook (1986).

As with the outlier problem, influential groups of cases may serve to mask each other and may not be found by examination of cases one at a time. In some problems, multiple-case methods may be desirable; see Cook and Weisberg (1982, Section 3.6).

9.3 NORMALITY ASSUMPTION

The assumption of normal errors plays only a minor role in regression analysis. It is needed primarily for inference with small samples, and even then the bootstrap outlined in Section 4.6 can be used for inference. Furthermore, nonnormality of the unobservable errors is very difficult to diagnose in small samples by examination of residuals. From (8.4), the relationship between the errors and the residuals is

$$\hat{\mathbf{e}} = (\mathbf{I} - \mathbf{H})\mathbf{Y}$$
$$= (\mathbf{I} - \mathbf{H})(\mathbf{X}\hat{\boldsymbol{\beta}} + \mathbf{e})$$
$$= (\mathbf{I} - \mathbf{H})\mathbf{e}$$

because $(\mathbf{I} - \mathbf{H})\mathbf{X} = \mathbf{0}$. In scalar form, the ith residual is

$$\hat{e}_i = e_i - \left(\sum_{j=1}^{m} h_{ij} e_j \right) \tag{9.9}$$

Thus, $\hat{e}_i$ is equal to e_i, adjusted by subtracting off a weighted sum of all the errors. By the central limit theorem, the sum in (9.9) will be nearly normal even if the errors are not normal. With a small or moderate sample size n, the second term can dominate the first, and the residuals can behave like a normal sample even if the errors are not normal. Gnanadesikan (1997) refers to this as the *supernormality* of residuals.

As n increases for fixed p', the second term in (9.9) has small variance compared to the first term, so it becomes less important, and residuals can be used to assess normality; but in large samples, normality is much less important. Should a test of normality be desirable, a *normal probability plot* can be used. A general treatment of probability plotting is given by Gnanadesikan (1997). Suppose we have a sample of n numbers $z_1, z_2, \ldots, z_n$, and we wish to examine the hypothesis that the z's are a sample from a normal distribution with unknown mean μ and variance σ^2. A useful way to proceed is as follows:

1. Order the z's to get $z_{(1)} \le z_{(2)} \le \ldots \le z_{(n)}$. The ordered zs are called the *sample order statistics*.

2. Now, consider a standard normal sample of size n. Let $u_{(1)} \le u_{(2)} \le \ldots \le$ $u_{(n)}$ be the mean values of the order statistics that would be obtained if we repeatedly took samples of size n from the standard normal. The $u_{(i)}$s are called the *expected order statistics*. The $u_{(i)}$ are available in printed tables, or can be well approximated using a computer program[2].

3. If the zs are normal, then

$$E(z_{(i)}) = \mu + \sigma u_{(i)}$$

so that the regression of $z_{(i)}$ on $u_{(i)}$ will be a straight line. If it is not straight, we have evidence against normality.

Judging whether a probability plot is sufficiently straight requires experience. Daniel and Wood (1980) provided many pages of plots to help the analyst learn to use these plots; this can be easily recreated using a computer package that allows one quickly to look at many plots. Atkinson (1985) used a variation of the bootstrap to calibrate probability plots.

Many statistics have been proposed for testing a sample for normality. One of these that works extremely well is the Shapiro and Wilk (1965) W statistic, which is essentially the square of the correlation between the observed order statistics and the expected order statistics. Normality is rejected if W is too small. Royston (1982abc) provides details and computer routines for the calculation of the test and for finding p-values.

Figure 9.5 shows normal probability plots of the residuals for the heights data (Section 1.1) and for the transactions data (Section 4.6.1). Both have large enough

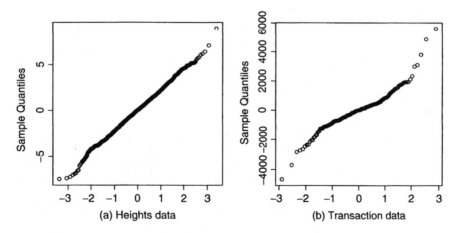

FIG. 9.5 Normal probability plots of residuals for (a) the heights data and (b) the transactions data.

[2]Suppose $\Phi(x)$ is a function that returns the area p to the left of x under a standard normal distribution, and $\Phi^{-1}(p)$ computes the inverse of the normal, so for a given value of p, it returns the associated value of x. Then the ith expected normal order statistic is approximately $\Phi^{-1}[(i - (3/8))/(n + (1/4))]$ (Blom, 1958).

samples for normal probability plots to be useful. For the heights data, the plot is very nearly straight, indicating no evidence against normality. For the transactions data, normality is in doubt because the plot is not straight. In particular, there are very large positive residuals well away from a fitted line. This supports the earlier claim that the errors for this problem are likely to be skewed with too many large values.

PROBLEMS

9.1. In an unweighted regression problem with $n = 54$, $p' = 5$, the results included $\hat{\sigma} = 4.0$ and the following statistics for four of the cases:

$\hat{e}_i$	h_{ii}
1.000	0.9000
1.732	0.7500
9.000	0.2500
10.295	0.185

For each of these four cases, compute r_i, D_i, and t_i. Test each of the four cases to be an outlier. Make a qualitative statement about the influence of each case on the analysis.

9.2. In the fuel consumption data, consider fitting the mean function

$$E(Fuel|X) = \beta_0 + \beta_1 Tax + \beta_2 Dlic + \beta_3 Income + \beta_4 \log(Miles)$$

For this regression, we find $\hat{\sigma} = 64.891$ with 46 df, and the diagnostic statistics for four states and the District of Columbia were:

	Fuel	$\hat{e}_i$	h_{ii}
Alaska	514.279	−163.145	0.256
New York	374.164	−137.599	0.162
Hawaii	426.349	−102.409	0.206
Wyoming	842.792	183.499	0.084
District of Columbia	317.492	−49.452	0.415

Compute D_i and t_i for each of these cases, and test for one outlier. Which is most influential?

9.3. The matrix $(X'_{(i)}X_{(i)})$ can be written as $(X'_{(i)}X_{(i)}) = X'X - x_i x'_i$, where x'_i is the ith row of X. Use this definition to prove that (A.37) holds.

9.4. The quantity $y_i - \mathbf{x}_i' \hat{\boldsymbol{\beta}}_{(i)}$ is the residual for the ith case when $\boldsymbol{\beta}$ is estimated without the ith case. Use (A.37) to show that

$$y_i - \mathbf{x}_i' \hat{\boldsymbol{\beta}}_{(i)} = \frac{\hat{e}_i}{1 - h_{ii}}$$

This quantity is called the *predicted residual*, or the *PRESS residual*.

9.5. Use (A.37) to verify (9.8).

9.6. Suppose that interest centered on $\boldsymbol{\beta}^*$ rather than $\boldsymbol{\beta}$, where $\boldsymbol{\beta}^*$ is the parameter vector excluding the intercept. Using (5.21) as a basis, define a distance measure D_i^* like Cook's D_i and show that (Cook, 1979)

$$D_i^* = \frac{r_i^2}{p} \left(\frac{h_{ii} - 1/n}{1 - h_{ii} + 1/n} \right)$$

where p is the number of terms in the mean function excluding the intercept.

9.7. Refer to the lathe data in Problem 6.2.

9.7.1. Starting with the full second-order model, use the Box–Cox method to show that an appropriate scale for the response is the logarithmic scale.

9.7.2. Find the two cases that are most influential in the fit of the quadratic mean function, and explain why they are influential. Delete these points from the data, refit the quadratic mean function, and compare to the fit with all the data.

9.8. Florida election 2000 In the 2000 election for US president, the counting of votes in Florida was controversial. In Palm Beach county in south Florida, for example, voters used a so-called butterfly ballot. Some believe that the layout of the ballot caused some voters to cast votes for Buchanan when their intended choice was Gore.

The data in the file florida.txt[3] has four variables, *County*, the county name, and *Gore*, *Bush*, and *Buchanan*, the number of votes for each of these three candidates. Draw the scatterplot of *Buchanan* versus *Bush*, and test the hypothesis that Palm Beach county is an outlier relative to the simple linear regression mean function for E(*Buchanan*|*Bush*). Identify another county with an unusual value of the Buchanan vote, given its Bush vote, and test that county to be an outlier. State your conclusions from the test, and its relevance, if any, to the issue of the butterfly ballot.

Next, repeat the analysis, but first consider transforming the variables in the plot to better satisfy the assumptions of the simple linear regression model. Again test to see if Palm Beach County is an outlier and summarize.

[3] Source: http://abcnews.go.com/sections/politics/2000vote/general/FL_county.html.

9.9. Refer to the United Nations data described in Problem 7.8 and consider the regression with response *ModernC* and predictors (log(*PPgdp*), *Change, Pop, Fertility, Frate, Purban*).

9.9.1. Examine added-variable plots for each of the terms in the regression model and summarize. Is it likely that any of the localities are influential for any of the terms? Which localities? Which terms?

9.9.2. Are there any outliers in the data?

9.9.3. Complete analysis of the regression of *ModernC* on the terms in the mean function.

9.10. The data in the data file `landrent.txt` were collected by Douglas Tiffany to study the variation in rent paid in 1977 for agricultural land planted to alfalfa. The variables are Y = average rent per acre planted to alfalfa, X_1 = average rent paid for all tillable land, X_2 = density of dairy cows (number per square mile), X_3 = proportion of farmland used as pasture, $X_4 = 1$ if liming is required to grow alfalfa; 0, otherwise.

The unit of analysis is a county in Minnesota; the 67 counties with appreciable rented farmland are included. Alfalfa is a high protein crop that is suitable feed for dairy cows. It is thought that rent for land planted to alfalfa relative to rent for other agricultural purposes would be higher in areas with a high density of dairy cows and rents would be lower in counties where liming is required, since that would mean additional expense. Use all the techniques learned so far to explore these data with regard to understanding rent structure. Summarize your results.

9.11. The data in the file `cloud.txt` summarize the results of the first Florida Area Cumulus Experiment, or FACE-1, designed to study the effectiveness of cloud seeding to increase rainfall in a target area (Woodley, Simpson, Biondini, and Berkeley, 1977). A fixed target area of approximately 3000 square miles was established to the north and east of Coral Gables, Florida. During the summer of 1975, each day was judged on its suitability for seeding. The decision to use a particular day in the experiment was based primarily on a suitability criterion S depending on a mathematical model for rainfall. Days with $S > 1.5$ were chosen as experimental days; there were 24 days chosen in 1975. On each day, the decision to seed was made by flipping a coin; as it turned out, 12 days were seeded, 12 unseeded. On seeded days, silver iodide was injected into the clouds from small aircraft. The predictors and the response are defined in Table 9.3.

The goal of the analysis is to decide if there is evidence that cloud seeding is effective in increasing rainfall. Begin your analysis by drawing appropriate graphs. Obtain appropriate transformations of predictors. Fit appropriate mean functions and summarize your results. (*Hint:* Be sure to check for influential observations and outliers.)

TABLE 9.3 The Florida Area Cumulus Experiment on Cloud Seeding

Variable	Description
A	Action, 1 = seed, 0 = do not seed
D	Days after the first day of the experiment (June 16, 1975=0)
S	Suitability for seeding
C	Percent cloud cover in the experimental area, measured using radar in Coral Gables, Florida
P	Prewetness, amount of rainfall in the hour preceding seeding in 10^7 cubic meters
E	Echo motion category, either 1 or 2, a measure of the type of cloud
Rain	Rainfall following the action of seeding or not seeding in 10^7 cubic meters

9.12. Health plans use many tools to try to control the cost of prescription medicines. For older drugs, *generic substitutes* that are equivalent to name-brand drugs are sometimes available at a lower cost. Another tool that may lower costs is restricting the drugs that physicians may prescribe. For example, if three similar drugs are available for treating the same symptoms, a health plan may require physicians to prescribe only one of them. Since the usage of the chosen drug will be higher, the health plan may be able to negotiate a lower price for that drug.

The data in the file `drugcost.txt`, provided by Mark Siracuse, can be used to explore the effectiveness of these two strategies in controlling drug costs. The response variable is *COST*, the average cost of drugs per prescription per day, and predictors include *GS* (the extent to which the plan uses generic substitution, a number between zero, no substitution, and 100, always use a generic substitute if available) and *RI* (a measure of the restrictiveness of the plan, from zero, no restrictions on the physician, to 100,

TABLE 9.4 The Drug Cost Data

Variable	Description
COST	Average cost to plan for one prescription for one day, dollars
RXPM	Average number of prescriptions per member per year
GS	Percent generic substitution used by the plan
RI	Restrictiveness index (0=none, 100=total)
COPAY	Average member copayment for prescriptions
AGE	Average member age
F	Percent female members
MM	Member months, a measure of the size of the plan
ID	An identifier for the name of the plan

the maximum possible restrictiveness). Other variables that might impact cost were also collected, and are described in Table 9.4. The data are from the mid-1990s, and are for 29 plans throughout the United States with pharmacies administered by a national insurance company.

Provide a complete analysis if these data, paying particular regard to possible outliers and influential cases. Summarize your results with regard to the importance of *GS* and *RI*. In particular, can we infer that more use of *GS* and *RI* will reduce drug costs?

CHAPTER 10

Variable Selection

We live in an era of cheap data but expensive information. A manufacturer studying the factors that impact the quality of its product, for example, may have many measures of quality, and possibly hundreds or even thousands of potential predictors of quality, including characteristics of the manufacturing process, training of employees, supplier of raw materials, and many others. In a medical setting, to model the size of tumor, we might have many potential predictors that describe the status of the patient, treatments given, and environmental factors thought to be relevant. In both of these settings, and in many others, we can have too many predictors.

One response to working with problems with many potential predictors is to try to identify the important or *active* predictors and the unimportant or *inactive* ones. *Variable selection methods* are often used for this purpose, and we will study them in this chapter. Estimates and predictions will generally be more precise from fitted models based only on relevant terms, although selection tends to *overestimate* significance. Sometimes, identifying the important predictors can be an end in itself. For example, learning if the supplier of raw materials impacts quality of a manufactured product may be more important than attempting to measure the exact form of the dependence.

Linear regression with variable selection is not the only approach to the problem of modeling a response as a function of a very large number of terms or predictors. The fields of *machine learning* and to some extent *data mining* provide alternative techniques for this problem, and in some circumstances, the methods developed in these areas can give superior answers. An introduction to these areas is given by Hastie, Tibshirani, and Friedman (2001). The methods described in this chapter are important in their own right because they are so widely used, and also because they can provide a basis for understanding newer methods.

10.1 THE ACTIVE TERMS

Given a response Y and a set of terms X derived from the predictors, the idealized goal of variable selection is to divide X into two pieces $X = (X_A, X_I)$, where X_A

is the set of active terms, while X_I is the set of inactive terms that are not relevant to the regression problem, in the sense that the two mean functions $E(Y|X_A, X_I)$ and $E(Y|X_A)$ would give exactly the same values. Identifying the active predictors can be surprisingly difficult.

Dividing $X = (X_A, X_I)$ into active and inactive terms, suppose the mean function

$$E(Y|X = \mathbf{x}) = \boldsymbol{\beta}'\mathbf{X} = \boldsymbol{\beta}'_A\mathbf{x}_A + \boldsymbol{\beta}'_I\mathbf{x}_I \tag{10.1}$$

was a correct specification. For example, using the methods of the previous chapters in this book, we might have selected transformations, added interactions, and possibly deleted a few outliers so that (10.1) holds at least to a reasonable approximation. For the inactive terms, we will have $\boldsymbol{\beta}_I = \mathbf{0}$. If we have a sufficiently large sample to estimate $\boldsymbol{\beta}$, then identifying X_A seems easy: the terms in X_A will have non zero corresponding elements in $\hat{\boldsymbol{\beta}}$, and the terms in X_I will correspond to elements in $\hat{\boldsymbol{\beta}}$ close to zero.

Simulated Data

To illustrate, we consider two cases based on artificial data, each with five terms in X, including the intercept. For both cases, the response is obtained as

$$y = 1 + x_1 + x_2 + 0x_3 + 0x_4 + \text{error}$$

where the error is $N(0, 1)$, so $\sigma^2 = 1$. For this mean function, X_A is the intercept plus first two components of X, and X_I is the remaining two components. In the first case we consider, $X_1 = (x_1, x_2, x_3, x_4)$ are independent standard normal random variables, and so the population covariance matrix for X_1 is

$$\text{Var}(X_1) = \begin{pmatrix} 1 & 0 & 0 & 0 \\ 0 & 1 & 0 & 0 \\ 0 & 0 & 1 & 0 \\ 0 & 0 & 0 & 1 \end{pmatrix}$$

In the second case, $X_2 = (x_1, x_2, x_3, x_4)$ are again normal with mean zero, but the population covariance matrix is

$$\text{Var}(X_2) = \begin{pmatrix} 1 & 0 & .95 & 0 \\ 0 & 1 & 0 & -.95 \\ .95 & 0 & 1 & 0 \\ 0 & -.95 & 0 & 1 \end{pmatrix} \tag{10.2}$$

so the first and third variables are highly positively correlated, and the second and fourth variables are highly negatively correlated.

Table 10.1 summarizes one set of simulated data for the first case and with $n = 100$. $\hat{\boldsymbol{\beta}}$ is reasonably close to the true value of 1 for the first three coefficients and 0 for the remaining two coefficients. The t-values for the first three terms are large, indicating that these are clearly estimated to be nonzero, while the t-values

TABLE 10.1 Regression Summary for the Simulated Data with No Correlation between the Predictors

	Estimate	Std. Error	t-value	$Pr(>|t|)$
(Intercept)	0.8022	0.0919	8.73	0.0000
x_1	0.9141	0.0901	10.14	0.0000
x_2	0.9509	0.0861	11.04	0.0000
x_3	−0.0842	0.1091	−0.77	0.4423
x_4	−0.2453	0.1109	−2.21	0.0294

$$\hat{\sigma} = 0.911, \ df = 95, \ R^2 = 0.714$$

$$\text{Var}(\hat{\beta}) = \frac{1}{100} \begin{pmatrix} 0.84 & 0.09 & 0.01 & -0.05 & 0.02 \\ 0.09 & 0.81 & -0.03 & -0.04 & -0.06 \\ 0.01 & -0.03 & 0.74 & -0.16 & -0.07 \\ -0.05 & -0.04 & -0.16 & 1.19 & 0.02 \\ 0.02 & -0.06 & -0.07 & 0.02 & 1.23 \end{pmatrix}$$

for the remaining two terms are much smaller. As it happens, the t-test for β_4 has a p-value of about 0.03, suggesting incorrectly that $\beta_3 \neq 0$. If we do tests at 5% level, then 5% of the time we will make errors like this. Also shown in Table 10.1 is $\text{Var}(\hat{\beta})$, which is approximately equal to $1/n$ times a diagonal matrix; ignoring the entries in the first row and column of this matrix that involve the intercept, the remaining 4×4 matrix should be approximately the inverse of $\text{Var}(X_1)$, which is the identity matrix. Apart from the intercept, all the estimates are equally variable and independent. If the sample size were increased to 1000, $\text{Var}(\hat{\beta})$ would be approximately the same matrix multiplied by 1/1000 rather than 1/100.

Table 10.2 gives the summary of the results when $n = 100$, and the covariance of the terms excluding the intercept is given by (10.2). X_A and X_I are not clearly identified. Since x_2 and x_4 are almost the same variable, apart from a sign change, identification of x_4 as more likely to be the active predictor is not surprising; the choice between x_2 and x_4 can vary from realization to realization of this simulation. All of the t-values are much smaller than in the first case, primarily because *with covariance between the terms, variances of the estimated coefficients are greatly inflated relative to uncorrelated terms*. To get estimates with about the same variances in case 2 as we got in case 1 requires about 11 times as many observations. The simulation for case 2 is repeated in Table 10.3 with $n = 1100$. Apart from the intercept, estimates and standard errors are now similar to those in Table 10.1, but the large correlations between some of the estimates, indicated by the large off-diagonal elements in the covariance matrix for $\hat{\beta}$, remain. Identification of the terms in X_A and X_I with correlation present can require huge sample sizes relative to problems with uncorrelated terms.

Selection methods try to identify the active terms and then refit, ignoring the terms thought to be inactive. Table 10.4a is derived from Table 10.2 by deleting the two terms with small t-values. This seems like a very good solution and summary of this problem, with one exception: it is the wrong answer, since x_4 is included

TABLE 10.2   Regression Summary for the Simulated Data with High Correlation between the Predictors

	Estimate	Std. Error	t-value	$Pr(>|t|)$
(Intercept)	0.8022	0.0919	8.73	0.0000
x_1	1.1702	0.3476	3.37	0.0011
x_2	0.2045	0.3426	0.60	0.5519
x_3	-0.2696	0.3494	-0.77	0.4423
x_4	-0.7856	0.3553	-2.21	0.0294

$$\hat{\sigma} = 0.911, \text{ df} = 95, \ R^2 = 0.702$$

$$\text{Var}(\hat{\boldsymbol{\beta}}) = \frac{1}{100} \begin{pmatrix} 0.84 & 0.25 & 0.08 & -0.17 & 0.07 \\ 0.25 & 12.08 & 0.14 & -11.73 & -0.34 \\ 0.08 & 0.14 & 11.73 & -0.36 & 11.78 \\ -0.17 & -11.73 & -0.36 & 12.21 & 0.17 \\ 0.07 & -0.34 & 11.78 & 0.17 & 12.63 \end{pmatrix}$$

TABLE 10.3   Regression Summary for the Simulated Data, Correlated Case But with $n = 1100$

	Estimate	Std. Error	t-value	$Pr(>|t|)$
(Intercept)	1.0354	0.0305	33.92	0.0000
x_1	1.0541	0.0974	10.83	0.0000
x_2	1.1262	0.0989	11.39	0.0000
x_3	-0.0106	0.0978	-0.11	0.9136
x_4	0.1446	0.1006	1.44	0.1511

$$\hat{\sigma} = 1.01, \text{ df} = 1095, \ R^2 = 0.68$$

$$\text{Var}(\hat{\boldsymbol{\beta}}) = \frac{1}{1100} \begin{pmatrix} 1.02 & -0.10 & 0.00 & 0.09 & -0.05 \\ -0.10 & 10.43 & -0.07 & -9.97 & -0.05 \\ 0.00 & -0.07 & 10.75 & 0.06 & 10.41 \\ 0.09 & -9.97 & 0.06 & 10.52 & 0.06 \\ -0.05 & -0.05 & 10.41 & 0.06 & 11.14 \end{pmatrix}$$

rather than x_2. Table 10.4b is the fit of the mean function using only the correct X_A as terms. The fit of this choice for the mean function is somewhat worse, with larger $\hat{\sigma}$ and smaller R^2.

10.1.1   Collinearity

Two terms X_1 and X_2 are exactly *collinear*, or linearly dependent, if there is a linear equation such as

$$c_1 X_1 + c_2 X_2 = c_0 \tag{10.3}$$

for some constants c_0, c_1 and c_2 that is true for all cases in the data. For example, suppose that X_1 and X_2 are amounts of two chemicals and are chosen so that

TABLE 10.4 Regression Summary for Two Candidate Subsets in the Simulated Data, Correlated Cases, with $n = 100$

	Estimate	Std. Error	t-value	$\Pr(>|t|)$
(a) Candidate terms are intercept, x_1 and x_4				
(Intercept)	0.7972	0.0912	8.74	0.0000
x_1	0.9146	0.0894	10.23	0.0000
x_4	−0.9796	0.0873	−11.22	0.0000

$$\hat{\sigma} = 0.906, \ \text{df} = 97, \ R^2 = 0.711$$

	Estimate	Std. Error	t-value	$\Pr(>|t|)$
(b) Candidate terms are intercept, x_1 and x_2				
(Intercept)	0.8028	0.0933	8.60	0.0000
x_1	0.9004	0.0915	9.84	0.0000
x_2	0.9268	0.0861	10.76	0.0000

$$\hat{\sigma} = 0.927, \ \text{df} = 97, \ R^2 = 0.691$$

$X_1 + X_2 = 50$ ml, then X_1 and X_2 are exactly collinear. Since $X_2 = 50 - X_1$, knowing X_1 is exactly the same as knowing both X_1 and X_2, and only one of X_1 or X_2 can be included in a mean function. Exact collinearities can occur by accident when, for example, weight in pounds and in kilograms are both included in a mean function, or with sets of dummy variables. *Approximate* collinearity is obtained if the equation (10.3) nearly holds for the observed data. For example, the variables *Dose* and *BodyWt* in the rat data in Section 9.2.3 are approximately collinear since *Dose* was approximately determined as a multiple of *BodyWt*. In the first of the two simulated cases in the last section, there is no collinearity because the terms are uncorrelated. In the second case, because of high correlation $x_1 \approx x_3$ and $x_2 \approx -x_4$, so these pairs of terms are collinear.

Collinearity between terms X_1 and X_2 is measured by the square of their sample correlation, r_{12}^2. Exact collinearity corresponds to $r_{12}^2 = 1$, and noncollinearity corresponds to $r_{12}^2 = 0$. As r_{12}^2 approaches 1, approximate collinearity becomes generally stronger. Most discussions of collinearity are really concerned with approximate collinearity.

The definition of approximate collinearity extends naturally to $p > 2$ terms. A set of terms, $X_1, X_2, \ldots, X_p$ are approximately collinear if, for constants $c_0, c_1, \ldots, c_p$,

$$c_1 X_1 + c_2 X_2 + \cdots + c_p X_p \approx c_0$$

with at least one $c_j \neq 0$. For that j, we can write

$$X_j \approx \frac{1}{c_j}\left(c_0 - \sum_{\ell \neq j} c_\ell X_\ell\right) = \frac{c_0}{c_j} + \sum_{\ell \neq j}\left(-\frac{c_\ell}{c_j}\right) X_\ell$$

which is similar to a linear regression mean function with intercept c_0/c_j and slopes $-c_\ell/c_j$. A simple diagnostic analogous to the squared correlation for the two-variable case is the square of the multiple correlation between X_j and the other X's, which we will call R_j^2. This number is computed from the regression of X_j on the other X's. If the largest R_j^2 is near 1, we would diagnose approximate collinearity.

When a set of predictors is exactly collinear, one or more predictors must be deleted, or else unique least squares estimates of coefficients do not exist. Since the deleted predictor contains no information after the others, no information is lost by this process although interpretation of parameters can be more complex. When collinearity is approximate, a usual remedy is again to delete variables from the mean function, with loss of information about fitted values expected to be minimal. The hard part is deciding which variables to delete.

10.1.2 Collinearity and Variances

We have seen in the initial example in this chapter that correlation between the terms increases the variances of estimates. In a mean function with two terms beyond the intercept,

$$E(Y|X_1 = x_1, X_2 = x_2) = \beta_0 + \beta_1 x_1 + \beta_2 x_2$$

suppose the sample correlation between X_1 and X_2 is r_{12}, and define the symbol $SX_j X_j = \sum (x_{ij} - \bar{x}_j)^2$ to be the sum of squares for the jth term in the mean function. It is an exercise (Problem 10.7) to show that, for $j = 1, 2$,

$$\text{Var}(\hat{\beta}_j) = \frac{\sigma^2}{1 - r_{12}^2} \frac{1}{SX_j X_j} \tag{10.4}$$

The variances of $\hat{\beta}_1$ and $\hat{\beta}_2$ are minimized if $r_{12}^2 = 0$, while as r_{12}^2 nears 1, these variances are greatly inflated; for example, if $r_{12}^2 = .95$, the variance of β_1 is 20 times as large as if $r_{12}^2 = 0$. The use of collinear predictors can lead to unacceptably variable estimated coefficients compared to problems with no collinearity.

When $p > 2$, the variance of the j-th coefficient is (Problem 10.7)

$$\text{Var}(\hat{\beta}_j) = \frac{\sigma^2}{1 - R_j^2} \frac{1}{SX_j X_j} \tag{10.5}$$

The quantity $1/(1 - R_j^2)$ is called the jth *variance inflation factor*, or *VIF*$_j$ (Marquardt, 1970). Assuming that the X_j's could have been sampled to make $R_j^2 = 0$ while keeping $SX_j X_j$ constant, the *VIF* represents the increase in variance due to the correlation between the predictors and, hence, collinearity.

In the first of the two simulated examples earlier in this section, all the terms are independent, so each of the R_j^2 should be close to zero, and the *VIF* are all close to their minimum value of one. For the second example, each of the $R_j^2 \approx 0.95^2$, and each *VIF* should be close to $1/(1 - .95^2) = 10.256$. Estimates in the second case are about 10 or 11 times as variable as estimates in the first case.

10.2   VARIABLE SELECTION

The goal of variable selection is to divide X into the set of active terms X_A and the set of inactive terms X_I. For this purpose, we assume that the mean function (10.1) is appropriate for the data at hand. If we have k terms in the mean function apart from the intercept, then there are potentially 2^k possible choices of X_A obtained from all possible subsets of the terms. If $k = 5$, there are only $2^5 = 32$ choices for X_A and all 32 possible can be fit and compared. If $k = 10$, there are 1024 choices, and fitting such a large number of models is possible but still an unpleasant prospect. For k as small as 30, the number of models possible is much too large to consider them all.

There are two basic issues. First, given a particular candidate X_C for the active terms, what criterion should be used to compare X_C to other possible choices for X_A? The second issue is computational: How do we deal with the potentially huge number of comparisons that need to be made?

10.2.1   Information Criteria

Suppose we have a particular candidate subset X_C. If $X_C = X_A$, then the fitted values from the fit of the mean function

$$E(Y|X_C = \mathbf{x}_C) = \boldsymbol{\beta}'_C\mathbf{x}_C \tag{10.6}$$

should be very similar to the fit of mean function (10.1), and the residual sum of squares for the fit of (10.6) should be similar to the residual sum of squares for (10.1). If X_C misses important terms, the residual sum of squares should be larger; see Problem 10.9.

Criteria for comparing various candidate subsets are based on the lack of fit of a model and its *complexity*. Lack of fit for a candidate subset X_C is measured by its residual sum of squares RSS_C. Complexity for multiple linear regression models is measured by the number of terms p_C in X_C, including the intercept[1]. The most common criterion that is useful in multiple linear regression and many other problems where model comparison is at issue is the *Akaike Information Criterion*, or *AIC*. Ignoring constants that are the same for every candidate subset, *AIC* is given by Sakamoto, Ishiguro, and Kitagawa (1987),

$$AIC = n\log(RSS_C/n) + p_C \tag{10.7}$$

---

[1] The complexity may also be defined as the number of parameters estimated in the regression as a whole, which is equal to the number of terms plus one for estimating σ^2.

Small values of *AIC* are preferred, so better candidate sets will have smaller *RSS* and a smaller number of terms p_C. An alternative to *AIC* is the *Bayes Information Criterion*, or *BIC*, given by Schwarz (1978),

$$BIC = n \log(RSS_C/n) + p_C \log(n) \tag{10.8}$$

which provides a different balance between lack of fit and complexity. Once again, smaller values are preferred.

Yet a third criterion that balances between lack of fit and complexity is *Mallows' C_p* (Mallows, 1973), where the subscript p is the number of terms in candidate X_C. This statistic is defined by

$$C_{pc} = \frac{RSS_C}{\hat{\sigma}^2} + 2p_C - n \tag{10.9}$$

where $\hat{\sigma}^2$ is from the fit of (10.1). As with many problems for which many solutions are proposed, there is no clear choice between the criteria for preferring a subset mean function. There is an important similarity between all three criteria: if we fix the complexity, meaning that we consider only the choices X_C with a fixed number of terms, then all three will agree that the choice with the smallest value of residual sum of squares is the preferred choice.

Highway Accident Data
We will use the highway accident data described in Section 7.2. The initial terms we consider include the transformations found in Section 7.2.2 and a few others and are described in Table 10.5. The response variable is, from Section 7.3, log(*Rate*). This mean function includes 14 terms to describe only $n = 39$ cases.

TABLE 10.5 Definition of Terms for the Highway Accident Data

Variable	Description
log(*Rate*)	Base-two logarithm of 1973 accident rate per million vehicle miles, the response
log(*Len*)	Base-two logarithm of the length of the segment in miles
log(*ADT*)	Base-two logarithm of average daily traffic count in thousands
log(*Trks*)	Base-two logarithm of truck volume as a percent of the total volume
Slim	1973 speed limit
Lwid	Lane width in feet
Shld	Shoulder width in feet of outer shoulder on the roadway
Itg	Number of freeway-type interchanges per mile in the segment
log(*Sigs1*)	Base-two logarithm of (number of signalized interchanges per mile in the segment + 1)/(length of segment)
Acpt	Number of access points per mile in the segment
Hwy	A factor coded 0 if a federal interstate highway, 1 if a principal arterial highway, 2 if a major arterial, and 3 otherwise

TABLE 10.6   Regression Summary for the Fit of All Terms in the Highway Accident Data[a]

	Estimate	Std. Error	t-value	$Pr(>|t|)$
Intercept	5.7046	2.5471	2.24	0.0342
logLen	−0.2145	0.1000	−2.15	0.0419
logADT	−0.1546	0.1119	−1.38	0.1792
logTrks	−0.1976	0.2398	−0.82	0.4178
logSigs1	0.1923	0.0754	2.55	0.0172
Slim	−0.0393	0.0242	−1.62	0.1172
Shld	0.0043	0.0493	0.09	0.9313
Lane	−0.0161	0.0823	−0.20	0.8468
Acpt	0.0087	0.0117	0.75	0.4622
Itg	0.0515	0.3503	0.15	0.8842
Lwid	0.0608	0.1974	0.31	0.7607
Hwy1	0.3427	0.5768	0.59	0.5578
Hwy2	−0.4123	0.3940	−1.05	0.3053
Hwy3	−0.2074	0.3368	−0.62	0.5437

$$\hat{\sigma} = 0.376 \text{ on 25 df, } R^2 = 0.791$$

[a]The terms *Hwy1*, *Hwy2*, and *Hwy3* are dummy variables for the highway factor.

The regression on all the terms is summarized in Table 10.6. Only two of the terms have t-values exceeding 2 in absolute value, in spite of the fact that $R^2 = 0.791$. Few of the predictors adjusted for the others are clearly important even though, taken as a group, they are useful for predicting accident rates. This is usually evidence that X_A is smaller than the full set of available terms.

To illustrate the information criteria, consider a candidate subset X_C consisting of the intercept and (log(Len), $Slim$, $Acpt$, log($Trks$), $Shld$). For this choice, $RSS_C = 5.016$ with $p_C = 6$. For the mean function with all the terms, $RSS_X = 3.537$ with $p_X = 14$, so $\hat{\sigma}^2 = 0.1415$. From these, we can compute the values of AIC, BIC, and C_p for the subset mean function,

$$AIC = 39 \log(5.016/39) + 2 \times 6 = -67.99$$

$$BIC = 39 \log(5.016/39) + 6 \log(39) = -58.01$$

$$C_p = \frac{5.016}{0.1415} + 2 \times 6 - 39 = 8.453$$

and for the full mean function,

$$AIC = 39 \log(3.5377/39) + 2 \times 14 = -65.611$$

$$BIC = 39 \log(3.5377/39) + 14 \log(39) = -42.322$$

$$C_p = \frac{3.537}{0.1415} + 2 \times 14 - 39 = 14$$

All three criteria are smaller for the subset, and so the subset is preferred over the mean function with all the terms. This subset need not be preferred to other subsets, however.

10.2.2 Computationally Intensive Criteria

Cross-validation can also be used to compare candidate subset mean functions. The most straightforward type of cross-validation is to split the data into two parts at random, a *construction set* and a *validation set*. The construction set is used to estimate the parameters in the mean function. Fitted values from this fit are then computed for the cases in the validation set, and the average of the squared differences between the response and the fitted values for the validation set is used as a summary of fit for the candidate subset. Good candidates for X_A will have small cross-validation errors. Correction for complexity is not required because different data are used for fitting and for estimating fitting errors.

Another version of cross-validation uses *predicted residuals* for the subset mean function based on the candidate X_C. For this criterion, compute the fitted value for each i from a regression that does not include case i. The sum of squares of these values is the predicted residual sum of squares, or *PRESS* (Allen, 1974; Geisser and Eddy, 1979),

$$PRESS = \sum_{i=1}^{n} (y_i - \mathbf{x}'_{Ci}\hat{\boldsymbol{\beta}}_{C(i)})^2 = \sum_{i=1}^{n} \left(\frac{\hat{e}_{Ci}}{1 - h_{Cii}} \right)^2 \qquad (10.10)$$

where $\hat{e}_{Ci}$ and h_{Cii} are, respectively, the residual and the leverage for the ith case in the subset model. For the subset mean function and the full mean function considered in the last section, the values of *PRESS* are computed to be 7.688 and 11.272, respectively, suggesting substantial improvement of this particular subset over the full mean function because it gives much smaller errors on the average.

The *PRESS* method for linear regression depends only on residuals and leverages, so it is fairly easy to compute. This simplicity does not carry over to problems in which the computation will require refitting the mean function n times. As a result, this method is not often used outside the linear regression model.

10.2.3 Using Subject-Matter Knowledge

The single most important tool in selecting a subset of variables is the analyst's knowledge of the area under study and of each of the variables. In the highway accident data, *Hwy* is a factor, so all of its levels should probably either be in the candidate subset or excluded. Also, the variable log(*Len*) should be treated differently from the others, since its inclusion in the active predictors may be required by the way highway segments are defined. Suppose that highways consist of "safe stretches" and "bad spots," and that most accidents occur at the bad spots. If we were to lengthen a highway segment in our study by a small amount, it is unlikely that we would add another bad spot to the section, assuming bad spots are rare, but the computed response, accidents per million vehicle miles on the section of roadway, will decrease. Thus, the response and log(*Len*) should be negatively correlated, and we should consider only subsets that include log(*Len*). Of the 14 terms, one is to be included in all candidate subsets, and three, the dummy variables for *Hwy*, are all to be included or excluded as a group. Thus, we have 10 terms (or groups of terms) that can be included or not, for a total of $2^{10} = 1024$ possible subset mean functions to consider.

10.3 COMPUTATIONAL METHODS

With linear regression, it is possible to find the few candidate subsets of each subset size that minimize the information criteria. Furnival and Wilson (1974) provided an algorithm called the *leaps and bounds* algorithm that uses information from regressions already computed to bound the possible value of the criterion function for regressions as yet not computed. This trick allows skipping the computation of most regressions. The algorithm has been widely implemented in statistical packages and in subroutine libraries. It cannot be used with factors, unless the factors are replaced by sets of dummy variables, and it cannot be used with computationally intensive criteria such as cross-validation or *PRESS*.

For problems other than linear least squares regression, or if cross-validation is to be used as the criterion function, exhaustive methods are generally not feasible, and computational compromise is required. *Stepwise methods* require examining only a few subsets of each subset size. The estimate of X_A is then selected from the few subsets that were examined. Stepwise methods are not guaranteed to find the candidate subset that is optimal according to any criterion function, but they often give useful results in practice.

Stepwise methods have three basic variations. For simplicity of presentation, we assume that no terms beyond the intercept are forced into the subsets considered. As before, let k be the number of terms, or groups of terms, that might be added to the mean function. *Forward selection* uses the following procedure:

[FS.1] Consider all candidate subsets consisting of one term beyond the intercept, and find the subset that minimizes the criterion of interest. If an information criterion is used, then this amounts to finding the term that is most highly correlated with the response because its inclusion in the subset gives the smallest residual sum of squares. Regardless of the criterion, this step requires examining k candidate subsets.

[FS.2] For all remaining steps, consider adding one term to the subset selected at the previous step. Using an information criterion, this will amount to adding the term with the largest partial correlation[2] with the response given the terms already in the subset, and so this is a very easy calculation. Using

[2]The *partial correlation* is the ordinary correlation coefficient between the two plotted quantities in an added-variable plot.

cross-validation, this will require fitting all subsets consisting of the subset selected at the previous step plus one additional term. At step ℓ, $k - \ell + 1$ subsets need to be considered.

[FS.3] Stop when all the terms are included in the subset, or when addition of another term increases the value of the selection criterion.

If the number of terms beyond the intercept is k, this algorithm will consider at most $k + (k - 1) + \cdots + 1 = k(k - 1)/2$ of the 2^k possible subsets. For $k = 10$, the number of subsets considered is 45 of the 1024 possible subsets. The subset among these 45 that has the best value of the criterion selected is tentatively selected as the candidate for X_A.

The algorithm requires modification if a group of terms is to be treated as all included or all not included, as would be the case with a factor. At each step, we would consider adding the term or the group of terms that produces the best value on the criterion of interest. Each of the information criteria can now give different best choices because at each step, as we are no longer necessarily examining mean functions with p_C fixed.

The *backward elimination algorithm* works in the opposite order:

[BE.1] Fit first with candidate subset $X_C = X$, as given by (10.1).

[BE.2] At the next step, consider all possible subsets obtained by removing one term other than those to be forced to be in all mean functions from the candidate subset selected at the last step. Using an information criterion, this amounts to removing the term with the smallest t-value in the regression summary because this will give the smallest increase in residual sum of squares. Using cross-validation, all subsets formed by deleting one term from the current subset must be considered.

[BE.3] Continue until all terms but those forced into all mean functions are deleted, or until the next deletion increases the value of the criterion.

As with the forward selection method, only $k(k - 1)/2$ subsets are considered, and the best among those considered is the candidate for X_A. The subsets considered by forward selection and by backward elimination may not be the same. If factors are included among the terms, as with backward elimination, the information criteria need not all select the same subset of fixed size as the best.

The forward and backward algorithms can be combined into a *stepwise* method, where at each step, a term is either deleted or added so that the resulting candidate mean function minimizes the criterion function of interest. This will have the advantage of allowing consideration of more subsets, without the need for examining all 2^k subsets.

Highway Accidents

Table 10.7 presents a summary of the 45 mean functions examined using forward selection for the highway accident data, using *PRESS* as the criterion function for selecting subsets. The volume of information in this table may seem overwhelming,

TABLE 10.7   Forward Selection for the Highway Accident Data. Subsets within a Step Are Ordered According to the Value of *PRESS*

Step 1:   Base terms: (logLen)

	df	RSS		p	C(p)	AIC	BIC	PRESS
Add: Slim	36	6.11216	|	3	10.20	-66.28	-61.29	6.93325
Add: Shld	36	7.86104	|	3	22.56	-56.46	-51.47	9.19173
Add: Acpt	36	7.03982	|	3	16.76	-60.77	-55.78	9.66532
Add: Hwy	34	8.62481	|	5	31.96	-48.85	-40.53	10.4634
Add: logSigs1	36	9.41301	|	3	33.53	-49.44	-44.45	10.8866
Add: logTrks	36	9.89831	|	3	36.96	-47.48	-42.49	11.5422
Add: logADT	36	10.5218	|	3	41.37	-45.09	-40.10	12.0428
Add: Itg	36	10.962	|	3	44.48	-43.50	-38.51	12.5544
Add: Lane	36	10.8671	|	3	43.81	-43.84	-38.84	12.5791
Add: Lwid	36	11.0287	|	3	44.95	-43.26	-38.27	15.3326

Step 2:   Base terms: (logLen Slim)

	df	RSS		p	C(p)	AIC	BIC	PRESS
Add: logTrks	35	5.5644	|	4	8.33	-67.94	-61.29	6.43729
Add: Hwy	33	5.41187	|	6	11.25	-65.02	-55.04	6.79799
Add: logSigs1	35	5.80682	|	4	10.04	-66.28	-59.62	6.94127
Add: Itg	35	6.10666	|	4	12.16	-64.31	-57.66	7.07987
Add: Lane	35	6.10502	|	4	12.15	-64.32	-57.67	7.15826
Add: logADT	35	6.05881	|	4	11.82	-64.62	-57.97	7.18523
Add: Shld	35	6.0442	|	4	11.72	-64.71	-58.06	7.28524
Add: Acpt	35	5.51181	|	4	7.96	-68.31	-61.66	7.77756
Add: Lwid	35	6.07752	|	4	11.96	-64.50	-57.85	8.9025

Step 3:   Base terms: (logLen Slim logTrks)

	df	RSS		p	C(p)	AIC	BIC	PRESS
Add: Hwy	32	4.82665	|	7	9.12	-67.49	-55.84	6.28517
Add: Itg	34	5.55929	|	5	10.29	-65.98	-57.66	6.54584
Add: logADT	34	5.46616	|	5	9.64	-66.63	-58.32	6.6388
Add: logSigs1	34	5.45673	|	5	9.57	-66.70	-58.38	6.65431
Add: Lane	34	5.56426	|	5	10.33	-65.94	-57.62	6.66387
Add: Shld	34	5.41802	|	5	9.30	-66.98	-58.66	6.71471
Add: Acpt	34	5.15186	|	5	7.41	-68.94	-60.63	7.341
Add: Lwid	34	5.51339	|	5	9.97	-66.30	-57.98	8.12161

Step 4:   Base terms: (logLen Slim logTrks Hwy)

	df	RSS		p	C(p)	AIC	BIC	PRESS
Add: logSigs1	31	3.97747	|	8	5.11	-73.03	-59.73	5.67779
Add: Itg	31	4.8187	|	8	11.06	-65.55	-52.24	6.49463
Add: Lane	31	4.8047	|	8	10.96	-65.66	-52.36	6.54448
Add: logADT	31	4.82664	|	8	11.12	-65.49	-52.18	6.73021
Add: Shld	31	4.82544	|	8	11.11	-65.50	-52.19	6.88205
Add: Acpt	31	4.61174	|	8	9.60	-67.26	-53.95	7.72218
Add: Lwid	31	4.80355	|	8	10.95	-65.67	-52.37	8.05348

Step 5:   Base terms: (logLen Slim logTrks Hwy logSigs1)

	df	RSS		p	C(p)	AIC	BIC	PRESS
Add: Itg	30	3.90937	|	9	6.63	-71.71	-56.74	5.70787
Add: Lane	30	3.91112	|	9	6.64	-71.69	-56.72	5.7465

(Continued overleaf)

TABLE 10.7 *(Continued)*

```
Step 5:  Base terms: (logLen Slim logTrks Hwy logSigs1)
Add: logADT   30   3.66683   |   9   4.92 -74.21 -59.23   5.86015
Add: Shld     30   3.97387   |   9   7.09 -71.07 -56.10   6.25166
Add: Acpt     30   3.92837   |   9   6.77 -71.52 -56.55   7.15377
Add: Lwid     30   3.97512   |   9   7.10 -71.06 -56.08   7.68299
Add: Itg      30   3.90937   |   9   6.63 -71.71 -56.74   5.70787
Add: Lane     30   3.91112   |   9   6.64 -71.69 -56.72   5.7465
Add: logADT   30   3.66683   |   9   4.92 -74.21 -59.23   5.86015
Add: Shld     30   3.97387   |   9   7.09 -71.07 -56.10   6.25166
Add: Acpt     30   3.92837   |   9   6.77 -71.52 -56.55   7.15377
Add: Lwid     30   3.97512   |   9   7.10 -71.06 -56.08   7.68299
```

```
Step 6:  Base terms: (logLen Slim logTrks Hwy logSigs1 Itg)
             df    RSS      |   p   C(p)    AIC    BIC    PRESS
Add: Lane    29   3.86586   |  10   8.32 -70.14 -53.51   5.78305
Add: logADT  29   3.66672   |  10   6.92 -72.21 -55.57   6.1424
Add: Shld    29   3.90652   |  10   8.61 -69.74 -53.10   6.30147
Add: Acpt    29   3.86515   |  10   8.32 -70.15 -53.52   7.17893
Add: Lwid    29   3.90718   |  10   8.62 -69.73 -53.09   7.73347
```

```
Step 7:  Base terms: (logLen Slim logTrks Hwy logSigs1 Itg Lane)
             df    RSS      |   p    C(p)    AIC    BIC    PRESS
Add: logADT  28   3.65494   |  11   8.83 -70.33 -52.03   6.31797
Add: Shld    28   3.86395   |  11  10.31 -68.16 -49.86   6.40822
Add: Acpt    28   3.8223    |  11  10.02 -68.59 -50.29   7.32972
Add: Lwid    28   3.86487   |  11  10.32 -68.15 -49.85   7.89833
```

```
Step 8: Base terms: (logLen Slim logTrks Hwy logSigs1 Itg Lane logADT)
            df    RSS      |   p    C(p)    AIC    BIC    PRESS
Add: Shld   27   3.654     |  12  10.83 -68.34 -48.38   6.93682
Add: Acpt   27   3.55292   |  12  10.11 -69.44 -49.47   8.2891
Add: Lwid   27   3.63541   |  12  10.70 -68.54 -48.58   8.3678
```

```
Step 9:  Base terms: (logLen Slim logTrks Hwy logSigs1 Itg Lane
                      logADT Shld)
            df    RSS      |   p    C(p)    AIC    BIC    PRESS
Add: Lwid   26   3.61585   |  13  12.56 -66.75 -45.12   8.85795
Add: Acpt   26   3.55037   |  13  12.09 -67.46 -45.84   9.33926
```

```
Step 10:  Base terms: (logLen Slim logTrks Hwy logSigs1 Itg Lane
                       logADT Shld Lwid)
            df    RSS      |   p    C(p)    AIC    BIC    PRESS
Add: Acpt   25   3.53696   |  14  14.00 -65.61 -42.32  11.2722
```

so some description is in order. At Step 1, the mean function consists of the single term log(*Len*) beyond the intercept because this term is to be included in all mean functions. Ten mean functions can be obtained by adding one of the remaining 10-candidate terms, counting *Hwy* as a single term. For each candidate mean function, the df, *RSS*, and number of terms $p = p_C$ in the mean function are printed, as are

PRESS and the three information criteria. If none of the terms were factors, then all three information criteria would order the terms identically. Since *Hwy* is a factor, the ordering need not be the same on all criteria. *PRESS* may choose a different ordering from the other three. All the criteria agree for the "best" term to add at the first step, since adding *Slim* gives the smallest value of each criterion.

Step 2 starts with the base mean function consisting of log(*Len*) and *Slim*, and *PRESS* selects log(*Trks*) at this step. Both C_p and *BIC* would select a different term at this step, leading to different results (see Problem 10.10). This process is repeated at each step.

The candidate mean function with the smallest value of *PRESS* is given by (log(*Len*), *Slim*, log(*Trks*), *Hwy*, log(*Sigs1*)), with a value of *PRESS* = 5.678. Several other subsets have values of *PRESS* that differ from this one by only a trivial amount, and, since the values of all the criteria are random variables, declaring this subset to the "best" needs to be tempered with a bit of common sense. The estimated active predictors should be selected from among the few essentially equivalent subsets on some other grounds, such as agreement with theory. The candidate for the active subset has $R^2 = 0.765$, as compared to the maximum possible value of 0.791, the R^2 for the mean function using all the predictors. Further analysis of this problem is left to homework problems.

10.3.1 Subset Selection Overstates Significance

All selection methods can overstate significance. Consider another simulated example. A data set of $n = 100$ cases with a response Y and 50 predictors $X = (X_1, \ldots, X_{50})$ was generated using standard normal random deviates, so there are no active terms, and the true multiple correlation between Y and X is also exactly zero. The regression of Y on X is summarized in the first line of Table 10.8. The value of $R^2 = 0.54$ may seem surprisingly large, considering that all the data are independent random numbers. The overall F-test, which is in a scale more easily calibrated, gives a p-value of .301 for the data; Rencher and Pun (1980) and Freedman (1983) report similar simulations with the overall p-value varying from near 0 to near 1, as it should since the null hypothesis of $\beta = 0$ is true. In the simulation reported here, 11 of 50 terms had $|t| > \sqrt{2}$, while 7 of 50 had $|t| > 2$. Line 2 of Table 10.8 displays summary statistics from the fit of the mean function that retains all the terms with $|t| > \sqrt{2}$. The value of R^2 drops to 0.28. The major change is in the perceived significance of the result. The overall F now has p-value of about .001, and the $|t|$-values for five terms exceed 2. The third line is similar to the second, except a more stringent $|t| > 2$ was used. Using seven terms, $R^2 = 0.20$, and four terms have $|t|$-values exceeding two.

This example demonstrates many lessons. Significance is overstated. The coefficients for the terms left in the mean function will generally be too large in absolute value, and have t- or F-values that are too large. Even if the response and the predictors are unrelated, R^2 can be large: when $\beta = 0$, the expected value of R^2 is $k/(n-1)$. With selection, R^2 can be much too large.

TABLE 10.8   Results of a Simulated Example with 50 Terms and $n = 100$

Method	Number of Terms	R^2	p-value of Overall F	Number $|t| > \sqrt{2}$	Number $|t| > 2$
No selection	50	0.48	0.301	11	7
$|t| > \sqrt{2}$	11	0.28	0.001	7	5
$|t| > 2$	5	0.20	0.003	5	4

10.4   WINDMILLS

The windmill data discussed in Problems 2.13, 4.6 and 6.11 provide another case study for model selection. In Problem 2.13, only the wind speed at the reference site was used in the mean function. In Problem 6.11, wind direction at the reference site was used to divide the data into 16 bins, and a separate regression was fit in each of the bins, giving a mean function with 32 parameters. We now consider several other potential mean functions.

10.4.1   Six Mean Functions

For this particular candidate site, we used as a reference site the closest site where the National Center for Environmental Modeling data is collected, which is southwest of the candidate. There are additional possible reference sites to the northwest, the northeast, and the southeast of the candidate site. We could use data from all four of these candidates to predict wind speed at the candidate site. In addition, we could consider the use of *lagged variables*, in which we use the wind speed at the reference site six hours before the current time to model the current wind speed at the candidate site. Lagged variables are commonly used with data collected at regularly spaced intervals and can help account for serial correlations between consecutive measurements.

In all, we will consider six very different mean functions for predicting *CSpd*, using the terms defined in Table 10.9:

[Model 1] $E(CSpd|Spd1) = \beta_0 + \beta_1 Spd1$. This was fit in Problem 2.13.

[Model 2] Fit as in Model 1, but with a separate intercept and slope for each of 16 bins determined by the wind direction at reference site 1.

[Model 3] This mean function uses the information about the wind directions in a different way. Writing θ for the wind direction at the reference site, the mean function is

$$E(CSpd|X) = \beta_0 + \beta_1 Spd1 + \beta_2 \cos(\theta) + \beta_3 \sin(\theta)$$
$$= +\beta_4 \cos(\theta)Spd1 + \beta_5 \sin(\theta)Spd1$$

This mean function uses four terms to include the information in the wind direction. The term $\cos(\theta)Spd1$ is the wind component in the east−west

TABLE 10.9 Description of Data in the Windmill Data in the File wm4.txt

Label	Description
Date	Date and time of measurement. "2002/3/4/12" means March 4, 2002 at 12 hours after midnight
Dir1	Wind direction θ at reference site 1 in degrees
Spd1	Wind speed at reference site 1 in meters per second. Site 1 is the closest site to the candidate site
Spd2	Wind speed at reference site 2 in m/s
Spd3	Wind speed at reference site 3 in m/s
Spd4	Wind speed at reference site 4 in m/s
Spd1Lag1	Wind speed at reference site 1 six hours previously
Spd2Lag1	Wind speed at reference site 2 six hours previously
Spd3Lag1	Wind speed at reference site 3 six hours previously
Spd4Lag1	Wind speed at reference site 4 six hours previously
Bin	Bin number
Spd1Sin1	$Spd1 \times \sin(\theta)$, site 1
Spd1Cos1	$Spd1 \times \cos(\theta)$, site 1
CSpd	Wind speed in m/s at the candidate site

direction, while $\sin(\theta)Spd1$ is the component in the north–south direction. The terms in sine and cosine alone are included to allow information from the wind direction alone.

[Model 4] This model uses the mean function

$$E(CSpd|X) = \beta_0 + \beta_1 Spd1 + \beta_2 Spd1Lag1$$

that ignores information from the angles but includes information from the wind speed at the previous period.

[Model 5] This model uses wind speed from all four candidate sites,

$$E(CSpd|X) = \beta_0 + \beta_1 Spd1 + \beta_2 Spd2 + \beta_3 Spd3 + \beta_4 Spd4$$

[Model 6] The final mean function starts with model 5 and then adds information on the lagged wind speeds:

$$E(CSpd|X) = \beta_0 + \beta_1 Spd1 + \beta_2 Spd2 + \beta_3 Spd3 + \beta_4 Spd4$$
$$+ \beta_5 Spd1Lag1 + \beta_6 Spd2Lag1 + \beta_7 Spd3Lag1$$
$$+ \beta_8 Spd4Lag1$$

All six of these mean functions were fit using the data in the file wm4.txt. The first case in the data does not have a value for the lagged variables, so it has been deleted from the file. Since the month of May 2002 is also missing, the first case in June 2002 was also deleted.

TABLE 10.10   Summary Criteria for the Fit of Six Mean Function to the Windmill Data

	df	AIC	BIC	PRESS
Model 1	2	2014.9	2024.9	6799.0
Model 2	32	1989.3	2149.8	6660.2
Model 3	6	2020.7	2050.8	6836.3
Model 4	3	1920.6	1935.6	6249.1
Model 5	5	1740.6	1765.7	5320.2
Model 6	9	1711.2	1756.3	5188.5

Table 10.10 gives the information criteria for comparing the fit of these six mean functions, along with *PRESS*. All three criteria agree on the ordering of the models. The simplest model 1 is preferred over model 3; evidently, the information in the sine and cosine of the direction is not helpful. Adding the lagged wind speed in Model 4 is clearly helpful, and apparently is more useful than the information from binning the directions used in Model 2. Adding information from four reference sites, as in models 5 and 6, gives a substantial improvement, with about a 15% decrease in the criterion statistics. Model 6, which includes lags but not information on angles, appears to be the most appropriate model here.

10.4.2   A Computationally Intensive Approach

The windmill data provides an unusual opportunity to look at model selection by examining the related problem of estimating the long-term average wind speed not at the candidate site but at the closest reference site. The data file wm5.txt[3] gives 55 years of data from all four candidate sites. We can simulate the original problem by estimating a regression model for predicting wind speed at site 1, given the data from the remaining three sites, and then see how well we do by comparing the prediction to the actual value, which is the known average over 55 years of data. We used the following procedure:

1. The year 2002 at the candidate site had $n = 1116$ data points. We begin by selecting n time points at random from the 55 years of data to comprise the "year" of complete data.
2. For the sample of times selected, fit the models we wish to compare. In this simulation, we considered only the model with one site as the predictor without binning; one site as predictor with wind directions binned into 16 bins, and using the wind speeds at all three remaining sites as predictors without using bins or lagged variables. Data from the simulated year were used to estimate the parameters of the model, predict the average wind speed

---

[3]Because this file is so large, it is not included with the other data files and must be downloaded separately from the web site for this book.

over the remaining time points, and also compute the standard error of the prediction, using the methodology outlined previously in this section, and in Problems 2.13 and 6.11.

3. Repeat the first two steps 1000 times, and summarize the results in histograms.

The summarizing histograms are shown in Figure 10.1. The first column shows the histograms for the estimates of the long-term average wind speed for the three mean functions. The vertical dashed lines indicate the true mean wind speed at site 1 over the 55 years of data collection. All three methods have distributions of estimates that are centered very close to the true value and appear to be more or less normally distributed. The second column gives the standard errors of the estimated mean for the 1000 simulations, and the dashed line corresponds to the

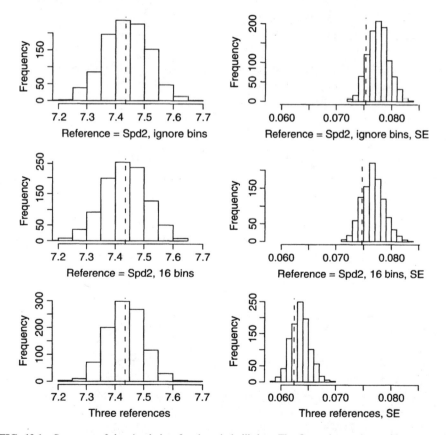

FIG. 10.1 Summary of the simulation for the windmill data. The first column gives a histogram of the estimated mean wind speed at reference site 1 for 1000 simulations using three mean functions. The second column gives a histogram of the 1000 standard errors. The dashed lines give the true values, the average of the wind speed measurements from 1948 to 2003 for the averages, and the standard deviation of the 1000 averages from the simulation for the standard errors.

"true value," actually the standard deviation of the 1000 means in the first column. In each case, most of the histogram is to the right of the dashed line, indicating that the standard formulas will generally overestimate the actual standard error, perhaps by 5%. Also, the mean functions that use only one reference, with or without binning, are extremely similar, suggesting only a trivial improvement due to binning. Using three references, however, shifts the distribution of the standard errors to the left, so this method is much more precise than the others.

Generalizing these results to the candidate site from reference site 1 does not seem to be too large a leap. This would suggest that we can do better with more reference sites than with one and that the information about wind direction, at least at this candidate site, is probably unimportant.

PROBLEMS

10.1. Generate data as described for the two simulated data sets in Section 10.1, and compare the results you get to the results given in the text.

10.2. Using the "data" in Table 10.11 with a response Y and three predictors X_1, X_2 and X_3 from Mantel (1970) in the file `mantel.txt`, apply the BE and FS algorithms, using C_p as a criterion function. Also, find AIC and C_p for all possible models, and compare results. What is X_A?

TABLE 10.11   Mantel's Data for Problem 10.2

	Y	X_1	X_2	X_3
1	5.00	1.00	1004.00	6.00
2	6.00	200.00	806.00	7.30
3	8.00	−50.00	1058.00	11.00
4	9.00	909.00	100.00	13.00
5	11.00	506.00	505.00	13.10

10.3. Use BE with the highway accident data and compare with the results in Table 10.7.

10.4. For the boys in the Berkeley Guidance Study in Problem 3.1, find a model for *HT18* as a function of the other variables for ages 9 or earlier. Perform a complete analysis, including selection of transformations and diagnostic analysis, and summarize your results.

10.5. An experiment was conducted to study *O2UP*, oxygen uptake in milligrams of oxygen per minute, given five chemical measurements shown in Table 10.12 (Moore, 1975). The data were collected on samples of dairy wastes kept in suspension in water in a laboratory for 220 days. All observations were on the

TABLE 10.12   Oxygen Update Experiment

Variable	Description
Day	Day number
BOD	Biological oxygen demand
TKN	Total Kjeldahl nitrogen
TS	Total solids
TVS	Total volatile solids
COD	Chemical oxygen demand
O2UP	Oxygen uptake

same sample over time. We desire an equation relating $\log(O2UP)$ to the other variables. The goal is to find variables that should be further studied with the eventual goal of developing a prediction equation; day cannot be used as a predictor. The data are given in the file `dwaste.txt`.

Complete the analysis of these data, including a complete diagnostic analysis. What diagnostic indicates the need for transforming *O2UP* to a logarithmic scale?

10.6. Prove the results (10.4)–(10.5). To avoid tedious algebra, start with an added-variable plot for X_j after all the other terms in the mean function. The estimated slope $\hat{\beta}_j$ is the OLS estimated slope in the added-variable plot. Find the standard error of this estimate, and show that it agrees with the given equations.

10.7. Galápagos Islands The Galápagos Islands off the coast of Ecuador provide an excellent laboratory for studying the factors that influence the development and survival of different life species. Johnson and Raven (1973) have presented data in the file `galapagos.txt`, giving the number of species and related variables for 29 different islands (Table 10.13). Counts are given for both the total number of species and the number of species that occur only on that one island (the endemic species).

Use these data to find factors that influence diversity, as measured by some function of the number of species and the number of endemic species, and summarize your results. One complicating factor is that elevation is not recorded for six very small islands, so some provision must be made for this. Four possibilities are: (1) find the elevations; (2) delete these six islands from the data; (3) ignore elevation as a predictor of diversity, or (4) substitute a plausible value for the missing data. Examination of large-scale maps suggests that none of these elevations exceed 200 m.

10.8. Suppose that (10.1) holds with $\beta_I = 0$, but we fit a subset model using the terms $X_C \neq X_A$; that is, X_C does not include all the relevant terms. Give general conditions under which the mean function $E(Y|X_C)$ is a linear mean function. (*Hint:* See Appendix A.2.4.)

TABLE 10.13   Galápagos Island Data

Variable	Description
Island	Island name
NS	Number of species
ES	Number of endemic species (occurs only on that island)
Area	Surface area of island, hectares
Anear	Area of closest island, hectares
Dist	Distance to closest island, kilometers
DistSC	Distance from Santa Cruz Island, kilometers
Elevation	Elevation in m, missing values given as zero
EM	1 if elevation is observed, 0 if missing

10.9. For the highway accident data, fit the regression model with active predictors given by the subset with the smallest value of *PRESS* in Table 10.7. The coefficient estimate of *Slim* is *negative*, meaning that segments with higher speed limits *lower* accident rates. Explain this finding.

10.10. Reëxpress C_p as a function of the F-statistic used for testing the null hypothesis (10.6) versus the alternative (10.1). Discuss.

10.11. In the windmill data discussed in Section 10.4, data were collected at the candidate site for about a year, for about 1200 observations. One issue is whether the collection period could be shortened to six months, about 600 observations, or three months, about 300 observations, and still give a reliable estimate of the long-term average wind speed.

Design and carry out a simulation experiment using the data described in Section 10.4.2 to characterize the increase in error due to shortening the collection period. For the purpose of the simulation, consider site #1 to be the "candidate" site and site #2 to be the reference site, and consider only the use of *Spd2* to predict *Spd1*. (*Hint:* The sampling scheme used in Section 10.4.2 may not be appropriate for time periods shorter than a year because of seasonal variation. Rather than picking 600 observations at random to make up a simulated six-month period, a better idea might be to pick a starting observation at random, and then pick 600 consecutive observations to comprise the simulated six months.)

CHAPTER 11

Nonlinear Regression

A regression mean function cannot always be written as a linear combination of the terms. For example, in the turkey diet supplement experiment described in Section 1.1, the mean function

$$E(Y|X = x) = \theta_1 + \theta_2(1 - \exp(-\theta_3 x)) \tag{11.1}$$

where Y was growth and X the amount of supplement added to the turkey diet, was suggested. This mean function has three parameters, θ_1, θ_2, and θ_3, but only one predictor, X. It is a nonlinear mean function because the mean function is not a linear combination of the parameters. In (11.1), θ_2 multiplies $1 - \exp(-\theta_3 x)$, and θ_3 enters through the exponent.

Another nonlinear mean function we have already seen was used in estimating transformations of predictors to achieve linearity, given by

$$E(Y|X = x) = \beta_0 + \beta_1 \psi_S(x, \lambda) \tag{11.2}$$

where $\psi_S(x, \lambda)$ is the scaled power transformation defined by (7.3), page 150. This is a nonlinear model because the slope parameter β_1 multiplies $\psi_S(x, \lambda)$, which depends on the parameter λ. In Chapter 7, we estimated λ visually and then estimated the βs from the linear model assuming λ is fixed at its estimated value. If we estimate all three parameters simultaneously, then the mean function is nonlinear.

Nonlinear mean functions usually arise when we have additional information about the dependence of the response on the predictor. Sometimes, the mean function is selected because the parameters of the function have a useful interpretation. In the turkey growth example, when $X = 0$, $E(Y|X = 0) = \theta_1$, so θ_1 is the expected growth with no supplementation. Assuming $\theta_3 > 0$, as X increases, $E(Y|X = x)$ will approach $\theta_1 + \theta_2$, so the sum of the first two parameters is the maximum growth possible for any dose called an *asymptote*, and θ_2 is the maximum additional growth due to supplementation. The final parameter θ_3 is a rate parameter; for larger

Applied Linear Regression, Third Edition, by Sanford Weisberg
ISBN 0-471-66379-4   Copyright © 2005 John Wiley & Sons, Inc.

values of θ_3, the expected growth approaches its maximum more quickly than it would if θ_3 were smaller.

11.1 ESTIMATION FOR NONLINEAR MEAN FUNCTIONS

Here is the general setup for nonlinear regression. We have a set of p terms X, and a vector $\boldsymbol{\theta} = (\theta_1, \ldots, \theta_k)'$ of parameters such that the mean function relating the response Y to X is given by

$$E(Y|X = \mathbf{x}) = \mathsf{m}(\mathbf{x}, \boldsymbol{\theta}) \tag{11.3}$$

We call the function m a *kernel mean function*. The two examples of m we have seen so far in this chapter are in (11.1) and (11.2) but there are of course many other choices, both simpler and more complex. The linear kernel mean function, $\mathsf{m}(\mathbf{x}, \boldsymbol{\theta}) = \mathbf{x}'\boldsymbol{\theta}$ is a special case of the nonlinear kernel mean function. Many nonlinear mean functions impose restrictions on the parameters, like $\theta_3 > 0$ in (11.1).

As with linear models, we also need to specify the variance function, and for this we will use exactly the same structure as for the linear model and assume

$$\text{Var}(Y|X = \mathbf{x}_i) = \sigma^2/w_i \tag{11.4}$$

where, as before, the w_i are known, positive weights and σ^2 is an unknown positive number. Equations (11.3) and (11.4) together with the assumption that observations are independent of each other define the nonlinear regression model. The only difference between the nonlinear regression model and the linear regression model is the form of the mean function, and so we should expect that there will be many parallels that can be exploited.

The data consist of observations $(\mathbf{x}_i, y_i)$, $i = 1, \ldots, n$. Because we have retained the assumption that observations are independent and that the variance function (11.4) is known apart from the scale factor σ^2, we can use least squares to estimate the unknown parameters, so we need to minimize over all permitted values of $\boldsymbol{\theta}$ the residual sum of squares function,

$$RSS(\boldsymbol{\theta}) = \sum_{i=1}^{n} w_i(y_i - \mathsf{m}(\mathbf{x}_i, \boldsymbol{\theta}))^2 \tag{11.5}$$

We have OLS if all the weights are equal and WLS if they are not all equal.

For linear models, there is a formula for the value $\hat{\boldsymbol{\theta}}$ of $\boldsymbol{\theta}$ that minimizes $RSS(\boldsymbol{\theta})$, given at (A.21) in the Appendix. For nonlinear regression, there generally is no formula, and minimization of (11.5) is a numerical problem. We present some theory now that will approximate (11.5) at each iteration of a computing algorithm by a nearby linear regression problem. Not only will this give one of the standard computing algorithms used for nonlinear regression but will also provide expressions for approximate standard errors and point out how to do approximate tests. The derivation uses some calculus.

We begin with a brief refresher on approximating a function using a Taylor series expansion[1]. In the scalar version, suppose we have a function $g(\beta)$, where β is a scalar. We want to approximate $g(\beta)$ for values of β close to some fixed value β^*. The Taylor series approximation is

$$g(\beta) = g(\beta^*) + (\beta - \beta^*)\frac{dg(\beta)}{d\beta} + \frac{1}{2}(\beta - \beta^*)^2\frac{d^2g(\beta)}{d\beta^2} + \text{Remainder} \qquad (11.6)$$

All the derivatives in equation (11.6) are evaluated at β^*, and so Taylor series approximates $g(\beta)$, the function on the left side of (11.6) using the polynomial in β on the right side of (11.6). We have only shown a two-term Taylor expansion and have collected all the higher-order terms into the remainder. By taking enough terms in the Taylor expansion, any function g can be approximated as closely as wanted. In most statistical applications, only one or two terms of the Taylor series are needed to get an adequate approximation. Indeed, in the application of the Taylor expansion here, we will mostly use a one-term expansion that includes the quadratic term in the remainder.

When $g(\boldsymbol{\theta})$ is a function of a vector-valued parameter $\boldsymbol{\theta}$, the two-term Taylor series is very similar,

$$g(\boldsymbol{\theta}) = g(\boldsymbol{\theta}^*) + (\boldsymbol{\theta} - \boldsymbol{\theta}^*)'\mathbf{u}(\boldsymbol{\theta}^*) + \tfrac{1}{2}(\boldsymbol{\theta} - \boldsymbol{\theta}^*)'\mathbf{H}(\boldsymbol{\theta}^*)(\boldsymbol{\theta} - \boldsymbol{\theta}^*) + \text{Remainder} \quad (11.7)$$

where we have defined two new quantities in (11.7), the *score vector* $\mathbf{u}(\boldsymbol{\theta}^*)$ and the *Hessian matrix* $\mathbf{H}(\boldsymbol{\theta}^*)$. If $\boldsymbol{\theta}^*$ has k elements, then $\mathbf{u}(\boldsymbol{\theta}^*)$ also has k elements, and its jth element is given by $\partial g(\mathbf{x}, \boldsymbol{\theta})/\partial\theta_j$, evaluated at $\boldsymbol{\theta} = \boldsymbol{\theta}^*$. The Hessian is a $k \times k$ matrix whose (ℓ, j) element is the partial second derivative $\partial^2 g(\mathbf{x}, \boldsymbol{\theta})/(\partial\theta_\ell\partial\theta_j)$, evaluated at $\boldsymbol{\theta} = \boldsymbol{\theta}^*$.

We return to the problem of minimizing (11.5). Suppose we have a current guess $\boldsymbol{\theta}^*$ of the value of $\boldsymbol{\theta}$ that will minimize (11.5). The general idea is to approximate $m(\boldsymbol{\theta}, \mathbf{x}_i)$ using a Taylor approximation around $\boldsymbol{\theta}^*$. Using a one-term Taylor series, ignoring the term with the Hessian in (11.7), we get

$$m(\boldsymbol{\theta}, \mathbf{x}_i) \approx m(\boldsymbol{\theta}^*, \mathbf{x}_i) + \mathbf{u}_i(\boldsymbol{\theta}^*)'(\boldsymbol{\theta} - \boldsymbol{\theta}^*) \qquad (11.8)$$

We have put the subscript i on the $\mathbf{u}$ because the value of the derivatives can be different for every value of $\mathbf{x}_i$. The $\mathbf{u}_i(\boldsymbol{\theta}^*)$ play the same role as the terms in the multiple linear regression model. There are as many elements of $\mathbf{u}_i(\boldsymbol{\theta}^*)$ as parameters in the mean function. The difference between nonlinear and linear models is that the $\mathbf{u}_i(\boldsymbol{\theta}^*)$ may depend on unknown parameters, while in multiple linear regression, the terms depend only on the predictors.

---

[1] Jerzy Neyman (1894–1981), one of the major figures in the development of statistics in the twentieth century, often said that arithmetic had five basic operations: addition, subtraction, multiplication, division, and Taylor series.

Substitute the approximation (11.8) into (11.5) and simplify to get

$$RSS(\theta) = \sum_{i=1}^{n} w_i \left[y_i - m(\theta, \mathbf{x}_i) \right]^2$$

$$\approx \sum_{i=1}^{n} w_i \left[y_i - m(\theta^*, \mathbf{x}_i) - \mathbf{u}_i(\theta^*)'(\theta - \theta^*) \right]^2$$

$$= \sum_{i=1}^{n} w_i \left[\hat{e}_i^* - \mathbf{u}_i(\theta^*)'(\theta - \theta^*) \right]^2 \qquad (11.9)$$

where $\hat{e}_i^* = y_i - m(\theta^*, \mathbf{x}_i)$ is the ith *working residual* that depends on the current guess θ^*. The approximate $RSS(\theta)$ is now in the same form as the residual sum of squares function for multiple linear regression (5.5), with response given by the $\hat{e}_i^*$, terms given by $\mathbf{u}_i(\theta^*)$, parameter given by $\theta - \theta^*$, and weights w_i. We switch to matrix notation and let $\mathbf{U}(\theta^*)$ be an $n \times k$ matrix with ith row $\mathbf{u}_i(\theta^*)'$, $\mathbf{W}$ is an $n \times n$ diagonal matrix of weights, and $\hat{\mathbf{e}}^* = (\hat{e}_1^*, \ldots, \hat{e}_n^*)'$. The least squares estimate is then

$$\widehat{\theta - \theta^*} = [\mathbf{U}(\theta^*)'\mathbf{W}\mathbf{U}(\theta^*)]^{-1}\mathbf{U}(\theta^*)'\mathbf{W}\hat{\mathbf{e}}^* \qquad (11.10)$$

$$\hat{\theta} = \theta^* + [\mathbf{U}(\theta^*)'\mathbf{W}\mathbf{U}(\theta^*)]^{-1}\mathbf{U}(\theta^*)'\mathbf{W}\hat{\mathbf{e}}^* \qquad (11.11)$$

We will use (11.10) in two ways, first to get a computing algorithm for estimating θ in the rest of this section and then as a basis for inference in the next section.

Here is the *Gausss–Newton algorithm* that is suggested by (11.10)–(11.11):

1. Select an initial guess $\theta^{(0)}$ for θ, and compute $RSS(\theta^{(0)})$.
2. Set the iteration counter at $j = 0$.
3. Compute $\mathbf{U}(\theta^{(j)})$ and $\hat{\mathbf{e}}^{(j)}$ with ith element $y_i - m(\mathbf{x}_i, \theta^{(j)})$. Evaluating (11.11) requires the estimate from a weighted linear least squares problem, with response $\hat{\mathbf{e}}^{(j)}$, predictors $\mathbf{U}(\theta^{(j)})$, and weights given by the w_i. The new estimator is $\theta^{(j+1)}$. Also, compute the residuals sum of squares $RSS(\theta^{(j+1)})$.
4. Stop if $RSS(\theta^{(j)}) - RSS(\theta^{(j+1)})$ is sufficiently small, in which case there is convergence. Otherwise, set $j = j + 1$. If j is too large, stop, and declare that the algorithm has failed to converge. If j is not too large, go to step 3.

The Gauss–Newton algorithm estimates the parameters of a nonlinear regression problem by a sequence of approximating linear WLS calculations.

Most statistical software for nonlinear regression uses the Gauss–Newton algorithm, or a modification of it, for estimating parameters. Some programs allow using a general function minimizer based on some other algorithm to minimize (11.5). We provide some references at the end of the chapter.

There appear to be two impediments to the use of the Gauss–Newton algorithm. First, the score vectors, which are the derivatives of m with respect to the parameters, are needed. Some software may require the user to provide expressions for the derivatives, but many packages compute derivatives using either symbolic or numeric differentiation. Also, the user must provide starting values $\theta^{(0)}$; there appears to be no general way to avoid specifying starting values. The optimization routine may also converge to a local minimum of the residuals sum of squares function rather than a global minimum, and so finding good starting values can be very important in some problems. With poor starting values, an algorithm may fail to converge to any estimate. We will shortly discuss starting values in the context of an example.

11.2 INFERENCE ASSUMING LARGE SAMPLES

We repeat (11.11), but now we reinterpret θ^* as the *true, unknown value of θ*. In this case, the working residuals $\hat{e}^*$ are now the actual errors e, the differences between the response and the true means. We write

$$\hat{\theta} = \theta^* + [U(\theta^*)'WU(\theta^*)]^{-1}U(\theta^*)'We \qquad (11.12)$$

This equation is based on the assumption that the nonlinear kernel mean function m can be accurately approximated close to θ^* by the linear approximation (11.8), and this can be guaranteed only if the sample size n is large enough. We then see that $\hat{\theta}$ is equal to the true value plus a linear combination of the elements of e, and by the central limit theorem $\hat{\theta}$ under regularity conditions will be approximately normally distributed,

$$\hat{\theta} \sim N(\theta^*, \sigma^2[U(\theta^*)'WU(\theta^*)]^{-1}) \qquad (11.13)$$

An estimate of the large-sample variance is obtained by replacing the unknown θ^* by $\hat{\theta}$ on the right side of (11.13),

$$\widehat{Var}(\hat{\theta}) = \hat{\sigma}^2[U(\hat{\theta})'WU(\hat{\theta})]^{-1} \qquad (11.14)$$

where the estimate of σ^2 is

$$\hat{\sigma}^2 = \frac{RSS(\hat{\theta})}{n-k} \qquad (11.15)$$

where k is the number of parameters estimated in the mean function.

These results closely parallel the results for the linear model, and consequently the inferential methods such as F- and t-tests and the analysis of variance for comparing nested mean functions, can be used for nonlinear models. One change that is recommended is to use the normal distribution rather than the t for inferences where the t would be relevant, but since (11.13) is really expected to be valid only in large samples, this is hardly important. *We emphasize that in small samples, large-sample inferences may be inaccurate.*

We can illustrate using these results with the turkey growth experiment. Methionine is an amino acid that is essential for normal growth in turkeys. Depending on the ingredients in the feed, turkey producers may need to add supplemental methionine for a proper diet. Too much methionine could be toxic. Too little methionine could result in malnourished birds.

An experiment was conducted to study the effects on turkey growth of different amounts A of methionine, ranging from a control with no supplementation to 0.44% of the total diet. The experimental unit was a pen of young turkeys, and treatments were assigned to pens at random so that 10 pens get the control (no supplementation) and 5 pens received each of the other five amounts used in the experiment, for a total of 35 pens. Pen weights, the average weight of the turkeys in the pen, were obtained at the beginning and the end of the experiment three weeks later. The response variable is *Gain*, the average weight gain in grams per turkey in a pen. The weight gains are shown in Table 11.1 and are also given in the file `turk0.txt` (Cook and Witmer, 1985). The primary goal of this experiment is to understand how expected weight gain $E(Gain|A)$ changes as A is varied. The data are shown in Figure 11.1.

In Figure 11.1, $E(Gain|A)$ appears to increase with A, at least over the range of values of A in the data. In addition, there is considerable pen-to-pen variation, reflected by the variability between repeated observations at the same value of A. The mean function is certainly not a straight line since the difference in the means when $A > 0.3$ is much smaller than the difference in means when $A < 0.2$. While a polynomial of degree two or three might well match the mean at the six values of A in the experiment, it will surely not match the data outside the range of A, and the parameters would have little physical meaning (see Problem 6.15). A nonlinear mean function is preferable for this problem.

For turkey growth as a function of an amino acid, the mean function

$$E(Gain|A) = \theta_1 + \theta_2(1 - \exp(-\theta_3 A)) \qquad (11.16)$$

TABLE 11.1   The Turkey Growth Data

Amount, A	Gain
0.00	644, 631, 661, 624, 633
	610, 615, 605, 608, 599
0.04	698, 667, 657, 685, 635
0.10	730, 715, 717, 709, 707
0.16	735, 712, 726, 760, 727
0.28	809, 796, 763, 791, 811
0.44	767, 771, 799, 799, 791

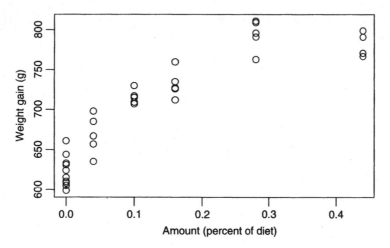

FIG. 11.1 Turkey data.

was suggested by Parks (1982). To estimate the parameters in (11.16), we need starting values for θ. While there is no absolute rule for selecting starting values, the following approaches are often useful:

Guessing Sometimes, starting values can be obtained by guessing values for the parameters. In the turkey data, from Figure 11.1, the intercept is about 620 and the asymptote is around 800. This leads to starting values $\theta_1^{(0)} = 620$ and $\theta_2^{(0)} = 800 - 620 = 180$. Guessing a value for the rate parameter θ_3 is harder.

Solving equations for a subset of the data Select as many distinct data points as parameters, and solve the equations for the unknown parameters. The hope is that the equations will be easy to solve. Selecting data points that are diverse often works well. In the turkey data, given $\theta_1^{(0)} = 620$ and $\theta_2^{(0)} = 180$ from the graph, we can get an initial estimate for θ_3 by solving only one equation in one unknown. For example, when $D = 0.16$, a plausible value of *Gain* is *Gain* = 750, so

$$750 = 620 + 180(1 - \exp(-\theta_3^{(0)}(.16)))$$

which is easily solved to give $\theta_3^{(0)} \approx 8$. Thus, we now have starting values for all three parameters.

Linearization If possible, transform to a multiple linear regression mean function, and fit it to get starting values. In the turkey data, we can move the parameters θ_1 and θ_2 to the left side of the mean function to get

$$\frac{(\theta_1 + \theta_2) - y_i}{\theta_2} = \exp(-\theta_3 D)$$

Taking logarithms of both sides,

$$\log\left(\frac{(\theta_1 + \theta_2) - y_i}{\theta_2}\right) = -\theta_3 D$$

Substituting initial guesses $\theta_1^{(0)} = 620$ and $\theta_2^{(0)} = 180$ on the left side of this equation, we can compute an initial guess for θ_3 by the linear regression of $\log[(y_i - 800)/180]$ on $-D$, through the origin. The OLS estimate in this approximate problem is $\theta_3^{(0)} \approx 12$.

Many computer packages for nonlinear regression require specification of the function m using an expression such as

```
y ~ th1 + th2*(1 - exp(-th3*A))
```

In this equation, the symbol ~ can be read "is modeled as," and the mathematical symbols +, -, * and / represent addition, subtraction, multiplication and division, respectively. Similarly, exp represents exponentiation, and ^, or, in some programs, **, is used for raising to a power. Parentheses are used according to the usual mathematical rules. Generally, the formula on the right of the model specification will be similar to an expression in a computer programming language such as Basic, Fortran, or C. This form of a computer model should be contrasted with the model statements described in Sections 6.2.1–6.2.2. Model statements for nonlinear models include explicit statement of both the parameters and the terms on the right side of the equation. For linear models, the parameters are usually omitted and only the terms are included in the model statement.

If the starting values are adequate and the nonlinear optimizer converges, output including the quantities in Table 11.2 will be produced. This table is very similar to the usual output for linear regression. The column marked "Estimate" gives $\hat{\theta}$. Since there is no necessary connection between terms and parameters, the lines of the table are labeled with the names of the parameters, not the names of the terms. The next column labeled "Std. Error" gives the square root of the diagonal entries of the matrix given at (11.14), so the standard errors are based on large-sample approximation. The column labeled "t-value" is the ratio of the estimate to its large-sample standard error and can be used for a test of the null hypothesis

TABLE 11.2   Nonlinear Least Squares Fit of (11.16)

```
Formula: Gain ~ th1 + th2 * (1 - exp(-th3 * A))

Parameters:
     Estimate Std. Error t-value Pr(>|t|)
th1  622.958       5.901  105.57  < 2e-16
th2  178.252      11.636   15.32 2.74e-16
th3    7.122       1.205    5.91 1.41e-06
---
Residual standard error: 19.66 on 32 degrees of freedom
```

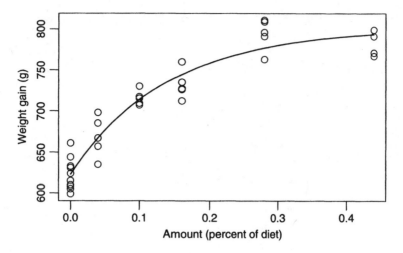

FIG. 11.2 Fitted mean function.

that a particular parameter is equal to zero against either a general or one-sided alternative. The column marked "P(> |t|)" is the significance level for this test, using a normal reference distribution rather than a t-distribution. Given at the foot of the table is the estimate $\hat{\sigma}$ and its df, which is the number of cases minus the number of elements in θ that were estimated, df $= 35 - 3 = 32$.

Since this example has only one predictor, Figure 11.1 is a summary graph for this problem. Figure 11.2 repeats this figure, but with the fitted mean function $\hat{E}(Gain|A = a) = 622.958 + 178.252(1 - \exp(-7.122a))$ added to the graph. The fitted mean function does not reproduce the possible decline of response for the largest value of A because it is constrained to increase toward an asymptote. For $A = 0.28$, the fitted function is somewhat less than the mean of the observed values, while at $A = 0.44$, it is somewhat larger than the mean of the observed values. If we believe that an asymptotic form is really appropriate for these data, then the fit of this mean function seems to be very good. Using the repeated observations at each level of A, we can perform a lack-of-fit test for the mean function, which is $F = 2.50$ with $(3, 29)$ df, for a significance level of 0.08, so the fit appears adequate.

Three Sources of Methionine

The purpose of this experiment was not only to estimate the weight gain response curve as a function of amount of methionine added but also to decide if the *source* of methionine was important. The complete experiment included three sources that we will call S_1, S_2, S_3. We can imagine a separate response curve such as Figure 11.2 for each of the three sources, and the goal might be to decide if the three response curves are different.

Suppose we create three dummy variables $S_i, i = 1, 2, 3$, so that S_i is equal to one if an observation is from source i, and it is zero otherwise. Assuming that the mean function (11.16) is appropriate for each source, the largest model we might contemplate is

$$E(Gain|A = a, S_1, S_2, S_3) = S_1[\theta_{11} + \theta_{21}(1 - \exp(-\theta_{31}a))]$$
$$+ S_2[\theta_{12} + \theta_{22}(1 - \exp(-\theta_{32}a))]$$
$$+ S_3[\theta_{13} + \theta_{23}(1 - \exp(-\theta_{33}a))] \quad (11.17)$$

This equation has a separate intercept, rate parameter, and asymptote for each group, and so has nine parameters. For this particular problem, this function has too many parameters because a dose of $A = 0$ from source 1 is the same as $A = 0$ with any of the sources, so the expected response at $A = 0$ must be the same for all three sources. This requires that $\theta_{11} = \theta_{12} = \theta_{13} = \theta_1$, which is a model of common intercepts, different asymptotes, and different slopes,

$$E(Gain|A = a, S_1, S_2, S_3) = \theta_1 + S_1[\theta_{21}(1 - \exp(-\theta_{31}a))]$$
$$+ S_2[\theta_{22}(1 - \exp(-\theta_{32}a))]$$
$$+ S_3[\theta_{23}(1 - \exp(-\theta_{33}a))] \quad (11.18)$$

Other reasonable mean functions to examine include common intercepts and asymptotes but separate rate parameters,

$$E(Gain|A = a, S_1, S_2, S_3) = \theta_1 +$$
$$\theta_2\{S_1[1 - \exp(-\theta_{31}a)] +$$
$$S_2[1 - \exp(-\theta_{32}a)] +$$
$$S_3[1 - \exp(-\theta_{33}a)]\} \quad (11.19)$$
$$= \theta_1 + \theta_2 \left(1 - \exp(-\sum \theta_{3i}S_i a)\right)$$

and finally the mean function of identical mean functions, given by (11.16).

The data from this experiment are given in the data file turkey.txt. This file is a little different because it does not give the response in each pen, but rather for each combination of A and source it gives the number m of pens with the combination, the mean response for those pens, and SD the standard deviation of the m pen responses. From Section 5.4, we can use these standard deviations to get a pure error estimate of σ^2 that can be used in lack-of-fit testing,

$$\hat{\sigma}_{pe}^2 = \frac{SS_{pe}}{df_{pe}} = \frac{\sum(m-1)SD^2}{\sum(m-1)} = \frac{19916}{70} = 284.5$$

The data are shown in Figure 11.3. A separate symbol was used for each of the three groups. Each point shown is an average over m pens, where $m = 5$ for every

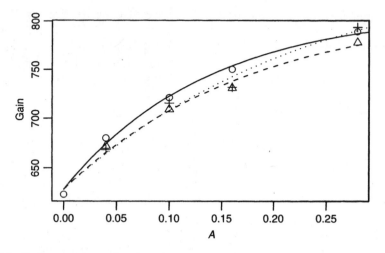

FIG. 11.3 Turkey growth as a function of methionine added for three sources of methionine. The lines shown on the figure are for the fit of (11.18), the most general reasonable mean function for these data.

point except at $A = 0$, where $m = 10$. The point at $A = 0$ is common to all three groups.

The four mean functions (11.16)–(11.19) can all be fit using nonlinear weighted least squares, with weights equal to the ms. Starting values for the estimates can be obtained in the same way as for fitting for one group. Table 11.3 summarizes the fit of the four mean functions, giving the RSS and df for each. For comparing the mean functions, we start with a lack-of-fit test for the most restrictive, common mean function, for which the F-test for lack of fit is

$$F = \frac{4326.1/10}{284.5} = 1.52$$

which, when compared with the $F(10, 70)$ distribution, gives a p-value of about 0.65, or no evidence of lack of fit of this mean function. Since the most restrictive

TABLE 11.3 Four Mean Functions Fit to the Turkey Growth Data with Three Sources of Methionine

			Change in	
Source	df	SS	df	SS
Common mean function, (11.16)	10	4326.1		
Equal intercept and asymptote, (11.19)	8	2568.4	2	1757.7
Common intercepts, (11.18)	6	2040.0	2	528.4
Separate regressions, (11.17)	4	1151.2	2	888.9
Pure error	70	19916.0		

mean function is adequate, we need not test the other mean functions for lack of fit.

We can perform tests to compare the mean functions using the general F-testing procedure of Section 5.4. For example, to compare as a null hypothesis the equal intercept and slope mean function (11.19) versus the common intercept mean function (11.18), we need to compute the change in RSS, given in the table as 528.4 with 2 df. For the F-test, we can use $\hat{\sigma}_{pe}^2$ in the denominator, and so

$$F = \frac{528.4/2}{\hat{\sigma}_{pe}^2} = 0.93 \sim F(2, 70)$$

for which $p = 0.61$, suggesting no evidence against the simpler mean function. Continued testing suggests that the simplest mean function of no difference between sources appears appropriate for these data, so we conclude that there is no evidence of a difference in response curve due to source. If we did not have a pure error estimate of variance, the estimate of variance from the most general mean function would be used in the F-tests.

11.3 BOOTSTRAP INFERENCE

The inference methods based on large samples introduced in the last section *may be inaccurate and misleading in small samples*. We cannot tell in advance if the large-sample inference will be accurate or not, as it depends not only on the mean function but also on the way we parameterize it, since there are many ways to write the same nonlinear mean function, and on the actual values of the predictors and the response. Because of this possible inaccuracy, computing inferences in some other way, at least as a check on the large-sample inferences, is a good idea.

One generally useful approach is to use the bootstrap introduced in Section 4.6. The case resampling bootstrap described in Section 4.6.1 can be applied in non-linear regression. Davison and Hinkley (1997) describe an alternative bootstrap scheme on the basis of resampling residuals, but we will not discuss it here.

We illustrate the use of the bootstrap with data in the file segreg.txt, which consists of measurements of electricity consumption in KWH and mean temperature in degrees F for one building on the University of Minnesota's Twin Cities campus for 39 months in 1988–1992, courtesy of Charles Ng. The goal is to model consumption as a function of temperature. Higher temperature causes the use of air conditioning, so high temperatures should mean high consumption. This building is steam heated, so electricity is not used for heating. Figure 11.4, a plot of $C =$ consumption in KWH/day versus *Temp*, the mean temperature in degrees F.

The mean function for these data is

$$\mathrm{E}(C|Temp) = \begin{cases} \theta_0 & Temp \le \gamma \\ \theta_0 + \theta_1(Temp - \gamma) & Temp > \gamma \end{cases}$$

This mean function has three parameters, the *level* θ_0 of the first phase; the *slope* θ_1 of the second phase, and the *knot*, γ, and assumes that energy consumption is

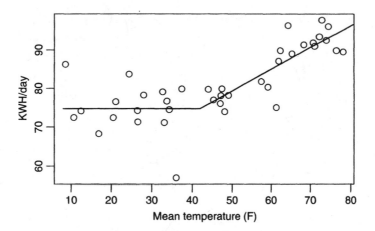

FIG. 11.4 Electrical energy consumption per day as a function of mean temperature for one building. The line shown on the graph is the least squares fit.

unaffected by temperature when the temperature is below the knot, but the mean increases linearly with temperature beyond the knot. The goal is to estimate the parameters.

The mean function can be combined into a single equation by writing

$$E(C|Temp) = \theta_0 + \theta_1(\max(0, Temp - \gamma))$$

Starting values can be easily obtained from the graph, with $\theta_0^{(0)} = 70$, $\theta_1^{(0)} = 0.5$ and $\gamma^{(0)} = 40$. The fitted model is summarized in Table 11.4. The baseline electrical consumption is estimated to be about $\hat{\theta}_0 \approx 75$ KWH per day. The knot is estimated to be at $\hat{\gamma} \approx 42°F$, and the increment in consumption beyond that temperature is about $\hat{\theta}_2 \approx 0.6$ KWH per degree increase.

From Figure 11.4, one might get the impression that information about the knot is asymmetric: γ could be larger than 42 but is unlikely to be substantially less than 42. We might expect that in this case confidence or test procedures based on asymptotic normality will be quite poor. We can confirm this using the bootstrap.

TABLE 11.4   Regression Summary for the Segmented Regression Example

```
Formula: C ~ th0 + th1 * (pmax(0, Temp - gamma))

Parameters:
        Estimate Std. Error t value Pr(>|t|)
th0      74.6953     1.3433  55.607  < 2e-16
th1       0.5674     0.1006   5.641 2.10e-06
gamma    41.9512     4.6583   9.006 9.43e-11

Residual standard error: 5.373 on 36 degrees of freedom
```

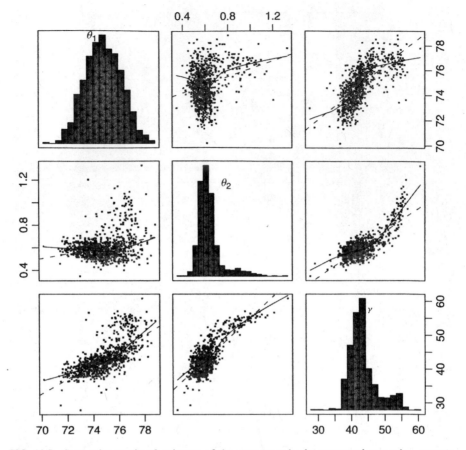

FIG. 11.5 Scatterplot matrix of estimates of the parameters in the segmented regression example, computed from $B = 999$ case bootstraps.

Figure 11.5 is a scatterplot matrix of $B = 999$ bootstrap replications. All three parameters are estimated on each replication. This scatterplot matrix is a little more complicated than the ones we have previously seen. The diagonals contain histograms of the 999 estimates of each of the parameters. If the normal approximation were adequate, we would expect that each of these histograms would look like a normal density function. While this may be so for θ_1, this is not the case for θ_2 and for γ. As expected, the histogram for γ is skewed to the right, meaning that estimates of γ much larger than about 40 occasionally occur but smaller values almost never occur. The univariate normal approximations are therefore poor.

The other graphs in the scatterplot matrix tell us about the distributions of the estimated parameters taken two at a time. If the normal approximation were to hold, these graphs should have approximately straight-line mean functions. The smoothers on Figure 11.5 are generally far from straight, and so the large-sample inferences are likely to be badly in error.

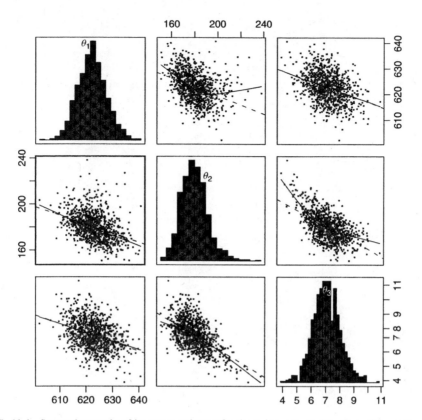

FIG. 11.6 Scatterplot matrix of bootstrap estimates for the turkey growth data. Two of the replicates were very different from the others and were deleted before graphing.

In contrast, Figure 11.6 is the bootstrap summary for the first source in the turkey growth data. Normality is apparent in histograms on the diagonal, and a linear mean function seems plausible for most of the scatterplots, and so the large-sample inference is adequate here.

Table 11.5 compares the estimates and confidence intervals produced by large-sample theory, and by the bootstrap. The bootstrap standard errors are the standard

TABLE 11.5 Comparison of Large-Sample and Bootstrap Inference for the Segmented Regression Data

	Large Sample				Bootstrap		
	θ_0	θ_1	γ		θ_0	θ_1	γ
Estimate	74.70	0.57	41.95	Mean	74.92	0.62	43.60
SE	1.34	0.10	4.66	SD	1.47	0.13	4.81
2.5%	72.06	0.37	32.82	2.5%	71.96	0.47	37.16
97.5%	77.33	0.76	51.08	97.5%	77.60	0.99	55.59

deviation of the bootstrap replicates, and the ends of the bootstrap 95% confidence interval are the 0.025 and 0.0975 quantiles of the bootstrap replicates. The large-sample theory confidence interval is given by the usual rule of estimate plus or minus 1.96 times the standard error computed from large-sample theory. Although the bootstrap SDs match the large-sample standard errors reasonably well, the confidence intervals for both θ_1 and for γ are shifted toward smaller values than the more accurate bootstrap estimates.

11.4 REFERENCES

Seber and Wild (1989) and Bates and Watts (1988) provide textbook-length treatments of nonlinear regression problems. Computational issues are also discussed in these references, and in Thisted (1988, Chapter 4). Ratkowsky (1990) provides an extensive listing of nonlinear mean functions that are commonly used in various fields of application.

PROBLEMS

11.1. Suppose we have a response Y, a predictor X, and a factor G with g levels. A generalization of the concurrent regression mean function given by Model 3 of Section 6.2.2, is, for $j = 1, \ldots, g$,

$$E(Y|X = x, G = j) = \beta_0 + \beta_{1j}(x - \gamma) \qquad (11.20)$$

for some point of concurrence γ.

11.1.1. Explain why (11.20) is a nonlinear mean function. Describe in words what this mean function specifies.

11.1.2. Fit (11.20) to the sleep data discussed in Section 6.2.2, so the mean function of interest is

$$E(TS|\log(BodyWt) = x, D = j) = \beta_0 + \beta_{1j}(x - \gamma)$$

(*Hint:* To get starting values, fit the concurrent regression model with $\gamma = 0$. The estimate of γ will be very highly variable, as is often the case with centering parameters like γ in this mean function.)

11.2. In fisheries studies, the most commonly used mean function for expected length of a fish at a given age is the *von Bertalanffy* function (von Bertalanffy, 1938; Haddon, 2001), given by

$$E(Length|Age = t) = L_\infty(1 - \exp(-K(t - t_0))) \qquad (11.21)$$

The parameter L_∞ is the expected value of *Length* for extremely large ages, and so it is the asymptotic or upper limit to growth, and K is a growth rate parameter that determines how quickly the upper limit to growth is reached. When $Age = t_0$, the expected length of the fish is 0, which allows fish to have nonzero length at birth if $t_0 < 0$.

11.2.1. The data in the file `lakemary.txt` give the *Age* in years and *Length* in millimeters for a sample of 78 bluegill fish from Lake Mary, Minnesota, in 1981 (courtesy of Richard Frie). *Age* is determined by counting the number of rings on a scale of the fish. This is a cross-sectional data set, meaning that all the fish were measured once. Draw a scatterplot of the data.

11.2.2. Use nonlinear regression to fit the von Bertalanffy function to these data. To get starting values, first guess at L_∞ from the scatterplot to be a value larger than any of the observed values in the data. Next, divide both sides of (11.21) by the initial estimate of L_∞, and rearrange terms to get just $\exp(-K(t - t_0))$ on the right of the equation. Take logarithms, to get a linear mean function, and then use OLS for the linear mean function to get the remaining starting values. Draw the fitted mean function on your scatterplot.

11.2.3. Obtain a 95% confidence interval for L_∞ using the large-sample approximation, and using the bootstrap.

11.3. The data in the file `walleye.txt` give the *length* in mm and the *age* in years of a sample of over 3000 male walleye, a popular game fish, captured in Butternut Lake in Northern Wisconsin (LeBeau, 2004). The fish are also classified according to the time *period* in which they were captured, with *period* = 1 for pre-1990, *period* = 2 for 1990–1996, and *period* = 3 for 1997–2000. Management practices on the lake were different in each of the periods, so it is of interest to compare the length at age for the three time periods.

Using the von Bertalanffy length at age function (11.21), compare the three time periods. If different, are all the parameters different, or just some of them? Which ones? Summarize your results.

11.4. A quadratic polynomial as a nonlinear model The data in the file `swan96.txt` were collected by the Minnesota Department of Natural Resources to study the abundance of black crappies, a species of fish, on Swan Lake, Minnesota in 1996. The response variable is *LCPUE*, the logarithm of the catch of 200 mm or longer black crappies per unit of fishing effort. It is believed that *LCPUE* is proportional to abundance. The single predictor is *Day*, the day on which the sample was taken, measured as the number of days after June 19, 1996. Some of the measurements were taken the following spring on the same population of fish before the young of the year are born in late June. No samples are taken during the winter months when the lake surface was frozen.

11.4.1. For these data, fit the quadratic polynomial

$$\mathrm{E}(LCPUE|Day = x) = \beta_0 + \beta_1 x + \beta_2 x^2$$

assuming $\mathrm{Var}(LCPUE|Day = x) = \sigma^2$. Draw a scatterplot of *LCPUE* versus *Day*, and add the fitted curve to this plot.

11.4.2. Using the delta method described in Section 6.1.2, obtain the estimate and variance for the value of *Day* that maximizes $E(LCPUE|Day)$.

11.4.3. Another parameterization of the quadratic polynomial is

$$E(Y|X) = \theta_1 - 2\theta_2\theta_3 x + \theta_3 x^2$$

where the θs can be related to the βs by

$$\theta_1 = \beta_0, \quad \theta_2 = -\beta_1/2\beta_2, \quad \theta_3 = \beta_2$$

In this parameterization, θ_1 is the intercept, θ_2 is the value of the predictor that gives the maximum value of the response, and θ_3 is a measure of curvature. This is a nonlinear model because the mean function is a nonlinear function of the parameters. Its advantage is that at least two of the parameters, the intercept θ_1 and the value of x that maximizes the response θ_2, are directly interpretable. Use nonlinear least squares to fit this mean function. Compare your results to the first two parts of this problem.

11.5. Nonlinear regression can be used to select transformations for a linear regression mean function. As an example, consider the highway accident data, described in Table 7.1, with response $\log(Rate)$ and two predictors $X_1 = Len$ and $X_2 = ADT$. Fit the nonlinear mean function

$$E(\log(Rate)|X_1 = x_1, X_2 = x_2, X_3 = x_3) = \beta_0 + \beta_1\psi_S(X_1, \lambda_1) + \beta_2\psi_S(X_2, \lambda_2)$$

where the scaled power transformations $\psi_S(X_j, \lambda_j)$ are defined at (7.3). Compare the results you get to results obtained using the transformation methodology in Chapter 7.

11.6. **POD models** Partial one-dimensional mean functions for problems with both factors and continuous predictors were discussed in Section 6.4. For the Australian athletes data discussed in that section, the mean function (6.26),

$$E(LBM|Sex, Ht, Wt, RCC) = \beta_0 + \beta_1 Sex + \beta_2 Ht + \beta_3 Wt + \beta_4 RCC$$

$$+ \eta_0 Sex + \eta_1 Sex \times (\beta_2 Ht + \beta_3 Wt + \beta_4 RCC)$$

was suggested. This mean function is nonlinear because η_1 multiplies each of the βs. Problem 6.21 provides a simple algorithm for finding estimates using only standard linear regression software. This method, however, will not produce the large-sample estimated covariance matrix that is available using nonlinear least squares.

11.6.1. Describe a reasonable method for finding starting values for fitting (6.26) using nonlinear least squares.

11.6.2. For the cloud seeding data, Problem 9.11, fit the partial one-dimensional model using the action variable A as the grouping variable, and summarize your results.

CHAPTER 12

Logistic Regression

A storm on July 4, 1999 with winds exceeding 90 miles per hour hit the Boundary Waters Canoe Area Wilderness (BWCAW) in northeastern Minnesota, causing serious damage to the forest. Roy Rich studied the effects of this storm using a very extensive ground survey of the area, determining for over 3600 trees the status, either alive or dead, species, and size. One goal of this study is to determine the dependence of survival on species, size of the tree, and on the local severity. Figure 12.1a shows a plot for 659 Balsam Fir trees, with the response variable Y coded as 1 for trees that were blown down and died and 0 for trees that survived, versus the single predictor $\log(D)$, the base-two logarithm of the diameter of the tree. To minimize overprinting, the plotted values of the variables were slightly jittered before plotting. Even with the jittering, this plot is much less informative than most of the plots of response versus a predictor that we have seen earlier in this book. Since the density of ink is higher in the lower-left and upper-right corners, the probability of blowdown is apparently higher for large trees than for small trees, but little more than that can be learned from this plot.

Figure 12.1b is an alternative to Figure 12.1a. This graph displays *density estimates*, which are like smoothed histograms for $\log(D)$, separately for the survivors, the solid line, and for the trees blown down, the dashed line[1]. Both densities are roughly shaped like a normal density. The density for the survivors, $Y = 1$, is shifted to the right relative to the density for the density for $Y = 0$, meaning that the trees that blew down are generally larger. If the histograms had no overlap, then the quantity on the horizontal axis, $\log(D)$, would be a perfect predictor of survival. Since there is substantial overlap of the densities, $\log(D)$ is not a perfect predictor of blowdown. For values of $\log(D)$ where the two densities have the same height, the probability of survival will be about 0.5. For values of $\log(D)$ where the height of the density for survivors is higher than the density for blowdown, then the probability of surviving exceeds 0.5; when the height of the density of

---

[1]Looking at overlapping histograms is much harder than looking at overlapping density estimates. Silverman (1986) and Bowman and Azzalini (1997), among others, provide discussions of density estimation.

Applied Linear Regression, Third Edition, by Sanford Weisberg
ISBN 0-471-66379-4 Copyright © 2005 John Wiley & Sons, Inc.

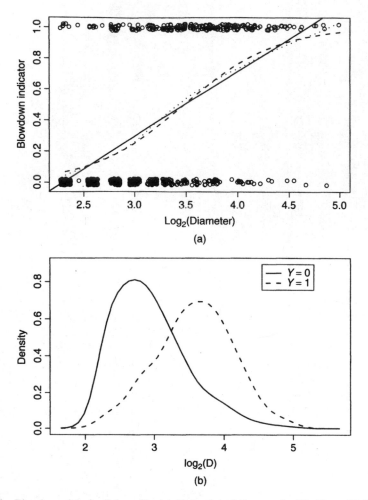

FIG. 12.1 Blowdown data for Balsam Fir. (a) Scatterplot of Y versus $\log(D)$. The solid line is the OLS line. The dotted line is the fit of a smoothing spline. The dashed line is the logistic regression fit. Data have been jittered in both variables to minimize overprinting. (b) Separate density estimates for $\log(D)$ for survivors, $Y = 0$, and blowdown, $Y = 1$.

blowdown is higher, the probability of surviving is less than 0.5. In Figure 12.1b, the probability of survival is greater than 0.5 if $\log(D)$ is less than about 3.3, and less than 0.5 if $\log(D)$ exceeds 3.3.

More generally, suppose in a problem with predictor X we let $\theta(x) = \Pr(Y = 1 | X = x)$ be the conditional probability that $Y = 1$ given the value of the predictor. This conditional probability plays the role of the mean function in regression problems when the response is either one or zero. For the blowdown data in Figure 12.1, the probability of blowdown increases from left to right.

We can visualize $\theta(\log(D))$ by adding a smoother to Figure 12.1a. The straight line on this graph is the OLS regression of Y on $\log(D)$. It includes estimated values of $\theta(\log(D))$ outside the range $(0, 1)$ for very small or large trees, and so the OLS line cannot be a good representation of $\theta(\log(D))$ for all values of $\log(D)$. The OLS line is often inappropriate for a bounded response because it will produce fitted values outside the permitted range. The dotted smoother in Figure 12.1 uses a smoother, so it estimates the mean function without a model. This estimate of the mean function has the characteristic shape of binary regression, with asymptotes at 0 and 1 for extreme values of the predictor. The logistic regression models we will study next also have this shape.

As with other regression problems, with a binary response we also have a set of terms or predictors X, and we are interested in the study of $\Pr(Y = 1|X = \mathbf{x})$ $= \theta(\mathbf{x})$ as $\mathbf{x}$ is varied. The response variable is really a category, like success or failure, alive or dead, passed or failed, and so on. In some problems, the ith value of the response y_i will be a count of the number of successes in m_i independent trials each with the same probability of success. If all the $m_i = 1$, then each element of Y has a Bernoulli distribution; if some of the $m_i > 1$, then each element of Y has a Binomial distribution if each of the trials has the same probability of "success" and all trials are independent. Bernoulli regression is a special case of binomial regression with all the $m_i = 1$.

12.1 BINOMIAL REGRESSION

We recall the basic facts about binomial random variables. Let y be the number of successes out of m independent trials, each with the same probability θ of success, so y can have any integer value between 0 and m. The random variable y has a *binomial distribution*. We write this as $y \sim \text{Bin}(m, \theta)$. The probability that Y equals a specific integer $j = 0, 1, \ldots, m$, is given by

$$\Pr(y = j) = \binom{m}{j} \theta^j (1 - \theta)^{(m-j)} \qquad (12.1)$$

where $\binom{m}{j} = m!/(j!(m - j)!)$ is the number of different orderings of j successes in m trials. Equation (12.1) is called the *probability mass function* for the binomial. The mean and variance of a binomial are

$$\text{E}(y) = m\theta; \quad \text{Var}(y) = m\theta(1 - \theta) \qquad (12.2)$$

Since m is known, both the mean and variance are determined by one parameter θ.

In the binomial regression problem, the response y_i counts the number of "successes" in m_i trials, and so $m_i - y_i$ of the trials were "failures." In addition, we have p' terms or predictors $\mathbf{x}_i$ possibly including a constant for the intercept, and assume that the probability of success for the ith case is $\theta(\mathbf{x}_i)$. We write this compactly as

$$(Y|X = \mathbf{x}_i) \sim \text{Bin}(m_i, \theta(\mathbf{x}_i)), i = 1, \ldots, n \qquad (12.3)$$

We use y_i/m_i, the observed fraction of successes at each i, as the response because the range of y_i/m_i is always between 0 and 1, whereas the range of y_i is between

0 and m_i and can be different for each i. Using (12.2), the mean and variance functions are

$$E(y_i/m_i|\mathbf{x}_i) = \theta(\mathbf{x}_i) \tag{12.4}$$

$$\text{Var}(y_i/m_i|\mathbf{x}_i) = \theta(\mathbf{x}_i)(1 - \theta(\mathbf{x}_i))/m_i \tag{12.5}$$

In the multiple linear regression model, the mean function and the variance function generally have completely separate parameters, but that is not so for binomial regression. The value of $\theta(\mathbf{x}_i)$ determines both the mean function and the variance function, so we need to estimate $\theta(\mathbf{x}_i)$. If the m_i are all large, we could simply estimate $\theta(\mathbf{x}_i)$ by y_i/m_i, the observed proportion of successes at $\mathbf{x}_i$. In many applications, the m_i are small—often $m_i = 1$ for all i—so this simple method will not always work.

12.1.1   Mean Functions for Binomial Regression

As with linear regression models, we assume that $\theta(\mathbf{x}_i)$ depends on $\mathbf{x}_i$ only through a linear combination $\boldsymbol{\beta}'\mathbf{x}_i$ for some unknown $\boldsymbol{\beta}$. This means that any two cases for which $\boldsymbol{\beta}'\mathbf{x}$ is equal will have the same probability of "success." We can write $\theta(\mathbf{x})$ as a function of $\boldsymbol{\beta}'\mathbf{x}$,

$$\theta(\mathbf{x}_i) = \mathsf{m}(\boldsymbol{\beta}'\mathbf{x}_i)$$

The quantity $\boldsymbol{\beta}'\mathbf{x}_i$ is called the *linear predictor*. As in nonlinear models, the function m is called a kernel mean function. $\mathsf{m}(\boldsymbol{\beta}'\mathbf{x}_i)$ should take values in the range $(0, 1)$ for all $\boldsymbol{\beta}'\mathbf{x}$. The most frequently used kernel mean function for binomial regression is the *logistic function*,

$$\theta(\mathbf{x}_i) = \mathsf{m}(\boldsymbol{\beta}'\mathbf{x}_i) = \frac{\exp(\boldsymbol{\beta}'\mathbf{x}_i)}{1 + \exp(\boldsymbol{\beta}'\mathbf{x}_i)} = \frac{1}{1 + \exp(-\boldsymbol{\beta}'\mathbf{x}_i)} \tag{12.6}$$

A graph of this kernel mean function is shown in Figure 12.2. The logistic mean function is always between 0 and 1, and has no additional parameters.

Most presentations of logistic regression work with the inverse of the kernel mean function called the *link function*. Solving (12.6) for $\boldsymbol{\beta}'\mathbf{x}$, we find

$$\log\left(\frac{\theta(\mathbf{x})}{1 - \theta(\mathbf{x})}\right) = \boldsymbol{\beta}'\mathbf{x} \tag{12.7}$$

The left side of (12.7) is called a *logit* and the right side is the linear predictor $\boldsymbol{\beta}'\mathbf{x}$. The logit is a linear function of the terms on the right side of (12.7). If we were to draw a graph of $\log(\theta(\mathbf{x})/(1 - \theta(\mathbf{x})))$ versus $\boldsymbol{\beta}'\mathbf{x}$, we would get a straight line.

The ratio $\theta(\mathbf{x})/(1 - \theta(\mathbf{x}))$ is the *odds of success*. For example, if the probability of success is 0.25, the odds of success are $.25/(1 - .25) = 1/3$, one success to each three failures. If the probability of success is 0.8, then the odds of success are $0.8/0.2 = 4$, or four successes to one failure. Whereas probabilities are bounded

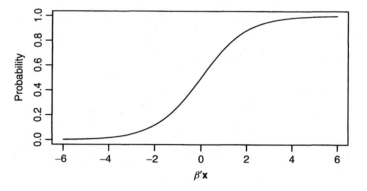

FIG. 12.2 The logistic kernel mean function.

between 0 and 1, odds can be any nonnegative number. The logit is the logarithm of the odds; natural logs are used in defining the logit. According to equation (12.7), the logit is equal to a linear combination of the terms.

In summary, the logistic regression model consists of the data and distribution specified by (12.3), and a fixed component that connects the response to the mean through (12.6).

12.2 FITTING LOGISTIC REGRESSION

Many standard statistical packages allow estimation for logistic regression models. The most common computational method is outlined in Section 12.3.2; for now, we return to our example.

12.2.1 One-Predictor Example

Consider first logistic regression with one predictor using the Balsam Fir data from the BWCAW blowdown shown in Figure 12.1. The data are given in the file blowBF.txt. The single predictor is $\log(D)$, using base-two logarithms. All the $m_i = 1$. We fit with two terms, the intercept and $\log(D)$. The results are summarized in Table 12.1.

TABLE 12.1 Logistic Regression Summary for the Balsam Fir Blowdown Data

```
Coefficients:
            Estimate Std. Error z value Pr(>|z|)
(Intercept)  -7.8923     0.6301  -12.53   <2e-16
logD          2.2626     0.1907   11.86   <2e-16
---
Residual deviance: 655.24  on 657  degrees of freedom
    Pearson's X^2: 677.44  on 657  degrees of freedom
```

The output reports estimate $\hat{\beta}_0 = -7.8923$ for the intercept and $\hat{\beta}_1 = 2.2626$ for the slope for $\log(D)$. The dashed curve drawn on Figure 12.1a corresponds to the fitted mean function

$$\hat{E}(Y|\log(D)) = \frac{1}{1 + \exp[-(-7.8923 + 2.2626\log(D))]}$$

The logistic fit matches the nonparametric spline fit fairly closely, except for the largest and smallest trees where smoothers and parametric fits are often in disagreement. It is not always easy to tell if the logistic fit is matching the data without comparing it to a nonparametric smooth.

Equation (12.7) provides a basis for understanding coefficients in logistic regression. In the example, the coefficient for $\log(D)$ is about 2.26. If $\log(D)$ were increased by one unit (since this is a base-two logarithm, increasing $\log(D)$ by one unit means that the value of D doubles), then the natural logarithm of the odds will increase by 2.26 and the odds will be *multiplied* by $\exp(2.26) = 9.6$. Thus, a tree with diameter 10 in. is 9.6 times as likely to blow down as a tree of diameter five inches, or a tree of 2-in. diameter is 9.6 times as likely to blow down as a tree of 1 in. diameter. In general, if $\hat{\beta}_j$ is an estimated coefficient in a logistic regression, then if x_j is increased by one unit, the odds of success, that is, the odds that $Y = 1$, are multiplied by $\exp(\hat{\beta}_j)$.

Table 12.1 also reports standard errors of the estimates, and the column marked z-value shows the ratio of the estimates to their standard errors. These values can be used for testing coefficients to be zero after adjusting for the other terms in the model, as in linear models, but the test should be compared to the standard normal distribution rather than a t-distribution. The deviance and Pearson's X^2 reported in the table will be discussed shortly. Since the variance of a binomial is determined by the mean, there is not a variance parameter that can be estimated separately. While an equivalent to an R^2 measure can be defined for logistic regression, its use is not recommended.

12.2.2 Many Terms

We introduce a second predictor into the blowdown data. The variable S is a local measure of severity of the storm that will vary from location to location, from near 0, with very few trees effected, to near 1, with nearly all trees blown down. Figure 12.3 shows two useful plots. Figure 12.3a gives the density for S for each of the two values of Y. In contrast to $\log(D)$, the two density estimates are much less nearly normal in shape, with one group somewhat skewed, and the other perhaps bimodal, or at least very diffuse. The two densities are less clearly separated, and this indicates that S is a weaker predictor of blowdown for these Balsam Fir trees. Figure 12.3b is a scatterplot of S versus $\log(D)$, with different symbols for $Y = 1$ and for $Y = 0$. This particular plot would be much easier to use with different colors indicating the two classes. In the upper-right of the plot, the symbol for $Y = 1$ predominates, while in the lower-left, the symbol for $Y = 0$ predominates.

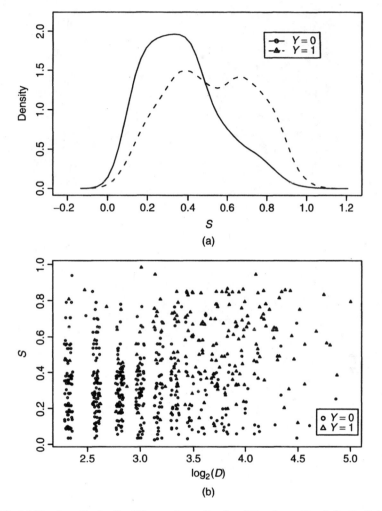

FIG. 12.3 (a) Density estimates for S for survivors, $Y = 0$ and blowdown, $Y = 1$, for the Balsam Fir data. (b) Plot of S versus $\log(D)$ with separate symbols for points with $Y = 1$ and $Y = 0$. The values of $\log(D)$ have been jittered.

This suggests that the two predictors have a joint effect because the prevalence of symbols for $Y = 1$ changes along a diagonal in the plot. If the prevalence changed from left to right but not down to up, then only the variable on the horizontal axis would have an effect on the probability of blowdown. If the prevalence of symbols for $Y = 1$ were uniform throughout the plot, then neither variable would be important.

Fitting logistic regression with the two predictors $\log(D)$ and S means that the probability of $Y = 1$ depends on these predictors only through a linear combination $\beta_1 \log(D) + \beta_2 S$ for some (β_1, β_2). This is like finding a sequence of parallel lines like those in Figure 12.4a so that the probability of blowdown is constant on the

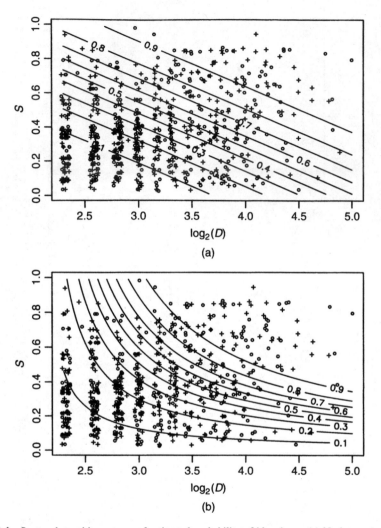

FIG. 12.4   Scatterplots with contours of estimated probability of blowdown. (a) No interaction mean function. (b) Mean function with interaction.

lines, and increases (or, in other problems, decreases) from lower left to upper right. The lines shown on Figure 12.4a come from the logistic regression fit that is summarized in Table 12.2a. From the table, we see that all points that have the same values of $2.2164\log(D) + 4.5086S$ have the same estimated probability of blowdown. For example, for the points near the line marked 0.5, we would expect about 50% symbols for $Y = 0$ and 50% symbols for $Y = 1$, but near the line 0.1, we would expect 90% symbols for $Y = 0$.

Figure 12.4b and Table 12.2b correspond to fitting with a mean function that includes the two terms and their interaction. The lines of constant estimated probability of $Y = 1$ shown on Figure 12.4b are now curves rather than straight lines,

TABLE 12.2   Logistic Regressions for the Balsam Fir Data

```
(a) No interaction
Coefficients:
             Estimate Std. Error z value Pr(>|z|)
(Intercept)  -9.5621     0.7499  -12.75   <2e-16 ***
logD          2.2164     0.2079   10.66   <2e-16 ***
S             4.5086     0.5159    8.74   <2e-16 ***
---

Residual deviance: 563.9  on 656  degrees of freedom
      Pearson's X^2: 715.3  on 656  degrees of freedom
```

```
(b) Mean function with interaction
Coefficients:
             Estimate Std. Error z value Pr(>|z|)
(Intercept)  -3.6788     1.4209  -2.589  0.00963
logD          0.4009     0.4374   0.916  0.35941
S           -11.2026     3.6143  -3.100  0.00194
logD:S        4.9098     1.1319   4.338 1.44e-05
---

Residual deviance: 541.75  on 655  degrees of freedom
      Pearson's X^2: 885.44  on 655  degrees of freedom
```

but otherwise the interpretation of this plot is the same as Figure 12.4a. Visually deciding which of these two mean functions matches the data more closely is difficult, and we will shortly develop a test for this comparison.

Interpretation of estimates of parameters in mean functions with no interaction works the same way with many predictors as it does with one predictor. For example, the estimated effect of increasing $\log(D)$ by one unit using Table 12.2a is to multiply the odds of blowdown by $\exp(2.2164) \approx 9.2$, similar to the estimated effect when S is ignored. Interpretation of estimates is complicated by interactions. Using the estimates in Table 12.2b, if $\log(D)$ is increased by one unit, then the odds of blowdown are multiplied by $\exp(0.4009 + 4.9098S)$, which depends on the value of S. For the effect of S, since S is bounded between 0 and 1, we cannot increase S by one unit, so we can summarize the S-effect by looking at an increase of 0.1 units. The odds multiplier for S is then $\exp(.1[-11.2026 + 4.9098\log(D)])$. These two functions are graphed in Figure 12.5. Big trees were much more likely to blow down in severe areas than in areas where severity was low. For fixed diameter, increasing severity by 0.1 has a relatively modest effect on the odds of blowdown.

As with logistic regression with a single term, the estimates are approximately normally distributed if the sample size is large enough, with estimated standard errors in Table 12.2. The ratios of the estimates to their standard errors are called *Wald tests*. For the no-interaction mean function in Table 12.2a, the *p*-values for all three Wald tests are very small, indicating all terms are important when adjusting for the other terms in the mean function. For the interaction mean function in Table 12.2b, the *p*-values for S and for the interaction are both small, but the main effect for $\log(D)$ has a large *p*-value. Using the *hierarchy principle*,

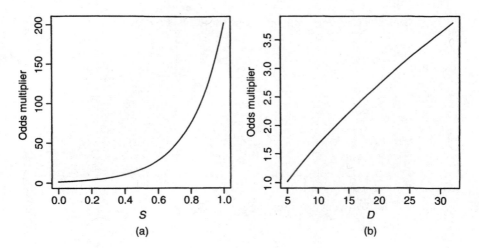

FIG. 12.5 Blowdown odds multiplier for (a) doubling the diameter of a Balsam Fir tree as a function of local severity, S, and (b) increasing S by 0.1 as a function of diameter.

however, we recommend that whenever an interaction is included in a mean function, all the main effects in that interaction be included as well, so in light of the significant interaction the test for $\log(D)$ is not relevant, and we would retain $\log(D)$.

In the next section, we derive tests analogous to F-tests for linear regression. Unlike the linear model where F-tests and Wald tests are equivalent, in logistic regression they can give conflicting results. The tests in the next section are to be preferred over the Wald tests.

12.2.3 Deviance

In multiple linear regression, the residual sum of squares provides the basis for tests for comparing mean functions. In logistic regression, the residual sum of squares is replaced by the *deviance*, which is often called G^2. The deviance is defined for logistic regression to be

$$G^2 = 2 \sum_{i=1}^{n} \left[y_i \log \left(\frac{y_i}{\hat{y}_i} \right) + (m_i - y_i) \log \left(\frac{m_i - y_i}{m_i - \hat{y}_i} \right) \right] \qquad (12.8)$$

where $\hat{y}_i = m_i \hat{\theta}(\mathbf{x}_i)$ are the fitted number of successes in m_i trials. The df associated with the deviance is equal to the number of cases n used in the calculation minus the number of elements of $\boldsymbol{\beta}$ that were estimated; in the example, $\mathrm{df} = 659 - 4 = 655$.

Methodology for comparing models parallels the results in Section 5.4. Write $\boldsymbol{\beta}'\mathbf{x} = \boldsymbol{\beta}'_1\mathbf{x}_1 + \boldsymbol{\beta}'_2\mathbf{x}_2$, and consider testing

$$\text{NH:} \quad \theta(\mathbf{x}) = m(\boldsymbol{\beta}'_1\mathbf{x}_1)$$
$$\text{AH:} \quad \theta(\mathbf{x}) = m(\boldsymbol{\beta}'_1\mathbf{x}_1 + \boldsymbol{\beta}'_2\mathbf{x}_2)$$

TABLE 12.3   Analysis of Deviance for Balsam Fir Blowdown Data

Terms	df	Deviance	Change in df	Deviance	$P(> |\text{Chi}|)$
1, $\log(D)$	657	655.24			
1, $\log(D)$, S, $S \times \log(D)$	655	541.75	2	113.50	0.0000

TABLE 12.4   Sequential Analysis of Deviance for Balsam Fir Blowdown Data

Terms	df	Deviance	Change in df	Deviance	$P(> |\text{Chi}|)$
Intercept	658	856.21			
Add $\log(D)$	657	655.24	1	200.97	0.0000
Add S	656	563.90	1	91.34	0.0000
Add interaction	655	541.75	1	22.16	0.0000

to see if the terms in $\mathbf{x}_2$ have zero coefficients. Obtain the deviance G^2_{NH} and degrees of freedom df_{NH} under the null hypothesis, and then obtain G^2_{AH} and df_{AH} under the alternative hypothesis. As with linear models, we will have evidence against the null hypothesis if $G^2_{NH} - G^2_{AH}$ is too large. To get p-values, we compare the difference $G^2_{NH} - G^2_{AH}$ with the χ^2 distribution with $df = df_{NH} - df_{AH}$, not with an F-distribution as was done for linear models.

If we set $\mathbf{x}_1 = (Ones, \log(D))$, where *Ones* is the vector of ones to fit the intercept, and $\mathbf{x}_2 = (S, S \times \log(D))$ to test that only $\log(D)$ and the intercept are required in the mean function. Fitting under the null hypothesis is summarized in Table 12.1, with $G^2_{NH} = 655.24$ with $df_{NH} = 657$. The alternative hypothesis is summarized in Table 12.2, where we see that $G^2_{AH} = 541.75$ with $df_{AH} = 655$. These results can be summarized in an analysis of deviance table, as in Table 12.3. The interpretation of this table parallels closely the results for multiple linear regression models in Section 5.4. The table includes the deviance and df for each of the models. The test statistic depends on the change in deviance and the change in df, as given in the table. The p-value is 0 to 4 decimals, and the larger model is preferred.

We could also have a longer sequence of models, for example, first fitting with an intercept only, then the intercept and $\log(D)$, then adding S, and finally the interaction. This would give an analysis of deviance table like Table 12.4 that parallels the sequential analysis of variance tables discussed in Section 3.5.6. The table displays the tests for comparing two adjacent mean functions.

12.2.4   Goodness-of-Fit Tests

When the number of trials $m_i > 1$, the deviance G^2 can be used to provide a goodness-of-fit test for a logistic regression model, essentially comparing the null hypothesis that the mean function used is adequate versus the alternative that a separate parameter needs to be fit for each value of i (this latter case is called the *saturated model*). When all the m_i are large enough, G^2 can be compared with the χ^2_{n-p} distribution to get an approximate p-value. The goodness-of-fit test is not applicable in the blowdown example because all the $m_i = 1$.

Pearson's X^2 is an approximation to G^2 defined for logistic regression by

$$
\begin{aligned}
X^2 &= \sum_{i=1}^n \left[(y_i - \hat{y}_i)^2 \left(\frac{1}{\hat{y}_i} + \frac{1}{m_i - \hat{y}_i} \right) \right] \\
&= \sum_{i=1}^n \frac{m_i (y_i/m_i - \hat{\theta}(\mathbf{x}_i))^2}{\hat{\theta}(\mathbf{x}_i)(1 - \hat{\theta}(\mathbf{x}_i))}
\end{aligned}
\tag{12.9}
$$

X^2 and G^2 have the same large-sample distribution and often give the same inferences. In small samples, there may be differences, and sometimes X^2 may be preferred for testing goodness-of-fit.

Titanic

The Titanic was a British luxury passenger liner that sank when it struck an iceberg about 640 km south of Newfoundland on April 14–15, 1912, on its maiden voyage to New York City from Southampton, England. Of 2201 known passengers and crew, only 711 are reported to have survived. The data in the file `titanic.txt` from Dawson (1995) classify the people on board the ship according to their *Sex* as Male or Female, *Age*, either child or adult, and *Class*, either first, second, third, or crew. Not all combinations of the three factors occur in the data, since no children were members of the crew. For each age/sex/class combination, the number of people M and the number surviving *Surv* are also reported. The data are shown in Table 12.5.

Table 12.6 gives the value of G^2 and Pearson's X^2 for the fit of five mean functions to these data. Since almost all the m_i exceed 1, we can use either G^2 or X^2 as a goodness-of-fit test for these models. The first two mean functions, the main effects only model, and the main effects plus the *Class* × *Sex* interaction, clearly do not fit the data because the values of G^2 and X^2 are both much larger then their df, and the corresponding p-values from the χ^2 distribution are

TABLE 12.5 Data from the Titanic Disaster of 1912. Each Cell Gives *Surv/M*, the Number of Survivors, and the Number of People in the Cell

Class	Female Adult	Female Child	Male Adult	Male Child
Crew	20/23	NA	192/862	NA
First	140/144	1/1	57/175	5/5
Second	80/93	13/13	14/168	11/11
Third	76/165	14/31	75/462	13/48

TABLE 12.6 Fit of Four Mean Functions for the Titanic Data. Each of the Mean Functions Treats *Age*, *Sex*, and *Class* as Factors, and Fits Different Main Effects and Interactions

Mean Function	df	G^2	X^2
Main effects only	8	112.57	103.83
Main effects + *Class* × *Sex*	5	45.90	42.77
Main effects + *Class* × *Sex* + *Class* × *Age*	3	1.69	1.72
Main effects + all two-factor interactions	2	0.00	0.00
Main effects, two-factor and three-factor interactions	0	0.00	0.00

0 to several decimal places. The third model, which adds the *Class* × *Age* interaction, has both G^2 and X^2 smaller than its df, with *p*-values of about 0.64, so this mean function seems to match the data well. Adding more terms can only reduce the value of G^2 and X^2, and adding the third interaction decreases these statistics to 0 to the accuracy shown. Adding the three-factor interaction fits one parameter for each cell, effectively estimating the probability of survival by the observed probability of survival in each cell. This will give an exact fit to the data.

The analysis of these data is continued in Problem 12.7.

12.3 BINOMIAL RANDOM VARIABLES

In this section, we provide a very brief introduction to maximum likelihood estimation and then provide a computing algorithm for finding maximum likelihood estimates for logistic regression.

12.3.1 Maximum Likelihood Estimation

Data can be used to estimate θ using *maximum likelihood estimation*. Suppose we have observed y successes in m independent trials, each with the same probability θ of success. The maximum likelihood estimate or MLE of θ is the value $\hat{\theta}$ of θ that maximizes the probability of observing y successes in m trials. This amounts to rewriting (12.1) as a function of θ, with y held fixed at its observed value,

$$L(\theta) = \left(\begin{array}{c} m \\ y \end{array} \right) \theta^y (1 - \theta)^{(m-y)} \qquad (12.10)$$

$L(\theta)$ is called the *likelihood function* for θ. Since the same value maximizes both $L(\theta)$ and $\log(L(\theta))$, we work with the more convenient log-likelihood, given by

$$\log(L(\theta)) = \log \left(\begin{array}{c} m \\ y \end{array} \right) + y \log(\theta) + (m - y) \log(1 - \theta) \qquad (12.11)$$

Differentiating (12.11) with respect to θ and setting the result to 0 gives

$$\frac{d\log(L(\theta))}{d\theta} = \frac{y}{\theta} - \frac{m-y}{1-\theta} = 0$$

Solving for θ gives the MLE,

$$\hat{\theta} = \frac{y}{m} = \frac{\text{Observed number of successes}}{\text{Observed fixed number of trials}}$$

which is the observed proportion of successes. Although we can find the variance of this estimator directly, we use a result that gives the large-sample variance of the MLE for most statistical problems. Suppose the parameter θ is a vector. Then in large samples,

$$\text{Var}(\hat{\theta}) = -\left[\text{E}\left(\frac{\partial^2 \log(L(\theta))}{\partial\theta(\partial\theta)'}\right)\right]^{-1} \tag{12.12}$$

For the binomial example, θ is a scalar, and

$$\left[-\text{E}\left(\frac{d^2 \log(L(\theta))}{d\theta^2}\right)\right]^{-1} = \left[-\text{E}\left(\frac{y}{\theta^2} - \frac{m-y}{(1-\theta)^2}\right)\right]^{-1}$$

$$= \left[\frac{m}{\theta(1-\theta)}\right]^{-1}$$

$$= \frac{\theta(1-\theta)}{m} \tag{12.13}$$

This variance is estimated by substituting $\hat{\theta}$ for θ. In large samples, the MLE $\hat{\theta}$ is approximately normally distributed with mean θ and variance given by (12.12).

12.3.2 The Log-Likelihood for Logistic Regression

Equation (12.10) provides the likelihood function for a single binomial random variable y with m trials and probability of success θ. We generalize now to having n independent random variables $(y_1, \ldots, y_n)$ with y_i a binomial random variable with m_i trials and probability of success $\theta(x_i)$ that depends on the value of x_i and so may be different for each i. The likelihood based on $(y_1, \ldots, y_n)$ is obtained by multiplying the likelihood for each observation,

$$L = \prod_{i=1}^{n} \binom{m_i}{y_i} (\theta(x_i))^{y_i} (1 - \theta(x_i))^{m_i - y_i}$$

$$\propto \prod_{i=1}^{n} (\theta(x_i))^{y_i} (1 - \theta(x_i))^{m_i - y_i}$$

In the last expression, we have dropped the binomial coefficients $\binom{m_i}{y_i}$ because they do not depend on parameters. After minor rearranging, the log-likelihood is

$$\log(L) \propto \sum_{i=1}^{n} \left[y_i \log \left(\frac{\theta(\mathbf{x}_i)}{1 - \theta(\mathbf{x}_i)} \right) + m_i \log(1 - \theta(\mathbf{x}_i)) \right]$$

Next, we substitute for $\theta(\mathbf{x}_i)$ using equation (12.7) to get

$$\log(L(\boldsymbol{\beta})) = \sum_{i=1}^{n} \left[(\boldsymbol{\beta}'\mathbf{x}_i)y_i - m_i \log(1 + \exp(\boldsymbol{\beta}'\mathbf{x}_i)) \right] \tag{12.14}$$

The log-likelihood depends on the regression parameters $\boldsymbol{\beta}$ explicitly, and we can maximize (12.14) to get estimates. An iterative procedure is required. The usual methods using either the *Newton–Raphson* algorithm or *Fisher scoring* attain convergence in just a few iterations, although problems can arise with unusual data sets, for example, if one or more of the predictors can determine the value of the response exactly; see Collett (2002, Section 3.12). Details of the computational method are provided by McCullagh and Nelder (1989, Section 2.5), Collett (2002), and Agresti (1996, 2002), among others.

The estimated covariance matrix of the estimates is given by

$$\text{Var}(\hat{\boldsymbol{\beta}}) = (\mathbf{X}'\widehat{\mathbf{W}}\mathbf{X})^{-1}$$

where $\widehat{\mathbf{W}}$ is a diagonal matrix with entries $m_i\hat{\theta}(\mathbf{x}_i)(1 - \hat{\theta}(\mathbf{x}_i))$, and $\mathbf{X}$ is a matrix with ith row $\mathbf{x}'$.

12.4 GENERALIZED LINEAR MODELS

Both the multiple linear regression model discussed earlier in this book and the logistic regression model discussed in this chapter are particular instances of a *generalized linear model*. Generalized linear models all share three basic characteristics:

1. The distribution of the response Y, given a set of terms X, is distributed according to an *exponential family distribution*. The important members of this class include the normal and binomial distributions we have already encountered, as well as the Poisson and gamma distributions. Generalized linear models based on the Poisson distributions are the basis of the most common models for contingency tables of counts; see Agresti (1996, 2002).

2. The response Y depends on the terms X only through the linear combination $\boldsymbol{\beta}'X$.

3. The mean $E(Y|X = \mathbf{x}) = \mathsf{m}(\boldsymbol{\beta}'\mathbf{x})$ for some kernel mean function m. For the multiple linear regression model, m is the identity function, and for logistic regression, it is the logistic function. There is considerable flexibility in selecting the kernel mean function. Most presentations of generalized linear models discuss the link function, which is the inverse of m rather than m itself.

These three components are enough to specify completely a regression problem along with methods for computing estimates and making inferences. The methodology for these models generally builds on the methods in this book, usually with only minor modification. Generalized linear models were first suggested by Nelder and Wedderburn (1972) and are discussed at length by McCullagh and Nelder (1989). Some statistical packages use common software to fit all generalized linear models, including the multiple linear regression model. Book-length treatments of binomial regression are given by Collett (2002) and by Hosmer and Lemeshow (2000).

PROBLEMS

12.1. Downer data For unknown reasons, dairy cows sometimes become recumbent—they lay down. Called *downers*, these cows may have a serious illness that may lead to death of the cow. These data are from a study of blood samples of over 400 downer cows studied at the Ruakura New Zealand Animal Health Laboratory during 1983–1984. A variety of blood tests were performed, and for many of the animals, the outcome (survived, died, or animal was killed) was determined. The goal is to see if survival can be predicted from the blood measurements. The variables in the data file downer.txt are described in Table 12.7. These data were collected from veterinary records, and not all variables were recorded for all cows.

12.1.1. Consider first predicting *Outcome* from *Myopathy*. Find the fraction of surviving cows of *Myopathy* = 0 and for *Myopathy* = 1.

12.1.2. Fit the logistic regression with response *Outcome*, and the single predictor *Myopathy*. Obtain a 95% confidence interval for coefficient for *Myopathy*, and compute the estimated decrease in odds of survival when *Myopathy* = 1. Obtain the estimated probability of survival when *Myopathy* = 0 and when *Myopathy* = 1, and compare with the observed survival fractions in Problem 12.1.1.

TABLE 12.7 The Downer Data

Variable	n	Description
AST	429	Serum asparate amino transferase (U/l at 30C)
Calving	431	0 if measured before calving, 1 if after
CK	413	Serum creatine phosphokinase (U/l at 30C)
Daysrec	432	Days recumbent when measurements were done
Inflamat	136	Is inflammation present? 0=no, 1=yes
Myopathy	222	Is muscle disorder present? 1=yes, 0=no
PCV	175	Packed cell volume (hematocrit), percent
Urea	266	Serum urea (mmol/l)
Outcome	435	1 if survived, 0 if died or killed

Source: Clark, Henderson, Hoggard, Ellison, and Young (1987).

12.1.3. Next, consider the regression problem with only *CK* as a predictor (*CK* is observed more often than is *Myopathy*, so this regression will be based on more cases than were used in the first two parts of this problem). Draw separate density estimates of *CK*, for *Outcome* = 0 and for *Outcome* = 1. Also, draw separate density estimates for log(*CK*) for the two groups. Comment on the graphs.

12.1.4. Fit the logistic regression mean function with log(*CK*) as the only term beyond the intercept. Summarize results.

12.1.5. Fit the logistic mean function with terms for log(*CK*), *Myopathy* and a *Myopathy* × log(*CK*) interaction. Interpret each of the coefficient estimates. Obtain a sequential deviance table for fitting the terms in the order given above, and summarize results. (Missing data can cause a problem here: if your computer program requires that you fit three separate mean functions to get the analysis of deviance, then you must be sure that each fit is based on the same set of observations, those for which *CK* and *Myopathy* are both observed.)

12.2. Starting with (12.6), prove (12.7).

12.3. Electric shocks A study carried out by R. Norell was designed to learn about the effect of small electrical currents on farm animals, with the eventual goal of understanding the effects of high-voltage power lines near farms. A total of $m = 70$ trials were carried out at each of six intensities, $0, 1, 2, 3, 4$, and 5 mA (shocks on the order of 15 mA are painful for many humans, Dalziel, Lagen, and Thurston 1941). The data are given in the file shocks.txt with columns *Intensity*, number of trials m, which is always equal to 70, and Y, the number of trials out of m for which the response, mouth movement, was observed.

Draw a plot of the fraction responding versus *Intensity*. Then, fit the logistic regression with predictor *Intensity*, and add the fitted curve to your plot. Test the hypothesis that the probability of response is independent of *Intensity*, and summarize your conclusions. Provide a brief interpretation of the coefficient for *Intensity*. (*Hint:* The response in the logistic regression is the number of successes in m trials. Unless the number of trials is one for every case, computer programs will require that you specify the number of trials in some way. Some programs will have an argument with a name like "trials" or "weights" for this purpose. Others, like R and JMP, require that you specify a bivariate response consisting of the number of successes Y and the number of failures $m - Y$.)

12.4. Donner party In the winter of 1846–1847, about 90 wagon train emigrants in the Donner party were unable to cross the Sierra Nevada Mountains of California before winter, and almost half of them starved to death. The data in file donner.txt from Johnson (1996) include some information about each of the members of the party. The variables include *Age*, the age of the person, *Sex*, whether male or female, *Status*, whether the person was a member of a family group, a hired worker for one of the family groups, or a single individual who did not appear to be a hired worker or a member of any of the larger family groups, and *Outcome*, coded 1 if the person survived and 0 if the person died.

12.4.1. How many men and women were in the Donner Party? What was the survival rate for each sex? Obtain a test that the survival rates were the same against the alternative that they were different. What do you conclude?

12.4.2. Fit the logistic regression model with response *Outcome* and predictor *Age*, and provide an interpretation for the fitted coefficient for *Age*.

12.4.3. Draw the graph of *Outcome* versus *Age*, and add both a smooth and a fitted logistic curve to the graph. The logistic regression curve apparently does not match the data: Explain what the differences are and how this failure might be relevant to understanding who survived this tragedy. Fit again, but this time, add a quadratic term in *Age*. Does the fitted curve now match the smooth more accurately?

12.4.4. Fit the logistic regression model with terms for an intercept, *Age*, Age^2, *Sex*, and a factor for *Status*. Provide an interpretation for the parameter estimates for *Sex* and for each of the parameter estimates for *Status*. Obtain tests on the basis of the deviance for adding each of the terms to a mean function that already includes the other terms, and summarize the results of each of the tests via a *p*-value and a one-sentence summary of the results.

12.4.5. Assuming that the logistic regression model provides an adequate summary of the data, give a one-paragraph written summary on the survival of members of the Donner Party.

12.5. Counterfeit banknotes The data in the file `banknote.txt` contains information on 100 counterfeit Swiss banknotes with $Y = 0$ and 100 genuine banknotes with $Y = 1$. Also included are six physical measurements of the notes, including the *Length*, *Diagonal* and the *Left* and *Right* edges of the note, all in millimeters, and the distance from the image to the *Top* edge and *Bottom* edge of the paper, all in millimeters (Flury and Riedwyl, 1988). The goal of the analysis is to estimate the probability or odds that a banknote is counterfeit, given the values of the six measurements.

12.5.1. Draw a scatterplot matrix of six predictors, marking the points different colors for the two groups (genuine or counterfeit). Summarize the information in the scatterplot matrix.

12.5.2. Use logistic regression to study the conditional distribution of *y*, given the predictors.

12.6. Challenger The file `challeng.txt` from Dalal, Fowlkes, and Hoadley (1989) contains data on O-rings on 23 U. S. space shuttle missions prior to the Challenger disaster of January 20, 1986. For each of the previous missions, the temperature at take-off and the pressure of a pre-launch test were recorded, along with the number of O-rings that failed out of six.

Use these data to try to understand the probability of failure as a function of temperature, and of temperature and pressure. Use your fitted model to estimate the probability of failure of an O-ring when the temperature was 31°F, the launch temperature on January 20, 1986.

12.7. Titanic Refer to the *Titanic* data, described in Section 12.2.4, page 262.

   12.7.1. Fit a logistic regression model with terms for factors *Sex*, *Age* and *Class*. On the basis of examination of the data in Table 12.5, explain why you expect that this mean function will be inadequate to explain these data.

   12.7.2. Fit a logistic regression model that includes all the terms of the last part, plus all the two-factor interactions. Use appropriate testing procedures to decide if any of the two-factor interactions can be eliminated. Assuming that the mean function you have obtained matches the data well, summarize the results you have obtained by interpreting the parameters to describe different survival rates for various factor combinations. (*Hint:* How does the survival of the crew differ from the passengers? First class from third class? Males from females? Children versus adults? Did children in first class survive more often than children in third class?)

12.8. BWCAW blowdown The data file `blowAPB.txt` contains the data for Rich's blowdown data, as introduced at the beginning of this chapter, but for the two species *SPP = A* for aspen, and *SPP = PB* for paper birch.

   12.8.1. Fit the same mean function used for Balsam Fir to each of these species. Is the interaction between *S* and *logD* required for these species?

   12.8.2. Ignoring the variable *S*, compare the two species, using the mean functions outlined in Section 6.2.2.

12.9. Windmill data For the windmill data in the data file `wm4.txt`, use the four-site data to estimate the probability that the wind speed at the candidate site exceeds six meters per second, and summarize your results.

Appendix

A.1 WEB SITE

The web address for material for this book is

<code>http://www.stat.umn.edu/alr</code>

The web site includes free text primers on how to do the computations described in the book with several standard computer programs, all the data files described in the book, errata for the book, and scripts for some of the packages that can reproduce the examples in the book.

A.2 MEANS AND VARIANCES OF RANDOM VARIABLES

Suppose we let $u_1, u_2, \ldots, u_n$ be random variables and also let $a_0, a_1, \ldots, a_n$ be $n + 1$ known constants.

A.2.1 E Notation

The symbol $E(u_i)$ is read as the expected value of the random variable u_i. The phrase "expected value" is the same as the phrase "mean value." Informally, the expected value of u_i is the average value of a very large sample drawn from the distribution of u_i. If $E(u_i) = 0$, then the average value we would get for u_i if we sampled its distribution repeatedly is 0. Since u_i is a random variable, any particular realization of u_i is likely to be nonzero.

The expected value is a *linear operator*, which means

$$E(a_0 + a_1 u_1) = a_0 + a_1 E(u_1)$$

$$E\left(a_0 + \sum a_i u_i\right) = a_0 + \sum a_i E(u_i) \tag{A.1}$$

Applied Linear Regression, Third Edition, by Sanford Weisberg
ISBN 0-471-66379-4   Copyright © 2005 John Wiley & Sons, Inc.

For example, suppose that $u_1, \ldots, u_n$ are a random sample from a population, and $E(u_i) = \mu$, $i = 1, \ldots, n$. The sample mean is $\bar{u} = \sum u_i/n = \sum (1/n)u_i$, and the expectation of the sample mean is

$$E(\bar{u}) = E\left(\sum \frac{1}{n}u_i\right) = \frac{1}{n}\sum E(u_i) = \frac{1}{n}(n\mu) = \mu$$

We say that $\bar{u}$ is an *unbiased* estimate of the population mean μ, since its expected value is μ.

A.2.2 Var Notation

The symbol $\text{Var}(u_i)$ is the variance of u_i. The variance is defined by the equation $\text{Var}(u_i) = E[u_i - E(u_i)]^2 =$, the expected squared difference between an observed value for u_i and its mean value. The larger $\text{Var}(u_i)$, the more variable observed values for u_i are likely to be. The symbol σ^2 is often used for a variance, or σ_u^2 might be used for the variance of the identically distributed u_i if several variances are being discussed.

The general rule for the variance of a sum of *uncorrelated* random variables is

$$\text{Var}\left(a_0 + \sum a_i u_i\right) = \sum a_i^2 \text{Var}(u_i) \tag{A.2}$$

The a_0 term vanishes because the variance of $a_0 + u$ is the same as the variance of u since the variance of a constant is 0. Assuming that $\text{Var}(u_i) = \sigma^2$, we can find the variance of the sample mean of independently, identically distributed u_i:

$$\text{Var}(\bar{u}) = \text{Var}\left(\sum \frac{1}{n}u_i\right) = \frac{1}{n^2}\sum E(u_i) = \frac{1}{n^2}(n\sigma^2) = \frac{\sigma^2}{n}$$

A.2.3 Cov Notation

The symbol $\text{Cov}(u_i, u_j)$ is read as the covariance between the random variables u_i and u_j and is defined by the equation

$$\text{Cov}(u_i, u_j) = E\left[(u_i - E(u_i))(u_j - E(u_j))\right] = \text{Cov}(u_j, u_i)$$

The covariance describes the way two random variables vary jointly. If the two variables are independent, then $\text{Cov}(u_i, u_j) = 0$, but zero correlation does not imply independence. The variance is a special case of covariance, since $\text{Cov}(u_i, u_i) = \text{Var}(u_i)$. The rule for covariance is

$$\text{Cov}(a_0 + a_1 u_1, a_3 + a_2 u_2) = a_1 a_2 \text{Cov}(u_1, u_2)$$

The *correlation coefficient* is defined by

$$\rho(u_i, u_j) = \frac{\text{Cov}(u_i, u_j)}{\sqrt{\text{Var}(u_i)\text{Var}(u_j)}}$$

The correlation does not depend on units of measurement and has a value between -1 and 1.

The general form for the variance of a linear combination of *correlated* random variables is

$$\text{Var}\left(a_0 + \sum_{i=1}^{n} a_i u_i\right) = \sum_{i=1}^{n} a_i^2 \text{Var}(u_i) + 2\sum_{i=1}^{n-1}\sum_{j=i+1}^{n} a_i a_j \text{Cov}(u_i, u_j) \qquad (\text{A.3})$$

A.2.4 Conditional Moments

Throughout the book, we use notation like $E(Y|X = x)$ to denote *the mean of the random variable Y in the population for which the value of X is fixed at the value $X = x$*. Similarly, $\text{Var}(Y|X = x)$ is the variance of the random variable Y in the population for which X is fixed at $X = x$.

There are simple relationships between the *conditional* mean and variance of Y given X and the *unconditional* mean and variances (see, for example, Casella and Berger, 1990):

$$E(Y) = E[E(Y|X = x)] \qquad (\text{A.4})$$

$$\text{Var}(Y) = E[\text{Var}(Y|X = x)] + \text{Var}(E(Y|X = x)) \qquad (\text{A.5})$$

For example, suppose that when we condition on the predictor X we have a simple linear regression mean function with constant variance, $E(Y|X = x) = \beta_0 + \beta_1 x$, $\text{Var}(Y|X = x) = \sigma^2$. In addition, suppose the unconditional moments of the predictor are $E(X) = \mu_x$ and $\text{Var}(X) = \tau_x^2$. Then for the unconditional random variable Y,

$$E(Y) = E[E(Y|X = x)]$$

$$= E[\beta_0 + \beta_1 x]$$

$$= \beta_0 + \beta_1 \mu_x$$

$$\text{Var}(Y) = E[\text{Var}(Y|X = x)] + \text{Var}[E(Y|X = x)]$$

$$= E[\sigma^2] + \text{Var}[\beta_0 + \beta_1 x]$$

$$= \sigma^2 + \beta_1^2 \tau_x^2$$

The mean of the unconditional variable Y is obtained by substituting the mean of the unconditional variable X into the conditional mean formula, and the unconditional variance of Y equals the conditional variance plus a second quantity that depends on both β_1^2 and on τ_x^2.

A.3 LEAST SQUARES FOR SIMPLE REGRESSION

The OLS estimates of β_0 and β_1 in simple regression are the values that minimize the residual sum of squares function,

$$RSS(\beta_0, \beta_1) = \sum_{i=1}^{n} (y_i - \beta_0 - \beta_1 x_i)^2 \tag{A.6}$$

One method of finding the minimizer is to differentiate with respect to β_0 and β_1, set the derivatives equal to 0, and solve

$$\frac{\partial RSS(\beta_0, \beta_1)}{\beta_0} = -2 \sum_{i=1}^{n} (y_i - \beta_0 - \beta_1 x_i) = 0$$

$$\frac{\partial RSS(\beta_0, \beta_1)}{\beta_1} = -2 \sum_{i=1}^{n} x_i (y_i - \beta_0 - \beta_1 x_i) = 0$$

Upon rearranging terms, we get

$$\beta_0 n + \beta_1 \sum x_i = \sum y_i$$
$$\beta_0 \sum x_i + \beta_1 \sum x_i^2 = \sum x_i y_i \tag{A.7}$$

Equations (A.7) are called the *normal equations* for the simple linear regression model (2.1). The normal equations depend on the data only through the sufficient statistics $\sum x_i$, $\sum y_i$, $\sum x_i^2$ and $\sum x_i y_i$. Using the formulas

$$SXX = \sum (x_i - \bar{x})^2 = \sum x_i^2 - n\bar{x}^2$$
$$SXY = \sum (x_i - \bar{x})(y_i - \bar{y}) = \sum x_i y_i - n\bar{x}\bar{y} \tag{A.8}$$

equivalent and numerically more stable sufficient statistics are given by $\bar{x}$, $\bar{y}$, SXX and SXY. Solving (A.7), we get

$$\hat{\beta}_0 = \bar{y} - \hat{\beta}_1 \bar{x}, \quad \hat{\beta}_1 = \frac{SXY}{SXX} \tag{A.9}$$

A.4 MEANS AND VARIANCES OF LEAST SQUARES ESTIMATES

The least squares estimates are linear combinations of the observed values $y_1, \ldots, y_n$ of the response, so we can apply the results of Appendix A.2 to the estimates found in Appendix A.3 to get the means, variances, and covariances of the estimates. Assume the simple regression model (2.1) is correct. The estimator $\hat{\beta}_1$ given at (A.9) can be written as $\hat{\beta}_1 = \sum c_i y_i$, where for each i, $c_i = (x_i - \bar{x})/SXX$.

Since we are conditioning on the values of X, the c_i are fixed numbers. By (A.1),

$$E(\hat{\beta}_1 | X) = E\left(\sum c_i y_i | X = x_i\right) = \sum c_i E(y_i | X = x_i)$$

$$= \sum c_i (\beta_0 + \beta_1 x_i)$$

$$= \beta_0 \sum c_i + \beta_1 \sum c_i x_i$$

By direct summation, $\sum c_i = 0$ and $\sum c_i x_i = 1$, giving

$$E(\hat{\beta}_1 | X) = \beta_1$$

which shows that $\hat{\beta}_1$ is unbiased for β_1. A similar computation will show that $\hat{\beta}_0 = \beta_0$.

Since the y_i are assumed independent, the variance of $\hat{\beta}_1$ is found by an application of (A.2),

$$Var(\hat{\beta}_1 | X) = Var\left(\sum c_i y_i | X = x_i\right)$$

$$= \sum c_i^2 Var(Y | X = x_i)$$

$$= \sigma^2 \sum c_i^2$$

$$= \sigma^2 / SXX$$

This computation also used $\sum c_i^2 = \sum (x_i - \bar{x})^2 / SXX^2 = 1/SXX$. Computing the variance of $\hat{\beta}_0$ requires an application of (A.3). We write

$$Var(\hat{\beta}_0) = Var(\bar{y} - \hat{\beta}_1 \bar{x} | X)$$

$$= Var(\bar{y} | X) + \bar{x}^2 Var(\hat{\beta}_1 | X) - 2\bar{x} Cov(\bar{y}, \hat{\beta}_1 | X) \qquad (A.10)$$

To complete this computation, we need to compute the covariance,

$$Cov(\bar{y}, \hat{\beta}_1 | X) = Cov\left(\frac{1}{n} \sum y_i, \sum c_i y_i\right)$$

$$= \frac{1}{n} \sum c_i Cov(y_i, y_i)$$

$$= \frac{\sigma^2}{n} \sum c_i$$

$$= 0$$

because the y_i are independent, and $\sum c_i = 0$. Substituting into (A.10) and simplifying,

$$\text{Var}(\hat{\beta}_0) = \sigma^2 \left(\frac{1}{n} + \frac{\bar{x}^2}{SXX} \right)$$

Finally,

$$
\begin{aligned}
\text{Cov}(\hat{\beta}_0, \hat{\beta}_1 | X) &= \text{Cov}(\bar{y} - \hat{\beta}_1 \bar{x}, \hat{\beta}_1 | X) \\
&= \text{Cov}(\bar{y}, \hat{\beta}_1) - \bar{x}\text{Cov}(\hat{\beta}_1, \hat{\beta}_1) \\
&= 0 - \sigma^2 \frac{\bar{x}}{SXX} \\
&= -\sigma^2 \frac{\bar{x}}{SXX}
\end{aligned}
$$

Further application of these results gives the variance of a fitted value, $\hat{y} = \hat{\beta}_0 + \hat{\beta}_1 x$:

$$
\begin{aligned}
\text{Var}(\hat{y}|X = x) &= \text{Var}(\hat{\beta}_0 + \hat{\beta}_1 x | X = x) \\
&= \text{Var}(\hat{\beta}_0 | X = x) + x^2 \text{Var}(\hat{\beta}_1 | X = x) + 2x\text{Cov}(\hat{\beta}_0, \hat{\beta}_1 | X = x) \\
&= \sigma^2 \left(\frac{1}{n} + \frac{\bar{x}^2}{SXX} \right) + \sigma^2 x^2 \frac{1}{SXX} - 2\sigma^2 x \frac{\bar{x}}{SXX} \\
&= \sigma^2 \left(\frac{1}{n} + \frac{(x - \bar{x})^2}{SXX} \right)
\end{aligned}
\tag{A.11}
$$

A prediction $\tilde{y}_*$ at the future value x_* is just $\hat{\beta}_0 + \hat{\beta}_1 x_*$. The variance of a prediction consists of the variance of the fitted value at x_* given by (A.11) plus σ^2, the variance of the error that will be attached to the future value,

$$\text{Var}(\tilde{y}_* | X = x_*) = \sigma^2 \left(\frac{1}{n} + \frac{(x - \bar{x})^2}{SXX} \right) + \sigma^2$$

as given by (2.25).

A.5 ESTIMATING E(Y|X) USING A SMOOTHER

For a 2D scatterplot of Y versus X, a scatterplot *smoother* provides an estimate of the mean function $E(Y|X = x)$ as x varies, without making parametric assumptions about the mean function, so we do not need to assume that the mean function is a straight line or any other particular form. We very briefly introduce one of many types of local smoothers and provide references to other approaches to smoothing.

The smoother we use most often in this book is the simplest case of the *loess* smoother (Cleveland, 1979; see also the first step in Algorithm 6.1.1 in Härdle, 1990, p. 192). This smoother estimates $E(Y|X = x_g)$ by $\tilde{y}_g$ at the point x_g via a weighted least squares simple regression, giving more weight to points close to x_g than to points distant from x_g. Here is the method:

1. Select a value for a *smoothing parameter* f, a number between 0 and 1. Values of f close to 1 will give curves that are too smooth and will be close to a straight line, while small values of f give curves that are too rough and match all the wiggles in the data. The value of f must be chosen to balance the bias of oversmoothing with the variability of undersmoothing. Remarkably, for many problems, $f \approx 2/3$ is a good choice. There is a substantial literature on the appropriate ways to estimate a smoothing parameter for *loess* and for other smoothing methods, but for the purposes of using a smoother to help us look at a graph, optimal choice of a smoothing parameter is not critical.

2. Find the fn closest points to x_g. For example, if $n = 100$, and $f = 0.6$, then find the $fn = 60$ closest points to x_g. Every time the value of x_g is changed, the points selected may change.

3. Among these fn *nearest neighbors* to x_g, compute the WLS estimates for the simple regression of Y on X, with weights determined so that points close to x_g have the highest weight, and the weights decline toward 0 for points farther from x_g. We use a triangular weight function that gives maximum weight to data at x_g, and weights that decrease linearly to 0 at the edge of the neighborhood. If a different weight function is used, answers are somewhat different.

4. The value of $\tilde{y}_g$ is the fitted value at x_g from the WLS regression using the nearest neighbors found at step 2 as the data, and the weights from step 3 as weights.

5. Repeat 1–4 for many values of x_g that form a grid of points that cover the interval on the x-axis of interest. Join the points.

Figure A.1 shows a plot of Y versus X, along with four smoothers. The first smoother is the OLS simple regression line, which does not match the data well because the mean function for the data in this figure is probably curved, not straight. The *loess* smooth with $f = 0.1$ is as expected very wiggly, matching the local variation rather than the mean. The line for $f = 2/3$ seems to match the data very well, while the *loess* fit for $f = .95$ is nearly the same as for $f = 2/3$, but it tends toward oversmoothing and attempts to match the OLS line. We would conclude from this graph that a straight-line mean function is likely to be inadequate because it does not match the data very well. Loader (2004) discusses a formal lack-of-fit test on the basis of comparing parametric and nonparametric estimates of the mean function that is presented in Problem 5.3.

The *loess* smoother is an example of a *nearest neighbor* smoother. Local polynomial regression smoothers and kernel smoothers are similar to *loess*, except they

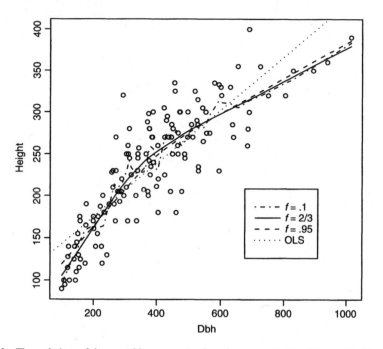

FIG. A.1 Three choices of the smoothing parameter for a *loess* smooth. The data used in this plot are discussed in Section 7.1.2.

give positive weight to all cases within a fixed distance of the point of interest rather than a fixed number of points.

There is a large literature on *nonparametric regression*, for which scatterplot smoothing is a primary tool. Recent reference on this subject include Simonoff (1996), Bowman and Azzalini (1997), and Loader (2004).

The literature on estimating a variance function from a scatterplot is much smaller than the literature on estimating the mean (but see Ruppert, Wand, Holst and Hössjer, 1997). Here is a simple algorithm that can produce a smoother that estimates the standard deviation function, which is the square root of the variance function:

1. Smooth the y_i on the x_i to get an estimate say $\tilde{y}_i$ for each value of $X = x_i$. Compute the squared residuals, $r_i = (y_i - \tilde{y}_i)^2$. Under normality of errors, the expectation $E(r_i|x_i) = \text{Var}(Y|X = x_i)$, so a mean smooth for the squared residuals estimates the variance smooth for Y.

2. Smooth the r_i on x_i to estimate $\text{Var}(Y|X = x_i)$ by s_i^2 at each value x_i. Then s_i^2 is the smoothed estimate of the variance, and s_i is a smoothed estimate of the standard deviation.

3. Add three lines to the scatterplot: The mean smooth $(x_i, \tilde{y}_i)$, the mean smooth plus one standard deviation, $(x_i, \tilde{y}_i + s_i)$ and the mean smooth minus one standard deviation, $(x_i, \tilde{y}_i - s_i)$.

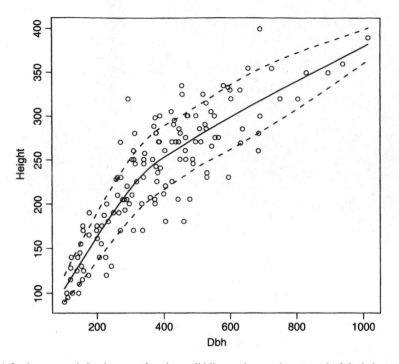

FIG. A.2 *loess* smooth for the mean function, solid line, and mean ± one standard deviation, dashed lines.

Figure A.2 shows the *loess* smooth for the mean function and the mean plus and minus one standard deviation for the same data as in Figure A.1. The variability appears to be a bit larger in the middle of the range than at the edges.

A.6 A BRIEF INTRODUCTION TO MATRICES AND VECTORS

We provide only a brief introduction to matrices and vectors. More complete references include Graybill (1969), Searle (1982), Schott (1996), or any good linear algebra book.

Boldface type is used to indicate matrices and vectors. We will say that $\mathbf{X}$ is an $r \times c$ matrix if it is an array of numbers with r rows and c columns. A specific 4×3 matrix $\mathbf{X}$ is

$$\mathbf{X} = \begin{pmatrix} 1 & 2 & 1 \\ 1 & 1 & 5 \\ 1 & 3 & 4 \\ 1 & 8 & 6 \end{pmatrix} = \begin{pmatrix} x_{11} & x_{12} & x_{13} \\ x_{21} & x_{22} & x_{23} \\ x_{31} & x_{32} & x_{33} \\ x_{41} & x_{42} & x_{43} \end{pmatrix} = (x_{ij}) \qquad (A.12)$$

The element x_{ij} of $\mathbf{X}$ is the number in the ith row and the jth column. For example, in the preceding matrix, $x_{32} = 3$.

A *vector* is a matrix with just one column. A specific 4×1 matrix $\mathbf{y}$, which is a vector of length 4, is given by

$$
\mathbf{y} = \begin{pmatrix} 2 \\ 3 \\ -2 \\ 0 \end{pmatrix} = \begin{pmatrix} y_1 \\ y_2 \\ y_3 \\ y_4 \end{pmatrix}
$$

The elements of a vector are generally singly subscripted; thus, $y_3 = -2$. A *row vector* is a matrix with one row. We do not use row vectors in this book. If a vector is needed to represent a row, a transpose of a column vector will be used (see below).

A *square matrix* has the same number of rows and columns, so $r = c$. A square matrix $\mathbf{Z}$ is *symmetric* if $z_{ij} = z_{ji}$ for all i and j. A square matrix is *diagonal* if all elements off the main diagonal are 0, $z_{ij} = 0$, unless $i = j$. The matrices $\mathbf{C}$ and $\mathbf{D}$ below are symmetric and diagonal, respectively:

$$
\mathbf{C} = \begin{pmatrix} 7 & 3 & 2 & 1 \\ 3 & 4 & 1 & -1 \\ 2 & 1 & 6 & 3 \\ 1 & -1 & 3 & 8 \end{pmatrix} \quad \mathbf{D} = \begin{pmatrix} 7 & 0 & 0 & 0 \\ 0 & 4 & 0 & 0 \\ 0 & 0 & 6 & 0 \\ 0 & 0 & 0 & 8 \end{pmatrix}
$$

The diagonal matrix with all elements on the diagonal equal to 1 is called the *identity matrix*, for which the symbol $\mathbf{I}$ is used. The 4×4 identity matrix is

$$
\mathbf{I} = \begin{pmatrix} 1 & 0 & 0 & 0 \\ 0 & 1 & 0 & 0 \\ 0 & 0 & 1 & 0 \\ 0 & 0 & 0 & 1 \end{pmatrix}
$$

A *scalar* is a 1×1 matrix, an ordinary number.

A.6.1 Addition and Subtraction

Two matrices can be added or subtracted only if they have the same number of rows and columns. The sum $\mathbf{C} = \mathbf{A} + \mathbf{B}$ of $r \times c$ matrices is also $r \times c$. Addition is done elementwise:

$$
\mathbf{C} = \mathbf{A} + \mathbf{B} = \begin{pmatrix} a_{11} & a_{12} \\ a_{21} & a_{22} \\ a_{31} & a_{32} \end{pmatrix} + \begin{pmatrix} b_{11} & b_{12} \\ b_{21} & b_{22} \\ b_{31} & b_{32} \end{pmatrix} = \begin{pmatrix} a_{11} + b_{11} & a_{12} + b_{12} \\ a_{21} + b_{21} & a_{22} + b_{22} \\ a_{31} + b_{31} & a_{32} + b_{32} \end{pmatrix}
$$

Subtraction works the same way, with the "+" signs changed to "−" signs. The usual rules for addition of numbers apply to addition of matrices, namely commutativity, $\mathbf{A} + \mathbf{B} = \mathbf{B} + \mathbf{A}$, and associativity, $(\mathbf{A} + \mathbf{B}) + \mathbf{C} = \mathbf{A} + (\mathbf{B} + \mathbf{C})$.

A.6.2 Multiplication by a Scalar

If k is a number and $\mathbf{A}$ is an $r \times c$ matrix with elements (a_{ij}), then $k\mathbf{A}$ is an $r \times c$ matrix with elements (ka_{ij}). For example, the matrix $\sigma^2\mathbf{I}$ has all diagonal elements equal to σ^2 and all off-diagonal elements equal to 0.

A.6.3 Matrix Multiplication

Multiplication of matrices follows rules that are more complicated than are the rules for addition and subtraction. For two matrices to be multiplied together in the order $\mathbf{AB}$, the number of columns of $\mathbf{A}$ must equal the number of rows of $\mathbf{B}$. For example, if $\mathbf{A}$ is $r \times c$, and $\mathbf{B}$ is $c \times q$, then $\mathbf{C} = \mathbf{AB}$ is $r \times q$. If the elements of $\mathbf{A}$ are (a_{ij}) and the elements of $\mathbf{B}$ are (b_{ij}), then the elements of $\mathbf{C} = (c_{ij})$ are given by the formula

$$c_{ij} = \sum_{k=1}^{c} a_{ik}b_{kj}$$

This formula says that c_{ij} is formed by taking the ith row of $\mathbf{A}$ and the jth column of $\mathbf{B}$, multiplying the first element of the specified row in $\mathbf{A}$ by the first element in the specified column in $\mathbf{B}$, multiplying second elements, and so on, and then adding the products together.

If $\mathbf{A}$ is $1 \times c$ and $\mathbf{B}$ is $c \times 1$, then the product $\mathbf{AB}$ is 1×1, an ordinary number. For example, if $\mathbf{A}$ and $\mathbf{B}$ are

$$\mathbf{A} = (1 \quad 3 \quad 2 \quad -1) \quad \mathbf{B} = \begin{pmatrix} 2 \\ 1 \\ -2 \\ 4 \end{pmatrix}$$

then the product $\mathbf{AB}$ is

$$\mathbf{AB} = (1 \times 2) + (3 \times 1) + (2 \times -2) + (-1 \times 4) = -3$$

$\mathbf{AB}$ is not the same as $\mathbf{BA}$. For the preceding matrices, the product $\mathbf{BA}$ will be a 4×4 matrix:

$$\mathbf{BA} = \begin{pmatrix} 2 & 6 & 4 & -2 \\ 1 & 3 & 2 & -1 \\ -2 & -6 & -4 & 2 \\ 4 & 12 & 8 & -4 \end{pmatrix}$$

The following small example illustrates what happens when all the dimensions are bigger than 1. A 3×2 matrix $\mathbf{A}$ times a 2×2 matrix $\mathbf{B}$ is given as

$$\begin{pmatrix} a_{11} & a_{12} \\ a_{21} & a_{22} \\ a_{31} & a_{32} \end{pmatrix} \begin{pmatrix} b_{11} & b_{12} \\ b_{21} & b_{22} \end{pmatrix} = \begin{pmatrix} a_{11}b_{11} + a_{12}b_{21} & a_{11}b_{12} + a_{12}b_{22} \\ a_{21}b_{11} + a_{22}b_{21} & a_{21}b_{12} + a_{22}b_{22} \\ a_{31}b_{11} + a_{32}b_{21} & a_{31}b_{12} + a_{32}b_{22} \end{pmatrix}$$

Using numbers, an example of multiplication of two matrices is

$$
\begin{pmatrix} 3 & 1 \\ -1 & 0 \\ 2 & 2 \end{pmatrix} \begin{pmatrix} 5 & 1 \\ 0 & 4 \end{pmatrix} = \begin{pmatrix} 15+0 & 3+4 \\ -5+0 & -1+0 \\ 10+0 & 2+8 \end{pmatrix} = \begin{pmatrix} 15 & 7 \\ -5 & -1 \\ 10 & 10 \end{pmatrix}
$$

In this example, $\mathbf{BA}$ is not defined because the number of columns of $\mathbf{B}$ is not equal to the number of rows of $\mathbf{A}$. However, the associative law holds: If $\mathbf{A}$ is $r \times c$, $\mathbf{B}$ is $c \times q$, and $\mathbf{C}$ is $q \times p$, then $\mathbf{A}(\mathbf{BC}) = (\mathbf{AB})\mathbf{C}$, and the result is an $r \times p$ matrix.

A.6.4 Transpose of a Matrix

The transpose of an $r \times c$ matrix $\mathbf{X}$ is a $c \times r$ matrix called $\mathbf{X}'$ such that if the elements of $\mathbf{X}$ are (x_{ij}), then the elements of $\mathbf{X}'$ are (x_{ji}). For the matrix $\mathbf{X}$ given at (A.12),

$$
\mathbf{X}' = \begin{pmatrix} 1 & 1 & 1 & 1 \\ 2 & 1 & 3 & 8 \\ 1 & 5 & 4 & 6 \end{pmatrix}
$$

The transpose of a column vector is a row vector. The transpose of a product $(\mathbf{AB})'$ is the product of the transposes, in *opposite order*, so $(\mathbf{AB})' = \mathbf{B}'\mathbf{A}'$.

Suppose that $\mathbf{a}$ is an $r \times 1$ vector with elements $a_1, \ldots, a_r$. Then the product $\mathbf{a}'\mathbf{a}$ will be a 1×1 matrix or scalar, given by

$$
\mathbf{a}'\mathbf{a} = a_1^2 + a_2^2 + \cdots + a_r^2 = \sum_{i=1}^{r} a_i^2 \tag{A.13}
$$

Thus, $\mathbf{a}'\mathbf{a}$ provides a compact notation for the sum of the squares of the elements of a vector $\mathbf{a}$. The square root of this quantity $(\mathbf{a}'\mathbf{a})^{1/2}$ is called the *norm* or *length* of the vector $\mathbf{a}$. Similarly, if $\mathbf{a}$ and $\mathbf{b}$ are both $r \times 1$ vectors, then we obtain

$$
\mathbf{a}'\mathbf{b} = a_1b_1 + a_2b_2 + \cdots + a_nb_n = \sum_{i=1}^{r} a_ib_i = \sum_{i=1}^{r} b_ia_i = \mathbf{b}'\mathbf{a}
$$

The fact that $\mathbf{a}'\mathbf{b} = \mathbf{b}'\mathbf{a}$ is often quite useful in manipulating the vectors used in regression calculations.

Another useful formula in regression calculations is obtained by applying the distributive law

$$
(\mathbf{a} - \mathbf{b})'(\mathbf{a} - \mathbf{b}) = \mathbf{a}'\mathbf{a} + \mathbf{b}'\mathbf{b} - 2\mathbf{a}'\mathbf{b} \tag{A.14}
$$

A.6.5 Inverse of a Matrix

For any scalar $c \neq 0$, there is another number called the *inverse* of c, say d, such that the product $cd = 1$. For example, if $c = 3$, then $d = 1/c = 1/3$, and the inverse of 3 is $1/3$. Similarly, the inverse of $1/3$ is 3. The number 0 does not have an inverse because there is no other number d such that $0 \times d = 1$.

Square matrices can also have an inverse. We will say that the inverse of a matrix $\mathbf{C}$ is another matrix $\mathbf{D}$, such that $\mathbf{CD} = \mathbf{I}$, and we write $\mathbf{D} = \mathbf{C}^{-1}$. Not all square matrices have an inverse. The collection of matrices that have an inverse are called *full rank*, *invertible*, or *nonsingular*. A square matrix that is not invertible is of less than full rank, or *singular*. If a matrix has an inverse, it has a unique inverse.

The inverse is easy to compute only in special cases, and its computation in general can require a very tedious calculation that is best done on a computer. High-level matrix and statistical languages such as Matlab, Maple, Mathematica, R and S-plus include functions for inverting matrices, or returning an appropriate message if the inverse does not exist.

The identity matrix $\mathbf{I}$ is its own inverse. If $\mathbf{C}$ is a diagonal matrix, say

$$\mathbf{C} = \begin{pmatrix} 3 & 0 & 0 & 0 \\ 0 & -1 & 0 & 0 \\ 0 & 0 & 4 & 0 \\ 0 & 0 & 0 & 1 \end{pmatrix}$$

then $\mathbf{C}^{-1}$ is the diagonal matrix

$$\mathbf{C} = \begin{pmatrix} \frac{1}{3} & 0 & 0 & 0 \\ 0 & -1 & 0 & 0 \\ 0 & 0 & \frac{1}{4} & 0 \\ 0 & 0 & 0 & 1 \end{pmatrix}$$

as can be verified by direct multiplication. For any diagonal matrix with nonzero diagonal elements, the inverse is obtained by inverting the diagonal elements. If any of the diagonal elements are 0, then no inverse exists.

A.6.6 Orthogonality

Two vectors $\mathbf{a}$ and $\mathbf{b}$ of the same length are *orthogonal* if $\mathbf{a}'\mathbf{b} = 0$. An $r \times c$ matrix $\mathbf{Q}$ has *orthonormal columns* if its columns, viewed as a set of $c \leq r$ different $r \times 1$ vectors, are orthogonal and in addition have length 1. This is equivalent to requiring that $\mathbf{Q}'\mathbf{Q} = \mathbf{I}$, the $r \times r$ identity matrix. A square matrix $\mathbf{A}$ is *orthogonal* if $\mathbf{A}'\mathbf{A} = \mathbf{AA}' = \mathbf{I}$, and so $\mathbf{A}^{-1} = \mathbf{A}'$. For example, the matrix

$$\mathbf{A} = \begin{pmatrix} \frac{1}{\sqrt{3}} & \frac{1}{\sqrt{2}} & \frac{1}{\sqrt{6}} \\ \frac{1}{\sqrt{3}} & 0 & -\frac{2}{\sqrt{6}} \\ \frac{1}{\sqrt{3}} & -\frac{1}{\sqrt{2}} & \frac{1}{\sqrt{6}} \end{pmatrix}$$

can be shown to be orthogonal by showing that $\mathbf{A}'\mathbf{A} = \mathbf{I}$, and therefore

$$\mathbf{A}^{-1} = \mathbf{A}' = \begin{pmatrix} \frac{1}{\sqrt{3}} & \frac{1}{\sqrt{3}} & \frac{1}{\sqrt{3}} \\ \frac{1}{\sqrt{2}} & 0 & -\frac{1}{\sqrt{2}} \\ \frac{1}{\sqrt{6}} & -\frac{2}{\sqrt{6}} & \frac{1}{\sqrt{6}} \end{pmatrix}$$

A.6.7 Linear Dependence and Rank of a Matrix

Suppose we have a $n \times p$ matrix $\mathbf{X}$, with columns given by the vectors $\mathbf{x}_1, \ldots, \mathbf{x}_p$; we consider only the case $p \leq n$. We will say that $\mathbf{x}_1, \ldots, \mathbf{x}_p$ are *linearly dependent* if we can find multipliers $a_1, \ldots, a_p$, not all of which are 0, such that

$$\sum_{i=1}^{p} a_i \mathbf{x}_i = \mathbf{0} \tag{A.15}$$

If no such multipliers exist, then we say that the vectors are *linearly independent*, and the matrix is *full rank*. In general, the *rank* of a matrix is the maximum number of $\mathbf{x}_i$ that form a linearly independent set.

For example, the matrix $\mathbf{X}$ given at (A.12) can be shown to have linearly independent columns because no a_i not all equal to zero can be found that satisfy (A.15). On the other hand, the matrix

$$\mathbf{X} = \begin{pmatrix} 1 & 2 & 5 \\ 1 & 1 & 4 \\ 1 & 3 & 6 \\ 1 & 8 & 11 \end{pmatrix} = (\mathbf{x}_1, \mathbf{x}_2, \mathbf{x}_3) \tag{A.16}$$

has linearly dependent columns and is singular because $\mathbf{x}_3 = 3\mathbf{x}_1 + \mathbf{x}_2$, or $3\mathbf{x}_1 + \mathbf{x}_2 - \mathbf{x}_3 = \mathbf{0}$. This matrix is of rank two because the linearly independent subset of the columns with the most elements, consisting of any two of the three columns, has two elements.

The matrix $\mathbf{X}'\mathbf{X}$ is a $p \times p$ matrix. If $\mathbf{X}$ has rank p, so does $\mathbf{X}'\mathbf{X}$. Full-rank square matrices always have an inverse. Square matrices of less than full rank never have an inverse.

A.7 RANDOM VECTORS

An $n \times 1$ vector $\mathbf{Y}$ is a *random vector* if each of its elements is a random variable. The mean of an $n \times 1$ random vector $\mathbf{Y}$ is also an $n \times 1$ vector whose elements are the means of the elements of $\mathbf{Y}$. The variance of an $n \times 1$ vector $\mathbf{Y}$ is an $n \times n$ square symmetric matrix, often called a *covariance matrix*, written Var($\mathbf{Y}$) with Var(y_i) as its (i, i) element and Cov(y_i, y_j) = Cov(y_j, y_i) as both the (i, j) and (j, i) element.

The rules for means and variances of random vectors are matrix equivalents of the scalar versions in Appendix A.2. If $\mathbf{a}_0$ is a vector of constants, and $\mathbf{A}$ is a matrix of constants,

$$E(\mathbf{a}_0 + \mathbf{AY}) = \mathbf{a}_0 + \mathbf{A}E(\mathbf{Y}) \tag{A.17}$$

$$\text{Var}(\mathbf{a}_0 + \mathbf{AY}) = \mathbf{A}\,\text{Var}(\mathbf{Y})\mathbf{A}' \tag{A.18}$$

A.8   LEAST SQUARES USING MATRICES

The multiple linear regression model can be written as

$$E(Y|X = \mathbf{x}) = \boldsymbol{\beta}'\mathbf{x}$$

$$\text{Var}(Y|X = \mathbf{x}) = \sigma^2$$

In matrix terms, we will write the model using errors as

$$\mathbf{Y} = \mathbf{X}\boldsymbol{\beta} + \mathbf{e}$$

where $\mathbf{Y}$ is the $n \times 1$ vector of response values and $\mathbf{X}$ is a $n \times p'$ matrix. If the mean function includes an intercept, then the first column of $\mathbf{X}$ is a vector of ones, and $p' = p + 1$. If the mean function does not include an intercept, then the column of one is not included in $\mathbf{X}$ and $p' = p$. The ith row of the $n \times p'$ matrix $\mathbf{X}$ is $\mathbf{x}_i'$, $\boldsymbol{\beta}$ is a $p' \times 1$ vector of parameters for the mean function, $\mathbf{e}$ is the $n \times 1$ vector of unobservable errors, and σ^2 is an unknown positive constant.

The OLS estimate $\hat{\boldsymbol{\beta}}$ of $\boldsymbol{\beta}$ is given by the arguments that minimize the residual sum of squares function,

$$RSS(\boldsymbol{\beta}) = (\mathbf{Y} - \mathbf{X}\boldsymbol{\beta})'(\mathbf{Y} - \mathbf{X}\boldsymbol{\beta})$$

Using (A.14), we obtain

$$RSS(\boldsymbol{\beta}) = \mathbf{Y}'\mathbf{Y} + \boldsymbol{\beta}'(\mathbf{X}'\mathbf{X})\boldsymbol{\beta} - 2\mathbf{Y}'\mathbf{X}\boldsymbol{\beta} \tag{A.19}$$

$RSS(\boldsymbol{\beta})$ depends on only three functions of the data: $\mathbf{Y}'\mathbf{Y}$, $\mathbf{X}'\mathbf{X}$, and $\mathbf{Y}'\mathbf{X}$. Any two data sets that have the same values of these three quantities will have the same least squares estimates. Using (A.8), the information in these quantities is equivalent to the information contained in the sample means of the terms plus the sample covariances of the terms and the response.

To minimize (A.19), differentiate with respect to $\boldsymbol{\beta}$ and set the result equal to 0. This leads to the matrix version of the *normal equations*,

$$\mathbf{X}'\mathbf{X}\boldsymbol{\beta} = \mathbf{X}'\mathbf{Y} \tag{A.20}$$

The OLS estimates are any solution to these equations. If the inverse of $(\mathbf{X}'\mathbf{X})$ exists, as it will if the columns of $\mathbf{X}$ are linearly independent, the OLS estimates are unique and are given by

$$\hat{\boldsymbol{\beta}} = (\mathbf{X}'\mathbf{X})^{-1}\mathbf{X}'\mathbf{Y} \tag{A.21}$$

If the inverse does not exist, then the matrix $(\mathbf{X}'\mathbf{X})$ is of less than full rank, and the OLS estimate is not unique. In this case, most computer programs will use a linearly independent subset of the columns of $\mathbf{X}$ in fitting the model, so that the reduced model matrix does have full rank. This is discussed in Section 4.1.4.

A.8.1 Properties of Estimates

Using the rules for means and variances of random vectors, (A.17) and (A.18), we find

$$
\begin{aligned}
E(\hat{\boldsymbol{\beta}}|\mathbf{X}) &= E((\mathbf{X}'\mathbf{X})^{-1}\mathbf{X}'\mathbf{Y}|\mathbf{X}) \\
&= (\mathbf{X}'\mathbf{X})^{-1}\mathbf{X}'E(\mathbf{Y}|\mathbf{X}) \\
&= (\mathbf{X}'\mathbf{X})^{-1}\mathbf{X}'\mathbf{X}\boldsymbol{\beta} \\
&= \boldsymbol{\beta}
\end{aligned}
\tag{A.22}
$$

so $\hat{\boldsymbol{\beta}}$ is unbiased for $\boldsymbol{\beta}$, as long as the mean function that was fit is the true mean function. The variance of $\hat{\boldsymbol{\beta}}$ is

$$
\begin{aligned}
\mathrm{Var}(\hat{\boldsymbol{\beta}}|\mathbf{X}) &= \mathrm{Var}((\mathbf{X}'\mathbf{X})^{-1}\mathbf{X}'\mathbf{Y}|\mathbf{X}) \\
&= (\mathbf{X}'\mathbf{X})^{-1}\mathbf{X}'\,[\mathrm{Var}(\mathbf{Y}|\mathbf{X})]\,\mathbf{X}(\mathbf{X}'\mathbf{X})^{-1} \\
&= (\mathbf{X}'\mathbf{X})^{-1}\mathbf{X}'\left[\sigma^2\mathbf{I}\right]\mathbf{X}(\mathbf{X}'\mathbf{X})^{-1} \\
&= \sigma^2(\mathbf{X}'\mathbf{X})^{-1}\mathbf{X}'\mathbf{X}(\mathbf{X}'\mathbf{X})^{-1} \\
&= \sigma^2(\mathbf{X}'\mathbf{X})^{-1}
\end{aligned}
\tag{A.23}
$$

The variances and covariances are compactly determined as σ^2 times a matrix whose elements are determined only by $\mathbf{X}$ and not by $\mathbf{Y}$.

A.8.2 The Residual Sum of Squares

Let $\hat{\mathbf{Y}} = \mathbf{X}\hat{\boldsymbol{\beta}}$ be the $n \times 1$ vector of fitted values corresponding to the n cases in the data, and $\hat{\mathbf{e}} = \mathbf{Y} - \hat{\mathbf{Y}}$ is the vector of residuals. One representation of the residual sum of squares, which is the residual sum of squares function evaluated at $\hat{\boldsymbol{\beta}}$, is

$$
RSS = (\mathbf{Y} - \hat{\mathbf{Y}})'(\mathbf{Y} - \hat{\mathbf{Y}}) = \hat{\mathbf{e}}'\hat{\mathbf{e}} = \sum_{i=1}^{n} \hat{e}_i^2
$$

which suggests that the residual sum of squares can be computed by squaring the residuals and adding them up. In multiple linear regression, it can also be computed more efficiently on the basis of summary statistics. Using (A.19) and the summary statistics $\mathbf{X}'\mathbf{X}$, $\mathbf{X}'\mathbf{Y}$ and $\mathbf{Y}'\mathbf{Y}$, we write

$$
RSS = RSS(\hat{\boldsymbol{\beta}}) = \mathbf{Y}'\mathbf{Y} + \hat{\boldsymbol{\beta}}'\mathbf{X}'\mathbf{X}\hat{\boldsymbol{\beta}} - 2\mathbf{Y}'\mathbf{X}\hat{\boldsymbol{\beta}}
$$

We will first show that $\hat{\beta}'\mathbf{X}'\mathbf{X}\hat{\beta} = \mathbf{Y}'\mathbf{X}\hat{\beta}$. Substituting for one of the $\hat{\beta}$s, we get

$$\hat{\beta}'\mathbf{X}'\mathbf{X}(\mathbf{X}'\mathbf{X})^{-1}\mathbf{X}'\mathbf{Y} = \hat{\beta}'\mathbf{X}'\mathbf{Y} = \mathbf{Y}'\mathbf{X}\hat{\beta}$$

the last result following because taking the transpose of a 1×1 matrix does not change its value. The residual sum of squares function can now be rewritten as

$$RSS = \mathbf{Y}'\mathbf{Y} - \hat{\beta}'\mathbf{X}'\mathbf{X}\hat{\beta}$$
$$= \mathbf{Y}'\mathbf{Y} - \hat{\mathbf{Y}}'\hat{\mathbf{Y}}$$

where $\hat{\mathbf{Y}} = \mathbf{X}\hat{\beta}$ are the fitted values. The residual sum of squares is the difference in the squares of the lengths of the two vectors $\mathbf{Y}$ and $\hat{\mathbf{Y}}$. Another useful form for the residual sum of squares is

$$RSS = SYY(1 - R^2)$$

where R^2 is the square of the sample correlation between $\hat{\mathbf{Y}}$ and $\mathbf{Y}$.

A.8.3   Estimate of Variance

Under the assumption of constant variance, the estimate of σ^2 is

$$\hat{\sigma}^2 = \frac{RSS}{d} \tag{A.24}$$

with d df, where d is equal to the number of cases n minus the number of terms with estimated coefficients in the model. If the matrix $\mathbf{X}$ is of full rank, then $d = n - p'$, where $p' = p$ for mean functions without an intercept, and $p' = p + 1$ for mean functions with an intercept. The number of estimated coefficients will be less than p' if $\mathbf{X}$ is not of full rank.

A.9   THE QR FACTORIZATION

Most of the formulas given in this book are convenient for derivations but can be inaccurate when used on a computer because inverting a matrix such as $(\mathbf{X}'\mathbf{X})$ leaves open the possibility of introducing significant rounding errors into calculations. Most statistical packages will use better methods of computing, and understanding how they work is useful.

   We start with the basic $n \times p'$ matrix $\mathbf{X}$ of terms. Suppose we could find an $n \times p'$ matrix $\mathbf{Q}$ and a $p' \times p'$ matrix $\mathbf{R}$ such that (1) $\mathbf{X} = \mathbf{QR}$; (2) $\mathbf{Q}$ has orthonormal columns, meaning that $\mathbf{Q}'\mathbf{Q} = \mathbf{I}_{p'}$ and (3) $\mathbf{R}$ is an upper triangular matrix, meaning that all the entries in $\mathbf{R}$ below the diagonal are equal to 0, but those on or above the diagonal can be nonzero.

Using the basic properties of matrices, we can write

$$\mathbf{X} = \mathbf{QR}$$

$$\mathbf{X'X} = (\mathbf{QR})'(\mathbf{QR}) = \mathbf{R'R}$$

$$(\mathbf{X'X})^{-1} = (\mathbf{R'R})^{-1} = \mathbf{R}^{-1}(\mathbf{R'})^{-1} \tag{A.25}$$

$$\hat{\beta} = \mathbf{X}(\mathbf{X'X})^{-1}\mathbf{X'Y} = \mathbf{R}^{-1}(\mathbf{Q'Y}) \tag{A.26}$$

$$\mathbf{H} = \mathbf{X}(\mathbf{X'X})^{-1}\mathbf{X'} = \mathbf{QQ'} \tag{A.27}$$

Equation (A.25) follows because $\mathbf{R}$ is a square matrix, and the inverse of the product of square matrices is the product of the inverses in opposite order. From (A.26), to compute $\hat{\beta}$, first compute $\mathbf{Q'Y}$, which is a $p' \times 1$ vector, and multiply on the left by $\mathbf{R}$ to get

$$\mathbf{R}\hat{\beta} = \mathbf{Q'Y} \tag{A.28}$$

This last equation is very easy to solve because $\mathbf{R}$ is a triangular matrix and so we can use *backsolving*. For example, to solve the equations

$$\begin{pmatrix} 7 & 4 & 2 \\ 0 & 2 & 1 \\ 0 & 0 & 1 \end{pmatrix} \hat{\beta} = \begin{pmatrix} 3 \\ 2 \\ 1 \end{pmatrix}$$

first solve the last equation, so $\hat{\beta}_3 = 1$, substitute into the equation above it, so $2\hat{\beta}_2 + 1 = 2$, so $\hat{\beta}_2 = 1/2$. Finally, the first equation is $7\hat{\beta}_1 + 2 + 2 = 3$, so $\hat{\beta}_3 = -1/7$.

Equation (A.27) shows how the elements of the $n \times n$ hat matrix $\mathbf{H}$ can be computed without inverting a matrix and without using all the storage needed to save $\mathbf{H}$ in full. If $\mathbf{q}_i$ is the ith column of $\mathbf{Q}$, then an element h_{ij} of the $\mathbf{H}$ matrix is simply computed as $h_{ij} = \mathbf{q}'_i \mathbf{q}_j$.

Golub and Van Loan (1996) provide a complete treatment on computing and using the $\mathbf{QR}$ factorization. Very high quality computer code for computing this and related quantities for statistics is provided in the publicly available Lapack package, described on the internet at www.netlib.org/lapack/lug/. This code is also used in many standard statistical packages.

A.10 MAXIMUM LIKELIHOOD ESTIMATES

Maximum likelihood estimation is probably the most frequently used method of deriving estimates in statistics. A general treatment is given by Casella and Berger (1990, Section 7.2.2); here we derive the maximum likelihood estimates for the linear regression model assuming normality, without proof or much explanation. Our goal is to establish notation and define quantities that will be used in the discussion of Box–Cox transformations, and estimation for generalized linear models in Chapter 12.

The normal multiple linear regression model specifies for the ith observation that

$$(y_i | \mathbf{x}_i) \sim N(\boldsymbol{\beta}' \mathbf{x}_i, \sigma^2)$$

Given this model, the density for the ith observation y_i is the normal density function,

$$f_{y_i}(y_i | \mathbf{x}_i, \boldsymbol{\beta}, \sigma^2) = \frac{1}{\sqrt{2\pi}\sigma} \exp\left(-\frac{(y_i - \boldsymbol{\beta}' \mathbf{x}_i)^2}{2\sigma^2}\right)$$

Assuming the observations are independent, the likelihood function is just the product of the densities for each of the n observations, viewed as a function of the parameters with the data fixed rather than a function of the data with the parameters fixed:

$$L(\boldsymbol{\beta}, \sigma^2 | Y) = \prod_{i=1}^{n} f_{y_i}(y_i | \mathbf{x}_i, \boldsymbol{\beta}, \sigma^2)$$

$$= \prod_{i=1}^{n} \frac{1}{\sqrt{2\pi}\sigma} \exp\left(-\frac{(y_i - \boldsymbol{\beta}' \mathbf{x}_i)^2}{2\sigma^2}\right)$$

$$= \left(\frac{1}{\sqrt{2\pi}\sigma}\right)^n \exp\left(-\frac{1}{\sigma^2} \sum_{i=1}^{n}(y_i - \boldsymbol{\beta}' \mathbf{x}_i)^2\right)$$

The maximum likelihood estimates are simply the values of $\boldsymbol{\beta}$ and σ^2 that maximize the likelihood function.

The values that maximize the likelihood will also maximize the logarithm of the likelihood

$$\log\left(L(\boldsymbol{\beta}, \sigma^2 | Y)\right) = -\frac{n}{2}\log(2\pi) - \frac{n}{2}\log(\sigma^2) - \frac{1}{2\sigma^2}\sum_{i=1}^{n}(y_i - \boldsymbol{\beta}' \mathbf{x}_i)^2 \quad \text{(A.29)}$$

The log-likelihood function (A.29) is a sum of three terms. Since $\boldsymbol{\beta}$ is included only in the third term and this term has a negative sign in front of it, we recognize that maximizing the log-likelihood over $\boldsymbol{\beta}$ is the same as minimizing the third term, which, apart from constants, is the same as the residual sum of squares function (see Section 3.4.3). We have just shown that the maximum likelihood estimate of $\boldsymbol{\beta}$ for the normal linear regression problem is the same as the OLS estimator. Fixing $\boldsymbol{\beta}$ at the OLS estimator $\hat{\boldsymbol{\beta}}$, (A.29) becomes

$$\log\left(L(\hat{\boldsymbol{\beta}}, \sigma^2 | Y)\right) = -\frac{n}{2}\log(2\pi) - \frac{n}{2}\log(\sigma^2) - \frac{1}{2\sigma^2}RSS \quad \text{(A.30)}$$

and differentiating (A.30) with respect to σ^2 and setting the result to 0 gives the maximum likelihood estimator for σ^2 as RSS/n, the same estimate we have been using, apart from division by n rather than $n - p'$.

Maximum likelihood estimation has many important properties that make them useful. These estimates are approximately normally distributed in large samples, and the large-sample variance achieves the lower bound for the variance of all unbiased estimates.

A.11 THE BOX–COX METHOD FOR TRANSFORMATIONS

A.11.1 Univariate Case

Box and Cox (1964) derived the Box–Cox method for selecting a transformation using a likelihood-like method. They supposed that, for some value of λ, $\psi_M(Y, \lambda)$ given by (7.6), page 153, is normally distributed. With n independent observations, therefore, the log-likelihood function for $(\beta, \sigma^2, \lambda)$ is given by (A.29), but with y_i replaced by $\psi_M(Y, \lambda)$[1],

$$\log\left(L(\beta, \sigma^2, \lambda | Y)\right) = -\frac{n}{2}\log(2\pi) - \frac{n}{2}\log(\sigma^2) - \frac{1}{2\sigma^2}\sum_{i=1}^{n}(\psi_M(y_i, \lambda) - \beta'x_i)^2$$

(A.31)

For a fixed value of λ, (A.31) is the same as (A.29), and so the maximum likelihood estimates for β and σ^2 are obtained from the regression of $\psi_M(Y, \lambda)$ on X, and the value of the log-likelihood evaluated at these estimates is

$$\log\left(L(\beta(\lambda), \sigma^2(\lambda), \lambda | Y)\right) = -\frac{n}{2}\log(2\pi) - \frac{n}{2}\log(RSS(\lambda)/n) - \frac{n}{2} \qquad (A.32)$$

where $RSS(\lambda)$ is the residual sum of squares in the regression of $\psi_M(Y, \lambda)$ on X, as defined in Section 7.1.4. Only the second term in (A.32) involves data, and so the global maximum likelihood estimate of λ minimizes $RSS(\lambda)$.

Standard likelihood theory can be applied to get a $(1 - \alpha) \times 100\%$ confidence interval for λ to be the set

$$\left\{\lambda \Big| 2\left[\log(L(\beta(\hat{\lambda}), \sigma^2(\hat{\lambda}), \hat{\lambda} | Y)) - \log(L(\beta(\lambda), \sigma^2(\lambda), \lambda | Y))\right] < \chi^2(1, 1 - \alpha)\right\}$$

Or, setting $\alpha = .05$ so $\chi^2(1, .95) = 3.84$, and using (A.32)

$$\left\{\lambda | (n/2)(\log(RSS(\lambda)) - \log(RSS(\hat{\lambda}))) < 1.92\right\} \qquad (A.33)$$

---

[1] As λ is varied, the *units* of $\psi_M(Y, \lambda)$ can change, and so the joint density of the transformed data should require a Jacobian term; see Casella and Berger (1990, Section 4.3). The modified power transformations are defined so the Jacobian of the transformation is always equal to 1, and it can therefore be ignored.

Many statistical packages will have routines that will provide a graph of $RSS(\lambda)$ versus λ, or of $(n/2)\log(RSS(\lambda))$ versus λ as shown in Figure 7.8, for the highway accident data. Equation (A.32) shows that the confidence interval for λ includes all values of λ for which the log-likelihood is within 1.92 units of the maximum value of the log-likelihood, or between the two vertical lines in the figure.

A.11.2 Multivariate Case

Although the material in this section uses more mathematical statistics than most of this book, it is included because the details of computing the multivariate extension of Box–Cox transformations are not published elsewhere. The basic idea was proposed by Velilla (1993).

Suppose X is a set of p variables we wish to transform and define

$$\psi_M(X, \lambda) = (\psi_M(X_1, \lambda_1), \ldots, \psi_M(X_k, \lambda_k))$$

Here, we have used the modified power transformations (7.6) for each element of X, but the same general idea can be applied using other transformations such as the Yeo–Johnson family introduced in Section 7.4. In analogy to the univariate case, we assume that for some λ, we will have

$$\psi_M(X, \lambda) \sim N(\mu, V)$$

where V is some unknown positive definite symmetric matrix that needs to be estimated. If $\mathbf{x}_i$ is the observed value of X for the ith observation, then the likelihood function is given by

$$L(\mu, V, \lambda | X) = \prod_{i=1}^{n} \frac{1}{(2\pi |V|)^{1/2}}$$
$$\times \exp\left(-\frac{1}{2}(\psi_M(\mathbf{x}_i, \lambda) - \mu)' V^{-1}(\psi_M(\mathbf{x}_i, \lambda) - \mu)\right) \text{ (A.34)}$$

where $|V|$ is the determinant[2]. After rearranging terms, the log-likelihood is given by

$$\log(L(\mu, V, \lambda | X)) = -\frac{n}{2}\log(2\pi) - \frac{n}{2}\log(|V|)$$
$$-\frac{1}{2}\sum_{i=1}^{n} V^{-1}(\psi_M(\mathbf{x}_i, \lambda) - \mu)(\psi_M(\mathbf{x}_i, \lambda) - \mu)' \text{ (A.35)}$$

If we fix λ, then (A.35) is the standard log-likelihood for the multivariate normal distribution. The values of V and μ that maximize (A.35) are the sample mean and

---

[2]The determinant is defined in any linear algebra textbook.

sample covariance matrix, the latter with divisor n rather than $n - 1$,

$$\mu(\lambda) = \frac{1}{n} \sum_{i=1}^{n} \psi_M(\mathbf{x}_i, \lambda)$$

$$V(\lambda) = \frac{1}{n} \sum_{i=1}^{n} (\psi_M(\mathbf{x}_i, \lambda) - \mu(\lambda))(\psi_M(\mathbf{x}_i, \lambda) - \mu(\lambda))'$$

Substituting these estimates into (A.35) gives the profile log-likelihood for λ,

$$\log(L(\mu(\lambda), V(\lambda), \lambda | X)) = -\frac{n}{2} \log(2\pi) - \frac{n}{2} \log(|V(\lambda)|) - \frac{n}{2} \qquad \text{(A.36)}$$

This equation will be maximized by minimizing the determinant of $V(\lambda)$ over values of λ. This is a numerical problem for which there is no closed-form solution, but it can be solved using a general-purpose function minimizer.

Standard theory for maximum likelihood estimates can provide tests concerning λ and standard errors for the elements of λ. To test the hypothesis that $\lambda = \lambda_0$ against a general alternative, compute

$$G^2 = 2\left[\log(L(\mu(\hat{\lambda}), V(\hat{\lambda}), \hat{\lambda})) - \log(L(\mu(\lambda_0), V(\lambda_0), \lambda_0))\right]$$

and compare G^2 to a Chi-squared distribution with k df. The standard error of $\hat{\lambda}$ is obtained from the inverse of the expected information matrix evaluated at $\hat{\lambda}$. The expected information for $\hat{\lambda}$ is just the matrix of second derivatives of (A.36) with respect to λ evaluated at $\hat{\lambda}$. Many optimization routines, such as `optim` in R, will return the matrix of estimated second derivatives if requested; all that is required is inverting this matrix, and then the square roots of the diagonal elements are the estimated standard errors.

A.12   CASE DELETION IN LINEAR REGRESSION

Suppose $\mathbf{X}$ is the $n \times p'$ matrix of terms with linearly independent columns. We use the subscript "(i)" to mean "without case i," so that $\mathbf{X}_{(i)}$ is an $(n - 1) \times p'$ matrix. We can compute $(\mathbf{X}_{(i)}'\mathbf{X}_{(i)})^{-1}$ from the remarkable formula

$$(\mathbf{X}_{(i)}'\mathbf{X}_{(i)})^{-1} = (\mathbf{X}'\mathbf{X})^{-1} + \frac{(\mathbf{X}'\mathbf{X})^{-1}\mathbf{x}_i\mathbf{x}_i'(\mathbf{X}'\mathbf{X})^{-1}}{1 - h_{ii}} \qquad \text{(A.37)}$$

where $h_{ii} = \mathbf{x}_i'(\mathbf{X}'\mathbf{X})^{-1}\mathbf{x}_i$ is the ith leverage value, a diagonal value from the hat matrix. This formula was used by Gauss (1821); a history of it and many variations is given by Henderson and Searle (1981). It can be applied to give all the results that one would want relating multiple linear regression with and without the ith case. For example,

$$\hat{\boldsymbol{\beta}}_{(i)} = \hat{\boldsymbol{\beta}} - \frac{(\mathbf{X}'\mathbf{X})^{-1}\mathbf{x}_i \hat{e}_i}{1 - h_{ii}} \tag{A.38}$$

Writing $r_i = \hat{e}_i / \hat{\sigma} \sqrt{1 - h_{ii}}$, the estimate of variance is

$$\hat{\sigma}^2_{(i)} = \hat{\sigma}^2 \left(\frac{n - p' - 1}{n - p' - r_i^2} \right)^{-1} \tag{A.39}$$

and the studentized residual t_i is

$$t_i = r_i \left(\frac{n - p' - 1}{n - p' - r_i^2} \right)^{1/2} \tag{A.40}$$

The diagnostic statistics examined in this book were first thought to be practical because of simple formulas used to obtain various statistics when cases are deleted that avoided recomputing estimates. Advances in computing in the last 20 years or so have made the computational burden of recomputing without a case much less onerous, and so diagnostic methods equivalent to those discussed here can be applied to problems other than linear regression where the updating formulas are not available.

References

Agresti, A. (1996). *An Introduction to Categorical Data Analysis*. New York: Wiley.

Agresti, A. (2002). *Categorical Data Analysis*, Second Edition. New York: Wiley.

Allen, D. M. (1974). The relationship between variable selection and prediction. *Technometrics*, 16, 125–127.

Allison, T. and Cicchetti, D. V. (1976). Sleep in mammals: Ecological and constitutional correlates. *Science*, 194, 732–734.

Anscombe, F. (1973). Graphs in statistical analysis. *Am. Stat.*, 27, 17–21.

Atkinson, A. C. (1985). *Plots, Transformations and Regression*. Oxford: Oxford University Press.

Baes, C. and Kellogg, H. (1953). Effects of dissolved sulphur on the surface tension of liquid copper. *J. Metals*, 5, 643–648.

Barnett, V. and Lewis, T. (2004). *Outliers in Statistical Data*, Third Edition. Chichester: Wiley.

Bates, D. and Watts, D. (1988). Relative curvature measures of nonlinearity (with discussions). *J. R. Stat. Soc. Ser. B*, 22, 41–88.

Beckman, R. and Cook, R. D. (1983). Outliers. *Technometrics*, 25, 119–149.

Bland, J. (1978). A comparison of certain aspects of ontogeny in the long and short shoots of McIntosh apple during one annual growth cycle. Unpublished Ph. D. Dissertation, University of Minnesota, St. Paul.

Blom, G. (1958). *Statistical Estimates and Transformed Beta Variates*. New York: Wiley.

Bowman, A. and Azzalini, A. (1997). *Applied Smoothing Techniques for Data Analysis*. Oxford: Oxford University Press.

Box, G. E. P. and Cox, D. R. (1964). An analysis of transformations. *J. R. Stat. Soc., Ser. B*, 26, 211–246.

Brillinger, D. (1983). A generalized linear model with "Gaussian" regression variables. In Bickel, P. J., Doksum, K. A., and Hodges Jr., J. L., eds., *A Festschrift for Erich L. Lehmann*. New York: Chapman & Hall, 97–114.

Brown, P. (1994). *Measurement, Regression and Calibration*. Oxford: Oxford University Press.

Applied Linear Regression, Third Edition, by Sanford Weisberg
ISBN 0-471-66379-4   Copyright © 2005 John Wiley & Sons, Inc.

Burt, C. (1966). The genetic determination of differences in intelligence: A study of monozygotic twins reared together and apart. *Br. J. Psychol.*, 57, 137–153.

Casella, G. and Berger, R. (1990). *Statistical Inference*. Pacific Grove: Wadsworth & Brooks-Cole.

Chen, C. F. (1983). Score tests for regression models. *J. Am. Stat. Assoc.*, 78, 158–161.

Clapham, A. W. (1934). *English Romanesque Architecture After the Conquest*. Oxford: Clarendon Press.

Clark, R., Henderson, H. V., Hoggard, G. K., Ellison, R., and Young, B. (1987). The ability of biochemical and haematological tests to predict recovery in periparturient recumbent cows. *N. Z. Vet. J.*, 35, 126–133.

Clausius, R. (1850). Über die bewegende Kraft der Wärme und die Gezetze welche sich daraus für die Wärmelehre selbst abeiten lassen. *Annalen der Physik*, 79, 500–524.

Cleveland, W. (1979). Robust locally weighted regression and smoothing scatterplots. *J. of the Amer. Stat. Assoc.*, 74, 829–836.

Collett, D. (2002). *Modelling Binary Data*, Second Edition. Boca Raton: CRC Press.

Cook, R. D. (1977). Detection of influential observations in linear regression. *Technometrics*, 19, 15–18.

Cook, R. (1979). Influential observations in linear regression. *J. Amer. Stat. Assoc.*, 74, 169–174.

Cook, R. D. (1986). Assessment of local influence (with discussion). *J. R. Stat. Soc., Ser. B*, 48, 134–169.

Cook, R. D. (1998). *Regression Graphics: Ideas for Studying Regressions through Graphics*. New York: Wiley.

Cook, R. and Jacobson, J. (1978). Analysis of 1977 West Hudson Bay snow goose surveys. Unpublished Report, Canadian Wildlife Services.

Cook, R. D. and Prescott, P. (1981). Approximate significance levels for detecting outliers in linear regression. *Technometrics*, 23, 59–64.

Cook, R. D. and Weisberg, S. (1982). *Residuals and Influence in Regression*. London: Chapman & Hall.

Cook, R. D. and Weisberg, S. (1983). Diagnostics for heteroscedasticity in regression. *Biometrika*, 70, 1–10.

Cook, R. D. and Weisberg, S. (1994). Transforming a response variable for linearity. *Biometrika*, 81, 731–737.

Cook, R. D. and Weisberg, S. (1997). Graphics for assessing the adequacy of regression models. *J. of the Amer. Stat. Assoc.*, 92, 490–499.

Cook, R. D. and Weisberg, S. (1999a). *Applied Regression Including Computing and Graphics*. New York: Wiley.

Cook, R. D. and Weisberg, S. (1999b). Graphs in statistical: Analysis: Is the medium the message? *Amer. Stat.*, 53, 29–37.

Cook, R. D. and Weisberg, S. (2004). Partial one-dimensional regression models. *Amer. Stat.*, 58, 110–116.

Cook, R. D. and Witmer, J. (1985). A note on parameter-effects curvature. *Journal of the American Statistical Association*, 80, 872–878.

Cox, D. R. (1958). *The Planning of Experiments*. New York: Wiley.

Cox, D. R. and Oakes, D. (1984). *Analysis of Survival Data*. London: Chapman & Hall.

Cunningham, R. and Heathcote, C. (1989). Estimating a non-Gaussian regression model with multicollinearity. *Australian Journal of Statistics*, 31, 12–17.

Dalziel, C., Lagen, J., and Thurston, J. (1941). Electric shocks. *Transactions of the IEEE*, 60, 1073–1079.

Daniel, C. and Wood, F. (1980). *Fitting Equations to Data*, Second Edition. New York: Wiley.

Davison, A. and Hinkley, D. (1997). *Bootstrap Methods and their Application*. Cambridge: Cambridge University Press.

Dawson, R. (1995). The "Unusual Episode" Data Revisited. *Journal of Statistical Education*, 3, an electronic journal available at www.stat.ncsu.edu/info/jse.

Dempster, A., Laird, N., and Rubin, D. (1977). Maximum likelihood estimation from incomplete data via the EM algorithm. *J. R. Stat. Soc., Ser. B*, 29, 1–38.

Derrick, A. (1992). Development of the measure-correlate-predict strategy for site assessment. *Proceedings of the 14th BWEA Conference*, Nottingham, 259–265.

Diggle, P., Heagerty, P., Liang, K. Y., and Zeger, S. (2002). *Analysis of Longitudinal Data*, Second Edition. Oxford: Oxford University Press.

Dodson, S. (1992), Predicting crustacean zooplankton species richness. *Limnology and Oceanography*, 37, 848–856.

Efron, B. (1979). Bootstrap methods: another look at the jackknife. *Ann. Stat.*, 7, 1–26.

Efron, B. and Tibshirani, R. (1993). *An Introduction to the Bootstrap*. Boca Raton: Chapman & Hall.

Ezekiel, M. and Fox, K. A. (1959). *Methods of Correlation Analysis, Linear and Curvilinear*. New York: Wiley.

Finkelstein, M. (1980). The judicial reception of multiple regression studies in race and sex discrimination cases. *Columbia Law Review*, 80, 734–757.

Flury, B. and Riedwyl, H. (1988). *Multivariate Statistics: A Practical Approach*. London: Chapman & Hall.

Forbes, J. (1857). Further experiments and remarks on the measurement of heights by boiling point of water. *Trans. R. Soc. Edinburgh*, 21, 235–243.

Freedman, D. (1983). A note on screening regression equations. *Amer. Stat.*, 37, 152–157.

Freeman, M. and Tukey, J. (1950). Transformations related to angular and the square root. *Ann. Math. Stat.*, 21, 607–677.

Fuller, W. (1987). *Measurement Error Models*. New York: Wiley.

Furnival, G. and Wilson, R. (1974). Regression by leaps and bounds. *Technometrics*, 16, 499–511.

Gauss, C. (1821–1826). Theoria Combinationis Observationum Erroribus Minimis Obnoxiae (Theory of the combination of observations which leads to the smallest errors). *Werke*, 4, 1–93.

Geisser, S. and Eddy, W. F. (1979). A predictive approach to model selection. *J. Am. Stat. Assoc.*, 74, 753–160.

Gnanadesikan, R. (1997). *Methods for Statistical Analysis of Multivariate Data*, Second Edition. New York: Wiley.

Golub, G. and van Loan, C. (1996). *Matrix Computations*, Third Edition. Baltimore: Johns Hopkins.

Gould, S. J. (1966). Allometry and size in ontogeny and phylogeny. *Biol. Rev.*, 41, 587–640.

Gould, S. J. (1973). The shape of things to come. *Syst. Zool.*, 22, 401–404.

Graybill, F. (1969). *Introduction to Matrices with Statistical Applications.* Belmont, CA: Wadsworth.

Green, P. and Silverman, B. (1994). *Nonparametric Regression and Generalized Linear Models: A Roughness Penalty Approach.* London: Chapman & Hall.

Haddon, M. (2001). *Modelling and Quantitative Methods in Fisheries.* Boca Raton: Chapman & Hall.

Hahn, M., ed. (1979). *Development and Evolution of Brain Size.* New York: Academic Press.

Hald, A. (1960). *Statistical Theory with Engineering Applications.* New York: Wiley.

Hall, P. and Li, K. C. (1993). On almost linearity of low dimensional projections from high dimensional data. *Ann. Stat.*, 21, 867–889.

Härdle, W. (1990). *Applied Nonparametric Regression.* Cambridge: Cambridge University Press.

Hastie, T., Tibshirani, R., and Friedman, J. (2001). *The Elements of Statistical Learning.* New York: Springer.

Hawkins, D. M. (1980). *Identification of Outliers.* London: Chapman & Hall.

Hawkins, D. M., Bradu, D., and Kass, G. (1984). Locations of several outliers in multiple regression using elemental sets. *Technometrics*, 26, 197–208.

Henderson, H. V. and Searle, S. R. (1981). On deriving the inverse of a sum of matrices. *SIAM Rev.*, 23, 53–60.

Hernandez, F. and Johnson, R. A. (1980). The large sample behavior of transformations to normality. *J. of the Amer. Stat. Assoc.*, 75, 855–861.

Hinkley, D. (1985). Transformation diagnostics for linear models. *Biometrika*, 72, 487–496.

Hoaglin, D. C. and Welsch, R. (1978). The hat matrix in regression and ANOVA. *Am. Stat.*, 32, 17–22.

Hosmer, D. and Lemeshow, S. (2000). *Applied Logistic Regression*, Second Edition. New York: Wiley.

Huber, P. (1981). *Robust Statistics.* New York: Wiley.

Ibrahim, J., Lipsitz, S., and Horton, N. (2001). Using auxiliary data for parameter estimation with non-ignorably missing outcomes. *Applied Statistics*, 50, 361–373.

Jevons, W. S. (1868). On the condition of the gold coinage of the United Kingdom, with reference to the question of international currency. *J. [R.] Stat. Soc.*, 31, 426–464.

Johnson, K. (1996). *Unfortunate Emigrants: Narratives of the Donner Party.* Logan: Utah State University Press.

Johnson, M. P. and Raven, P. H. (1973). Species number and endemism: The Galápagos Archipelago revisted. *Science*, 179, 893–895.

Joiner, B. (1981). Lurking variables: Some examples. *Amer. Stat.*, 35, 227–233.

Kalbfleisch, J. D. and Prentice, R. L. (1980). *The Statistical Analysis of Failure Time Data.* New York: Wiley.

Kennedy, W. and Bancroft, T. (1971). Model building for prediction in regression based on repeated significance tests. *Ann. Math. Stat.*, 42, 1273–1284.

LeBeau, M. (2004). Evaluation of the intraspecific effects of a 15-inch minimum size limit on walleye populations in Northern Wisconsin. Unpublished Ph. D. Dissertation, University of Minnesota.

Li, K. C. and Duan, N. (1989). Regression analysis under link violation. *Ann. Stat.*, 17, 1009–1052.

Lindgren, B. L. (1993). *Statistical Theory*, Fourth Edition. New York: Macmillan.

Loader, C. (2004). Smoothing: Local regression techniques. In Gentle, J., Härdle, W., Mori, Y., *Handbook of Computational Statistics*. New York: Springer-Verlag.

Littell, R., Milliken, G., Stroup, W.Wolfinger, R. (1996). *SAS System for Mixed Models*. Cary: SAS Institute.

Little, R. and Rubin, D. (1987). *Statistical Analysis with Missing Data*. New York: Wiley.

Longley, J. (1967). An appraisal of least squares programs for the electronic computer from the point of view of the user. *J. Am. Stat. Assoc.*, 62, 819–841.

Mallows, C. (1973). Some comments on C_p. *Technometrics*, 15, 661–676.

Mantel, N. (1970). Why stepdown procedures in variable selection? *Technometrics*, 12, 621–625.

Marquardt, D. W. (1970). Generalized inverses, ridge regression and biased linear estimation. *Technometrics*, 12, 591–612.

McCullagh, P. and Nelder, J. (1989). *Generalized Linear Models*, Second Edition. London: Chapman & Hall.

Miller, R. (1981). *Simultaneous Inference*, Second Edition. New York: Springer.

Moore, J. A. (1975). Total biomedical oxygen demand of animal manures. Unpublished Ph. D. Dissertation, University of Minnesota.

Mosteller, F. and Wallace, D. (1964). *Inference and Disputed Authorship: The Federalist*. Reading: Addison-Wesley.

Nelder, J. A. and Wedderburn, R. W. M. (1972). Generalized linear models. *Journal of the Royal Statistical Society, Series A*, 135, 370–384.

Noll, S., Weibel, P., Cook, R. D., and Witmer, J. (1984). Biopotency of methionine sources for young turkeys. *Poultry Science*, 63, 2458–2470.

Oehlert, G. (2000). *A First Course in the Design and Analysis of Experiments*. New York: W.H. Freeman.

Pan, Wei, Connett, J., Porzio, G., and Weisberg, S. (2001). Graphical model checking using marginal model plots with longitudinal data. *Statistics in Medicine*, 20, 2935–2945.

Pardoe, I. and Weisberg, S. (2001). An Introduction to bootstrap methods using Arc. Unpublished Report available at www.stat.umn.edu/arc/bootmethREV.pdf.

Parks, J. (1982). *A Theory of Feeding and Growth of Animals*. New York: Springer.

Pearson, K. (1930). *Life and Letters and Labours of Francis Galton*, Vol IIIa. Cambridge: Cambridge University Press.

Pinheiro, J. and Bates, D. (2000). *Mixed-Effects Models in S and S-plus*. New York: Springer.

Pearson, E. S. and Lee, S. (1903). On the laws of inheritance in man. *Biometrika*, 2, 357–463.

Porzio, G. (2002). A simulated band to check binary regression models. *Metron*, 60, 83–95.

Ratkowsky, D. A. (1990). *Handbook of Nonlinear Regression Models*. New York: Marcel Dekker.

Rencher, A. and Pun, F. (1980). Inflation of R^2 in best subset regression. *Technometrics*, 22, 49–53.

Royston, J. P. (1982a). An extension of Shapiro and Wilk's W test for normality to large samples. *Appl. Stat.*, 31, 115–124.

Royston, J. P. (1982b). Expected normal order statistics (exact and approximate), algorithm AS177. *Appl. Stat.*, 31, 161–168.

Royston, J. P. (1982c). The *W* test for normality, algorithm AS181. *Appl. Stat.*, 31, 176–180.

Royston, P. and Altman, D. (1994). Regression using fractional polynomials of continuous covariates: parsimonious parametric modeling. *Applied Statistics*, 43, 429–467.

Rubin, D. (1976). Inference and missing data. *Biometrika*, 63, 581–592.

Ruppert, D., Wand, M., Holst, U., and Hössjer, O. (1997). Local polynomial variance-function estimation. *Technometrics*, 39, 262–273.

Sakamoto, Y., Ishiguro, M., and Kitagawa, G. (1987). *Akaike Information Criterion Statistics*. Dordrecht: Reidel.

Saw, J. (1966). A conservative test for concurrence of several regression lines and related problems. *Biometrika*, 53, 272–275.

Schafer, J. (1997). *Analysis of Incomplete Multivariate Data*. Boca Raton: Chapman & Hall/CRC.

Scheffé, H. (1959). *The Analysis of Variance*. New York: Wiley.

Sclove, S. (1968). Improved estimators for coefficients in linear regression. *J. Am. Stat. Assoc.*, 63, 596–606.

Schott, J. (1996). *Matrix Analysis for Statistics*. New York: Wiley.

Schwarz, G. (1978). Estimating the dimension of a model. *Annals of Statistics*, 6, 461–464.

Searle, S. R. (1971). *Linear Models*. New York: Wiley.

Searle, S. R. (1982). *Matrix Algebra Useful for Statistics*. New York: Wiley.

Seber, G. A. F. (1977). *Linear Regression Analysis*. New York: Wiley.

Seber, G. A. F. (2002). *The Estimation of Animal Abundance*. Caldwell, NJ: Blackburn Press.

Seber, G. and Wild, C. (1989). *Nonlinear Regression*. New York: Wiley.

Silverman, B. (1986). *Density Estimation for Statistics and Data Analysis*. Boca Raton: CRC/Chapman & Hall.

Simonoff, J. (1996). *Smoothing Methods in Statistics*. New York: Springer-Verlag.

Shapiro, S. S. and Wilk, M. B. (1965). An analysis of variance test for normality (complete samples). *Biometrika*, 52, 591–611.

Staudte, R. and Sheather, S. (1990). *Robust Estimation and Testing*. New York: Wiley.

Taff, S., Tiffany, D., and Weisberg, S. (1996). Measured effects of feedlots on residential property values in Minnesota: A report to the legislature. Technical Report, Department of Applied Economics, University of Minnesota, agecon.lib.umn.edu/mn/p96-12.pdf.

Tang, G., Little, R., and Raghunathan, T. (2003). Analysis of multivariate missing data with nonignorable nonresponse. *Biometrika*, 90, 747–764.

Thisted, R. (1988). *Elements of Statistical Computing*. New York: Chapman & Hall.

Tuddenham, R. and Snyder, M. (1954). Physical growth of California boys and girls from birth to age 18. *Calif. Publ . Child Dev.*, 1, 183–364.

Tukey, J. (1949). One degree of freedom for nonadditivity. *Biometrics*, 5, 232–242.

Velilla, S. (1993). A note on the multivariate Box-Cox transformation to normality. *Stat. Probab. Lett.*, 17, 259–263.

Verbeke, G. and Molenberghs, G. (2000). *Linear Mixed Models for Longitudinal Data*. New York: Springer-Verlag.

von Bertalanffy, L. (1938). A quantitative theory of organic growth. *Human Biology*, 10, 181–213.

Weisberg, H., Beier, H., Brody, H., Patton, R., Raychaudhari, K., Takeda, H., Thern, R., and Van Berg, R. (1978). *s*-dependence of proton fragmentation by hadrons. II. Incident laboratory momenta, 30–250 GeV/c. *Phys. Rev. D*, 17, 2875–2887.

Wilm, H. (1950). Statistical control in hydrologic forecasting. Research Notes, 71, Pacific Northwest Forest Range Experiment Station, Oregon.

Woodley, W. L., Simpson, J., Biondini, R., and Berkeley, J. (1977). Rainfall results 1970–75: Florida area cumulus experiment. *Science*, 195, 735–742.

Yeo, I. and Johnson, R. (2000). A new family of power transformations to improve normality or symmetry. *Biometrika*, 87, 954–959.

Zipf, G. (1949). *Human Behavior and the Principle of Least-Effort*. Cambridge: Addison-Wesley.

Author Index

Agresti, A., 100, 265, 293
Ahlstrom, M., 45
Allen, D. M., 220, 293
Allison, T., 86, 148, 293
Altman, D., 122, 298
Anscombe, F., 26, 12, 293
Atkinson, A. C., 205, 293
Azzalini, A., 112, 251, 277, 293

Baes, C., 161, 293
Bancroft, T., 117, 296
Barnett, V., 197, 293
Bates, D., 100, 136, 177, 185, 248, 293, 297
Beckman, R., 197, 293
Beier, H., 299
Berger, R., 27, 80, 287, 289, 294
Berkeley, J., 209, 299
Bickel, P. J., 293
Biondini, R., 209, 299
Bland, J., 104, 293
Blom, G., 205, 293
Bowman, A., 112, 251, 277, 293
Box, G. E. P., 130, 153, 289, 293
Bradu, D., 197, 296
Brillinger, D., 155, 293
Brody, H., 299
Brown, P., 143, 293
Burnside, O., 17
Burt, C., 138, 294

Campeyron, E., 39
Casella, G., 27, 80, 287, 289, 294
Chen, C. F., 180, 294
Cicchetti, D., 86, 148, 293

Clapham, A. W., 140, 294
Clark, R., 266, 294
Clausius, R., 39
Cleveland, W., 14, 276, 294
Collett, D., 265–266, 294
Connett, J., 191, 297
Cook, R. D., 8, 49, 113, 131, 152, 156–157, 159, 172–173, 180, 186, 197–198, 204, 238, 293–294, 297
Cox, D. R., 294, 78, 85, 130, 153, 289, 293–294
Cunningham, R., 88, 132, 294

Dalal, S., 268
Dalziel, C., 267, 295
Daniel, C., 205, 295
Davison, A., 87, 244, 295
Dawson, R., 262, 295
Dempster, A., 85, 295
Derrick, A., 140, 295
Diggle, P., 137, 295
Dodson, S., 193, 295
Doksum, K. A., 293
Duan, N., 155–156, 296

Eddy, W., 220, 295
Efron, B., 87, 295
Ellison, R., 266, 294
Ezekiel, M., 162, 295

Finkelstien, M., 295
Flury, B., 268, 295
Forbes, J., 4, 6, 295
Fowlkes, E. B., 268
Fox, K. A., 162, 295

Applied Linear Regression, Third Edition, by Sanford Weisberg
ISBN 0-471-66379-4   Copyright © 2005 John Wiley & Sons, Inc.

Freedman, D., 225, 295
Freeman, M., 179, 295
Frie, R., 249
Friedman, J., 211, 296
Fuller, W., 90, 295
Furnival, G., 221, 295

Galatowitsch, S., 134
Galto, F., 110
Gauss, C., 291, 295
Geisser, S., 220, 295
Gentle, J., 297
Gnanadesikan, R., 204, 295
Golub, G., 287, 295
Gosset, W. S., 196
Gould, S. J., 140, 150, 295
Graybill, F., 278, 296
Green, P., 14, 296

Haddon, M., 248, 296
Hahn, M., 150, 296
Hald, A., 162, 296
Hall, P., 156, 296
Härdle, W., 14, 276, 296–297
Hastie, T., 211, 296
Hawkins, D. M., 197, 296
Heagerty, P., 137, 295
Heathcote, C., 88, 294
Henderson, H. V., 266, 291, 294, 296
Hernandez, F., 153, 296
Hinkley, D., 87, 184, 244, 295–296
Hoadley, B., 268
Hoaglin, D. C., 169, 296
Hodges Jr., J. L., 293
Hoffstedt, C., 153
Hoggard, G. K., 266, 294
Holst, U., 277, 298
Horton, N., 86, 296
Hosmer, D., 266, 296
Hössjer, O., 277, 298
Hubbard, K., 17
Huber, P., 27, 296
Hutchinson, R., 18

Ibrahim, J., 86, 296
Ishiguro, M., 217, 298

Jacobson, J., 113, 294
Jevons, W. S., 114, 296
Johnson, K., 267, 296
Johnson, M. P., 231
Johnson, R. A., 153, 296, 299
Joiner, B., 79, 296

Kalbfleisch, J. D., 85, 296
Kass, G., 197, 296
Kellogg, H., 161, 293
Kennedy, W., 117, 296
Kitagawa, G., 217, 298

Lagen, J., 295
Laird, N., 85, 295
LeBeau, M., 249, 296
Lee, S., 2, 297
Lemeshow, S., 266, 296
Lewis, T., 197, 293
Li, K. C., 155–156, 296
Liang, K. Y., 137, 295
Lindgren, B., 80, 297
Lipsitz, S., 86, 296
Littell, R., 136, 297
Little, R., 84, 86, 297–298
Loader, C., 111–112, 276–277, 297
Longley, J., 94, 297

Mallows, C., 218, 297
Mantel, N., 230, 297
Marquardt, D. W., 216, 297
McCullagh, P., 100, 265–266, 297
Miklovic, S., 134
Miller, R., 45, 196, 297
Milliken, G., 136, 297
Molenberghs, G., 136, 298
Moore, J., 231, 297
Mori, Y., 297
Mosteller, F., 44, 297

Nelder, J. A., 100, 265–266, 297
Neyman, J., 235
Ng, C., 244
Noll, S., 8, 297
Norell, R., 267

Oakes, D., 85, 294
Oehlert, G., 78, 117, 131, 297

Pan, W., 191, 297
Pardoe, I., 89, 297
Parks, J., 238, 297
Patton, R., 299
Pearson, E. S., 2, 297
Pearson, K., 110, 297
Pierce, R., 90
Pinheiro, J., 100, 136, 177, 185, 297
Porzio, G., 191, 297
Prentice, R. L., 85, 296

Prescott, P., 197, 294
Pun, F., 225, 297

Raghunathan, T., 86, 298
Ratkowsky, D. A., 248, 297
Raven, P. H., 231, 296
Raychaudhari, K., 299
Rencher, A., 225, 297
Rice, J., 182
Rich, R., 251
Riedwyl, H., 268, 295
Robinson, A., 112, 151
Royston, J. P., 122, 205, 297–298
Rubin, D., 84–85, 295, 297–298
Ruppert, D., 277, 298

Sakamoto, Y., 217, 298
Saw, J., 130, 298
Schafer, J., 84, 298
Scheffé, H., 179, 298
Schott, J., 278, 298
Schwarz, G., 218, 298
Sclove, S., 162, 298
Searle, S. R., 74, 278, 291, 296, 298
Seber, G. A. F., 31, 91, 108, 116, 248, 298
Shapiro, S. S., 205, 298
Sheather, S., 197, 298
Silverman, B., 14, 251, 296, 298
Simonoff, J., 14, 112, 277, 298
Simpson, J., 209, 299
Siracuse, M., 210
Snyder, M., 65, 298
Staudte, R., 197, 298
Stigler, S., 114
Stroup, W., 136–137, 297

Taff, S., 78, 298
Takeda, H., 299
Tang, G., 86, 298

Telford, R., 132
Thern, R., 299
Thisted, R., 87, 248, 298
Thurston, J., 295
Tibshirani, R., 87, 211, 295–296
Tiffany, D., 78, 141, 208, 298
Tuddenham, R., 65, 298
Tukey, J., 176, 179, 295, 298

Van Berg, R., 299
Van Loan, C., 287, 295
Velilla, S., 157, 290, 298
Verbeke, G., 136, 298
Von Bertalanffy, L., 248, 299

Wallace, D., 44, 297
Wand, M., 277, 298
Watts, D., 248, 293
Wedderburn, R. W. M., 100, 266, 297
Weibel, P., 8, 297
Weisberg, H., 98, 102, 299
Weisberg, S., 49, 78, 89, 131, 152, 157, 159, 172–173, 180, 186, 191, 197, 204, 294, 297–298
Welsch, R., 169, 296
Wild, C., 248, 298
Wilk, M. B., 205, 298
Wilm, H., 42, 299
Wilson, R., 221, 295
Witmer, J., 8, 238, 294, 297
Wolfinger, R., 136–137, 297
Wood, F., 205, 295
Woodley, W. L., 209, 299

Yeo, I., 160, 299
Young, B., 266, 294

Zeger, S., 137, 295
Zipf, G., 43, 299

Subject Index

Added-variable plot, 49–50, 63, 66, 73, 203, 221
AIC, 217
Akaike Information Criterion, 217
Aliased, 73
Allometric, 150
Analysis of variance, 28–29, 131
  overall, 61
  sequential, 62, 64
anova, 28–29
  overall, 61
  sequential, 62,64
Arcsine square-root, 179
Asymptote, 233

Backsolving, 287
Backward elimination algorithm, 222
Bayes Information Criterion, 218
Bernoulli distribution, 253
BIC, 218
Binary regression, 251
Binomial distribution, 253, 263
Binomial regression, 251, 253
Bonferroni inequality, 196
Bootstrap, 87, 110, 112, 143, 178, 244
  bias, 89
  measurement error, 90
  nonlinear regression, 244
  ratio of estimators, 89
Box–Cox, 153, 157, 159–160, 289

Cases, 2
Causation, 69, 78
Censored regression, 85
Central composite design, 138

Class variable, 124
Coefficient of determination, 31, 62
Collinearity, 211, 214
Comparing groups
  linear regression, 126
  nonlinear regression, 241
  random coefficient model, 134
Complexity, 217
Computer packages, 270
Computer simulation, 87
Confidence interval, 32
  intercept, 32
  slope, 33
Confidence regions, 108
Cook's distance, 198
Correlation, 31, 54, 272
Covariance matrix, 283
Cross-validation, 220
Curvature
  in residual plot, 176

Data
  cross-sectional, 7
  longitudinal, 7
Data files, 270
  ais.txt, 132
  allshoots.txt, 104, 139
  anscombe.txt, 12
  baeskel.txt, 161
  banknote.txt, 268
  BGSall.txt, 65–66, 138
  BGSboys.txt, 65–66
  BGSgirls.txt, 65–66
  BigMac2003.txt, 164

Applied Linear Regression, Third Edition, by Sanford Weisberg
ISBN0-471-66379-4   Copyright © 2005 John Wiley & Sons, Inc.

Data files *(Continued)*
  blowAPB.txt, 269
  blowBF.txt, 255
  brains.txt, 148
  cakes.txt, 118, 137
  cathedral.txt, 139
  caution.txt, 172
  challeng.txt, 268
  chloride.txt, 134
  cloud.txt, 208
  domedata.txt, 145
  domedata1.txt, 146
  donner.txt, 267
  downer.txt, 266
  drugcost.txt, 209
  dwaste.txt, 231
  florida.txt, 207
  forbes.txt, 5
  ftcollinssnow.txt, 7
  fuel2001.txt, 15, 52
  galapagos.txt, 231
  galtonpeas.txt, 110
  heights.txt, 2, 41
  highway.txt, 153–154
  hooker.txt, 40
  htwt.txt, 38
  jevons.txt, 114
  lakemary.txt, 249
  lakes.txt, 193
  landrent.txt, 208
  lathe1.txt, 137
  longley.txt, 94
  longshoots.txt, 104
  mantel.txt, 230
  mile.txt, 143
  Mitchell.txt, 18
  MWwords.txt, 44
  npdata.txt, 91
  oldfaith.txt, 18
  physics.txt, 98
  physics1.txt, 114
  pipeline.txt, 191
  prodscore.txt, 141
  rat.txt, 200
  salary.txt, 141
  salarygov.txt, 163
  segreg.txt, 244
  shocks.txt, 267
  shortshoots.txt, 104
  sleep1.txt, 122
  snake.txt, 42
  sniffer.txt, 182
  snowgeese.txt, 113
  stopping.txt, 162
  swan96.txt, 249
  titanic.txt, 262
  transact.txt, 88
  turk0.txt, 142, 238
  turkey.txt, 8, 242
  twins.txt, 138
  ufcgf.txt, 112
  ufcwc.txt, 151
  UN1.txt, 18
  UN2.txt, 47
  UN3.txt, 166
  walleye.txt, 249
  water.txt, 18, 163
  wblake.txt, 7, 41
  wblake2.txt, 41
  wm1.txt, 45, 93
  wm2.txt, 140
  wm3.txt, 140
  wm4.txt, 227, 269
  wm5.txt, 228
  wool.txt, 130, 165
Data mining, 211
Degrees of freedom, 25
Delta method, 120, 143
Density estimate, 251
Dependence, 1
Deviance, 260
Diagnostics, 167–187, 194–204
Dummy variable, 52, 122

EM algorithm, 85
Errors, 19
Examples
  Alaska pipeline faults, 191
  Anscombe, 12, 198
  Apple shoot, 139
  Apple shoots, 104, 111
  Australian athletes, 132, 250
  Banknote, 268
  Berkeley guidance study, 65, 70, 92, 138–139, 230
  Big Mac data, 164–165
  Black crappies, 249
  Blowdown, 251, 255, 269
  Brain weight, 148
  Cake data, 137
  Cakes, 117
  California water, 18, 162, 192
  Cathedrals, 139
  Challenger data, 268
  Cloud seeding, 209, 250
  Donner party, 267
  Downer, 266
  Drug costs, 209

Electric shocks, 267
Feedlots, 78
Florida election 2000, 207
Forbes, 4, 22–24, 29, 33–34, 37, 39, 138
Ft. Collins snowfall, 7, 33, 44
Fuel consumption, 15, 52, 69, 93, 166, 173, 206
Galápagos species, 231
Galton's peas, 110
Government salary, 163
Heights, 2, 19, 23, 34–36, 41, 51, 205
Highway accidents, 153, 159, 218, 222, 230, 232, 250
Hooker, 40, 138
Jevons' coins, 114, 143
Lake Mary, 248
Land productivity, 140
Land rent, 208
Lathe, 137
Lathe data, 207
Longley, 94
Mantel, 230
Metrodome, 145
Mitchell, 17
Northern Pike, 90
Old Faithful, 18, 44
Oxygen uptake, 230
Physics data, 116
Rat, 200, 203
Segmented regression, 244
Sex discrimination, 141
Sleep, 86, 122, 126, 138, 248
Smallmouth bass, 6, 17, 41
Snake River levels, 42
Sniffer, 182
Snow geese, 113, 181
Stopping distances, 162
Strong interaction, 98, 114
Surface tension, 161
Titanic, 262
Transactions, 88, 143, 205
Turkey growth, 8, 142–143, 233, 238, 247
Twin study, 138
United Nations, 18, 41, 47, 51, 66, 166, 170, 176, 187, 198
Upper Flat Creek, 151, 112
Walleye growth, 249
Windmills, 45, 93, 140, 226, 232, 269
Wool data, 130, 165
World cities, 164
Zipf's Law, 43
Zooplankton species, 193
Expected information matrix, 291
Expected order statistics, 205

Experiment, 77
Exponential family distribution, 265

F-test, 28
  overall *anova*, 30
  partial, 63
  power, 107
  robustness, 108
Factor, 52, 122–136
Factor rule, 123
Fisher scoring, 265
Fitted value, 21, 35, 57, 65
Fixed significance level, 31
Forward selection, 221
Full rank matrix, 283

Gauss–Markov theorem, 27
Gauss–Newton algorithm, 236
Generalized least squares, 100, 178
Generalized linear models, 100, 178, 265
Geometric mean, 153
Goodness of fit, 261

Hat matrix, 168–169
Hat notation, 21
Hessian matrix, 235
Histogram, 251
Hyperplane, 51

Independent, 2, 19
Information matrix, 291
Influence, 167, 198
Inheritance, 2
Interaction, 52, 117
Intercept, 9, 19, 51
Intra-class correlation, 136
Inverse fitted value plot, 152, 159, 165
Inverse regression, 143

Jittering, 2

Kernel mean function, 234, 254
  logistic function, 254

Lack of fit
  F test, 103
  for nonlinear models, 241
  nonparametric, 111
  sum of squares, 103
  variance known, 100
  variance unknown, 102
Lagged variables, 226
Lapack, 287

Leaps and bounds, 221
Least squares
 nonlinear, 234
 ordinary, 6
 weighted, 96
Leverage, 4, 169, 196
Li–Duan theorem, 156
Likelihood function, 263
Likelihood ratio test, 108, 158, 260
Linear
 dependence, 73, 214, 283
 independence, 73, 283
 mixed model, 136
 operator, 270
 predictor, 156, 254
 regression, 1
Link function, 254
Local influence, 204
loess, 14–15, 111–112, 149, 181, 185, 187,
  276–277
Logarithms, 70
 and coefficient estimates, 76
 choice of base, 23
 log rule, 150
Logistic function, 254
Logistic regression, 100, 251–265
 log-likelihood, 264
 model comparisons, 261
Logit, 254
Lurking variable, 79

Machine learning, 211
Main effects, 130
Mallows' C_p, 218
Marginal model plot, 185–190
Matrix
 diagonal, 279
 full rank, 282
 identity, 279
 inverse, 282
 invertible, 282
 nonsingular, 282
 norm, 281
 notation, 54
 orthogonal, 282
 orthogonal columns, 282
 singular, 282
 square, 279
 symmetric, 279
Maximum likelihood, 27, 263–264, 287–288
Mean function, 9–11
Mean square, 25
Measure, correlate, predict, 140
Measurement error, 90

Median, 87
Missing
 at random, 85
 data, 84
 multiple imputation, 85
 values in data files, 270
Modified power family, 153
Multiple correlation coefficient, 62
Multiple linear regression, 47
Multivariate normal, 80
Multivariate transformations, 157

Nearest neighbor, 276
Newton–Raphson, 265
NID, 27
Noncentral χ^2, 108
Noncentral F, 31
Nonconstant variance, 177, 180
Nonlinear least squares, 152, 234
Nonlinear regression, 233–250
 comparing groups, 241
 large-sample inference, 237
 starting values, 237
Nonparametric, 10
Nonparametric regression, 277
Normal distributions, 80
Normal equations, 273, 284
Normality, 20, 58, 204
 large sample, 27
Normal probability plot, 204
Null plot, 13, 36

Observational study, 69, 77
Odds of success, 254
OLS, *see* Ordinary least squares
One-dimensional estimation result, 156
One-way analysis of variance, 124
Order statistics, 205
Ordinary least squares, 6–7, 10, 21–28, 116, 144,
  231, 234, 240, 284–285, 288
 same as maximum likelihood estimate, 288
Orthogonal, 192
 polynomials, 116
 projection, 169
Outlier, 4, 36, 194–197
 mean shift, 194
Over-parameterized, 73
Overplotting, 2

p', 59
p-value, 268
 interpreting, 31

Parameters, 9, 21
  interpreting, 69
  not the same as estimates, 21, 24
  terms in log scale, 70
Parametric bootstrap, 112
Partial correlation, 221
Partial one-dimensional, 131, 144, 250
Pearson residual, 171
Pearson's X^2, 262
POD model, 131–133, 144, 250
Polynomial regression, 115
Power, 31, 108
Power family, 148
Power transformations, 14
Predicted residual, 207, 220
Prediction, 34, 65
Predictor, 1, 51
PRESS, 220
  residual, 207
Probability mass function, 253
Profile log-likelihood, 291
Pure error, 103, 242

Quadratic regression, 115
  estimated minimum or maximum, 115

R^2, 31–32, 62, 81
  interpretation, 83
Random coefficients model, 135
Random intercepts model, 136
Random sampling, 81
Random vector, 283
Range rule, 150
Rectangles, 74
Regression, 1, 10
  binomial, 251
  multiple linear, 47
  nonlinear, 233
  sum of squares, 58
  through the origin, 42, 84
Removable nonadditivity, 166
Residual, 21, 23, 36, 57
  degrees of freedom, 25
  mean square, 25
  Pearson, 171
  properties, 168
  standardized, 195
  studentized, 196
  sum of squares, 21, 24, 57
  sum of squares function, 234
  weighted, 170–171
Residual plots, 171
  curvature, 176
  Tukey's test, 176

Response, 1

Sample correlations, 54
Sample covariance matrix, 57
Sample order statistics, 204
Saturated model, 262
Scalar, 279
Scaled power transformation, 233
Scatterplot, 1
Scatterplot matrix, 15–17
Score test, 180
  nonconstant variance, 180
Score vector, 235
se, 27
sefit, 35
sepred, 34
Second-order mean function, 117
Segmented regression, 244
Separated points, 3
Significance level
  fixed, 31
Simple regression
  model, 19
  deviations form, 41
  matrix version, 58
Slices, 3
Slope, 9, 19
Smoother, 10, 14, 275
Smoothing parameter, 276
Standard error, 27
  of fitted value, 2
  of prediction, 2
  of regression, 25
Standardized residual, 195
Statistical error, 19
Stepwise methods, 221–226
  backward elimination, 222
  forward selection, 221
Straight lines, 19
Studentized residual, 196
Studentized statistic, 196
Sum of squares
  due to regression, 29
  for lack of fit, 103
  regression, 58
Summary graph, 1–2, 11–12, 175
  multiple regression, 84
Supernormality, 204

Taylor series, 120, 235
Terms, 47, 51–54
Three-dimensional plot, 48

Transformations, 52, 147–160
  arcsine square-root, 179
  Box–Cox, 153
  families, 148
  for linearity, 233
  log rule, 150
  many predictors, 153–158
  modified power, 153
  multivariate Box–Cox, 157
  nonpositive variables, 160
  power family, 148
  predictor, 150
  range rule, 150
  response, 159
  scaled, 150
  using a nonlinear model, 233
  Yeo–Johnson, 160
t-tests, 32
Tukey's test for nonadditivity, 176

Unbiased, 271
Uncorrelated, 8

Variability explained, 83
Variable selection methods, 211
Variance estimate
  pure error, 103
Variance function, 11, 96
Variance inflation factor, 216
Variance stabilizing transformation, 100, 177
Vector, 279
  length, 281

Web site, 270
  www.stat.umn.edu/alr, 270
Weighted least squares, 96–99, 114, 234
  outliers, 196
Weighted residual, 171
Wind farm, 45
WLS, *see* Weighted least squares

Yeo–Johnson transformation family, 160, 290

Zipf's law, 43

WILEY SERIES IN PROBABILITY AND STATISTICS

ESTABLISHED BY WALTER A. SHEWHART AND SAMUEL S. WILKS

Editors: *David J. Balding, Noel A. C. Cressie, Nicholas I. Fisher,*
Iain M. Johnstone, J. B. Kadane, Geert Molenberghs. Louise M. Ryan,
David W. Scott, Adrian F. M. Smith, Jozef L. Teugels
Editors Emeriti: *Vic Barnett, J. Stuart Hunter, David G. Kendall*

The *Wiley Series in Probability and Statistics* is well established and authoritative. It covers many topics of current research interest in both pure and applied statistics and probability theory. Written by leading statisticians and institutions, the titles span both state-of-the-art developments in the field and classical methods.

Reflecting the wide range of current research in statistics, the series encompasses applied, methodological and theoretical statistics, ranging from applications and new techniques made possible by advances in computerized practice to rigorous treatment of theoretical approaches.

This series provides essential and invaluable reading for all statisticians, whether in academia, industry, government, or research.

ABRAHAM and LEDOLTER · Statistical Methods for Forecasting
AGRESTI · Analysis of Ordinal Categorical Data
AGRESTI · An Introduction to Categorical Data Analysis
AGRESTI · Categorical Data Analysis, *Second Edition*
ALTMAN, GILL, and McDONALD · Numerical Issues in Statistical Computing for the
    Social Scientist
AMARATUNGA and CABRERA · Exploration and Analysis of DNA Microarray and
    Protein Array Data
ANDĚL · Mathematics of Chance
ANDERSON · An Introduction to Multivariate Statistical Analysis, *Third Edition*
* ANDERSON · The Statistical Analysis of Time Series
ANDERSON, AUQUIER, HAUCK, OAKES, VANDAELE, and WEISBERG ·
    Statistical Methods for Comparative Studies
ANDERSON and LOYNES · The Teaching of Practical Statistics
ARMITAGE and DAVID (editors) · Advances in Biometry
ARNOLD, BALAKRISHNAN, and NAGARAJA · Records
* ARTHANARI and DODGE · Mathematical Programming in Statistics
* BAILEY · The Elements of Stochastic Processes with Applications to the Natural
    Sciences
BALAKRISHNAN and KOUTRAS · Runs and Scans with Applications
BARNETT · Comparative Statistical Inference, *Third Edition*
BARNETT and LEWIS · Outliers in Statistical Data, *Third Edition*
BARTOSZYNSKI and NIEWIADOMSKA-BUGAJ · Probability and Statistical Inference
BASILEVSKY · Statistical Factor Analysis and Related Methods: Theory and
    Applications
BASU and RIGDON · Statistical Methods for the Reliability of Repairable Systems
BATES and WATTS · Nonlinear Regression Analysis and Its Applications
BECHHOFER, SANTNER, and GOLDSMAN · Design and Analysis of Experiments for
    Statistical Selection, Screening, and Multiple Comparisons
BELSLEY · Conditioning Diagnostics: Collinearity and Weak Data in Regression

*Now available in a lower priced paperback edition in the Wiley Classics Library.
†Now available in a lower priced paperback edition in the Wiley–Interscience Paperback Series.

† BELSLEY, KUH, and WELSCH · Regression Diagnostics: Identifying Influential Data and Sources of Collinearity

BENDAT and PIERSOL · Random Data: Analysis and Measurement Procedures, *Third Edition*

BERRY, CHALONER, and GEWEKE · Bayesian Analysis in Statistics and Econometrics: Essays in Honor of Arnold Zellner

BERNARDO and SMITH · Bayesian Theory

BHAT and MILLER · Elements of Applied Stochastic Processes, *Third Edition*

BHATTACHARYA and WAYMIRE · Stochastic Processes with Applications

† BIEMER, GROVES, LYBERG, MATHIOWETZ, and SUDMAN · Measurement Errors in Surveys

BILLINGSLEY · Convergence of Probability Measures, *Second Edition*

BILLINGSLEY · Probability and Measure, *Third Edition*

BIRKES and DODGE · Alternative Methods of Regression

BLISCHKE AND MURTHY (editors) · Case Studies in Reliability and Maintenance

BLISCHKE AND MURTHY · Reliability: Modeling, Prediction, and Optimization

BLOOMFIELD · Fourier Analysis of Time Series: An Introduction, *Second Edition*

BOLLEN · Structural Equations with Latent Variables

BOROVKOV · Ergodicity and Stability of Stochastic Processes

BOULEAU · Numerical Methods for Stochastic Processes

BOX · Bayesian Inference in Statistical Analysis

BOX · R. A. Fisher, the Life of a Scientist

BOX and DRAPER · Empirical Model-Building and Response Surfaces

* BOX and DRAPER · Evolutionary Operation: A Statistical Method for Process Improvement

BOX, HUNTER, and HUNTER · Statistics for Experimenters: An Introduction to Design, Data Analysis, and Model Building

BOX and LUCEÑO · Statistical Control by Monitoring and Feedback Adjustment

BRANDIMARTE · Numerical Methods in Finance: A MATLAB-Based Introduction

BROWN and HOLLANDER · Statistics: A Biomedical Introduction

BRUNNER, DOMHOF, and LANGER · Nonparametric Analysis of Longitudinal Data in Factorial Experiments

BUCKLEW · Large Deviation Techniques in Decision, Simulation, and Estimation

CAIROLI and DALANG · Sequential Stochastic Optimization

CASTILLO, HADI, BALAKRISHNAN, and SARABIA · Extreme Value and Related Models with Applications in Engineering and Science

CHAN · Time Series: Applications to Finance

CHATTERJEE and HADI · Sensitivity Analysis in Linear Regression

CHATTERJEE and PRICE · Regression Analysis by Example, *Third Edition*

CHERNICK · Bootstrap Methods: A Practitioner's Guide

CHERNICK and FRIIS · Introductory Biostatistics for the Health Sciences

CHILÈS and DELFINER · Geostatistics: Modeling Spatial Uncertainty

CHOW and LIU · Design and Analysis of Clinical Trials: Concepts and Methodologies, *Second Edition*

CLARKE and DISNEY · Probability and Random Processes: A First Course with Applications, *Second Edition*

* COCHRAN and COX · Experimental Designs, *Second Edition*

CONGDON · Applied Bayesian Modelling

CONGDON · Bayesian Statistical Modelling

CONOVER · Practical Nonparametric Statistics, *Third Edition*

COOK · Regression Graphics

COOK and WEISBERG · Applied Regression Including Computing and Graphics

COOK and WEISBERG · An Introduction to Regression Graphics

*Now available in a lower priced paperback edition in the Wiley Classics Library.
†Now available in a lower priced paperback edition in the Wiley–Interscience Paperback Series.

CORNELL · Experiments with Mixtures, Designs, Models, and the Analysis of Mixture Data, *Third Edition*

COVER and THOMAS · Elements of Information Theory

COX · A Handbook of Introductory Statistical Methods

* COX · Planning of Experiments

CRESSIE · Statistics for Spatial Data, *Revised Edition*

CSÖRGŐ and HORVÁTH · Limit Theorems in Change Point Analysis

DANIEL · Applications of Statistics to Industrial Experimentation

DANIEL · Biostatistics: A Foundation for Analysis in the Health Sciences, *Eighth Edition*

* DANIEL · Fitting Equations to Data: Computer Analysis of Multifactor Data, *Second Edition*

DASU and JOHNSON · Exploratory Data Mining and Data Cleaning

DAVID and NAGARAJA · Order Statistics, *Third Edition*

* DEGROOT, FIENBERG, and KADANE · Statistics and the Law

DEL CASTILLO · Statistical Process Adjustment for Quality Control

DeMARIS · Regression with Social Data: Modeling Continuous and Limited Response Variables

DEMIDENKO · Mixed Models: Theory and Applications

DENISON, HOLMES, MALLICK and SMITH · Bayesian Methods for Nonlinear Classification and Regression

DETTE and STUDDEN · The Theory of Canonical Moments with Applications in Statistics, Probability, and Analysis

DEY and MUKERJEE · Fractional Factorial Plans

DILLON and GOLDSTEIN · Multivariate Analysis: Methods and Applications

DODGE · Alternative Methods of Regression

* DODGE and ROMIG · Sampling Inspection Tables, *Second Edition*

* DOOB · Stochastic Processes

DOWDY, WEARDEN, and CHILKO · Statistics for Research, *Third Edition*

DRAPER and SMITH · Applied Regression Analysis, *Third Edition*

DRYDEN and MARDIA · Statistical Shape Analysis

DUDEWICZ and MISHRA · Modern Mathematical Statistics

DUNN and CLARK · Basic Statistics: A Primer for the Biomedical Sciences, *Third Edition*

DUPUIS and ELLIS · A Weak Convergence Approach to the Theory of Large Deviations

* ELANDT-JOHNSON and JOHNSON · Survival Models and Data Analysis

ENDERS · Applied Econometric Time Series

ETHIER and KURTZ · Markov Processes: Characterization and Convergence

EVANS, HASTINGS, and PEACOCK · Statistical Distributions, *Third Edition*

FELLER · An Introduction to Probability Theory and Its Applications, Volume I, *Third Edition,* Revised; Volume II, *Second Edition*

FISHER and VAN BELLE · Biostatistics: A Methodology for the Health Sciences

FITZMAURICE, LAIRD, and WARE · Applied Longitudinal Analysis

* FLEISS · The Design and Analysis of Clinical Experiments

FLEISS · Statistical Methods for Rates and Proportions, *Third Edition*

FLEMING and HARRINGTON · Counting Processes and Survival Analysis

FULLER · Introduction to Statistical Time Series, *Second Edition*

FULLER · Measurement Error Models

GALLANT · Nonlinear Statistical Models

GHOSH, MUKHOPADHYAY, and SEN · Sequential Estimation

GIESBRECHT and GUMPERTZ · Planning, Construction, and Statistical Analysis of Comparative Experiments

GIFI · Nonlinear Multivariate Analysis

GLASSERMAN and YAO · Monotone Structure in Discrete-Event Systems

*Now available in a lower priced paperback edition in the Wiley Classics Library.
†Now available in a lower priced paperback edition in the Wiley–Interscience Paperback Series.

GNANADESIKAN · Methods for Statistical Data Analysis of Multivariate Observations, *Second Edition*

GOLDSTEIN and LEWIS · Assessment: Problems, Development, and Statistical Issues

GREENWOOD and NIKULIN · A Guide to Chi-Squared Testing

GROSS and HARRIS · Fundamentals of Queueing Theory, *Third Edition*

† GROVES · Survey Errors and Survey Costs

* HAHN and SHAPIRO · Statistical Models in Engineering

HAHN and MEEKER · Statistical Intervals: A Guide for Practitioners

HALD · A History of Probability and Statistics and their Applications Before 1750

HALD · A History of Mathematical Statistics from 1750 to 1930

HAMPEL · Robust Statistics: The Approach Based on Influence Functions

HANNAN and DEISTLER · The Statistical Theory of Linear Systems

HEIBERGER · Computation for the Analysis of Designed Experiments

HEDAYAT and SINHA · Design and Inference in Finite Population Sampling

HELLER · MACSYMA for Statisticians

HINKELMAN and KEMPTHORNE: · Design and Analysis of Experiments, Volume 1: Introduction to Experimental Design

HOAGLIN, MOSTELLER, and TUKEY · Exploratory Approach to Analysis of Variance

HOAGLIN, MOSTELLER, and TUKEY · Exploring Data Tables, Trends and Shapes

* HOAGLIN, MOSTELLER, and TUKEY · Understanding Robust and Exploratory Data Analysis

HOCHBERG and TAMHANE · Multiple Comparison Procedures

HOCKING · Methods and Applications of Linear Models: Regression and the Analysis of Variance, *Second Edition*

HOEL · Introduction to Mathematical Statistics, *Fifth Edition*

HOGG and KLUGMAN · Loss Distributions

HOLLANDER and WOLFE · Nonparametric Statistical Methods, *Second Edition*

HOSMER and LEMESHOW · Applied Logistic Regression, *Second Edition*

HOSMER and LEMESHOW · Applied Survival Analysis: Regression Modeling of Time to Event Data

† HUBER · Robust Statistics

HUBERTY · Applied Discriminant Analysis

HUNT and KENNEDY · Financial Derivatives in Theory and Practice

HUSKOVA, BERAN, and DUPAC · Collected Works of Jaroslav Hajek— with Commentary

HUZURBAZAR · Flowgraph Models for Multistate Time-to-Event Data

IMAN and CONOVER · A Modern Approach to Statistics

† JACKSON · A User's Guide to Principle Components

JOHN · Statistical Methods in Engineering and Quality Assurance

JOHNSON · Multivariate Statistical Simulation

JOHNSON and BALAKRISHNAN · Advances in the Theory and Practice of Statistics: A Volume in Honor of Samuel Kotz

JOHNSON and BHATTACHARYYA · Statistics: Principles and Methods, *Fifth Edition*

JOHNSON and KOTZ · Distributions in Statistics

JOHNSON and KOTZ (editors) · Leading Personalities in Statistical Sciences: From the Seventeenth Century to the Present

JOHNSON, KOTZ, and BALAKRISHNAN · Continuous Univariate Distributions, Volume 1, *Second Edition*

JOHNSON, KOTZ, and BALAKRISHNAN · Continuous Univariate Distributions, Volume 2, *Second Edition*

JOHNSON, KOTZ, and BALAKRISHNAN · Discrete Multivariate Distributions

JOHNSON, KOTZ, and KEMP · Univariate Discrete Distributions, *Second Edition*

JUDGE, GRIFFITHS, HILL, LÜTKEPOHL, and LEE · The Theory and Practice of Econometrics, *Second Edition*

JUREČKOVÁ and SEN · Robust Statistical Procedures: Aymptotics and Interrelations

JUREK and MASON · Operator-Limit Distributions in Probability Theory

KADANE · Bayesian Methods and Ethics in a Clinical Trial Design

KADANE AND SCHUM · A Probabilistic Analysis of the Sacco and Vanzetti Evidence

KALBFLEISCH and PRENTICE · The Statistical Analysis of Failure Time Data, *Second Edition*

KASS and VOS · Geometrical Foundations of Asymptotic Inference

KAUFMAN and ROUSSEEUW · Finding Groups in Data: An Introduction to Cluster Analysis

KEDEM and FOKIANOS · Regression Models for Time Series Analysis

KENDALL, BARDEN, CARNE, and LE · Shape and Shape Theory

KHURI · Advanced Calculus with Applications in Statistics, *Second Edition*

KHURI, MATHEW, and SINHA · Statistical Tests for Mixed Linear Models

* KISH · Statistical Design for Research

KLEIBER and KOTZ · Statistical Size Distributions in Economics and Actuarial Sciences

KLUGMAN, PANJER, and WILLMOT · Loss Models: From Data to Decisions, *Second Edition*

KLUGMAN, PANJER, and WILLMOT · Solutions Manual to Accompany Loss Models: From Data to Decisions, *Second Edition*

KOTZ, BALAKRISHNAN, and JOHNSON · Continuous Multivariate Distributions, Volume 1, *Second Edition*

KOTZ and JOHNSON (editors) · Encyclopedia of Statistical Sciences: Volumes 1 to 9 with Index

KOTZ and JOHNSON (editors) · Encyclopedia of Statistical Sciences: Supplement Volume

KOTZ, READ, and BANKS (editors) · Encyclopedia of Statistical Sciences: Update Volume 1

KOTZ, READ, and BANKS (editors) · Encyclopedia of Statistical Sciences: Update Volume 2

KOVALENKO, KUZNETZOV, and PEGG · Mathematical Theory of Reliability of Time-Dependent Systems with Practical Applications

LACHIN · Biostatistical Methods: The Assessment of Relative Risks

LAD · Operational Subjective Statistical Methods: A Mathematical, Philosophical, and Historical Introduction

LAMPERTI · Probability: A Survey of the Mathematical Theory, *Second Edition*

LANGE, RYAN, BILLARD, BRILLINGER, CONQUEST, and GREENHOUSE · Case Studies in Biometry

LARSON · Introduction to Probability Theory and Statistical Inference, *Third Edition*

LAWLESS · Statistical Models and Methods for Lifetime Data, *Second Edition*

LAWSON · Statistical Methods in Spatial Epidemiology

LE · Applied Categorical Data Analysis

LE · Applied Survival Analysis

LEE and WANG · Statistical Methods for Survival Data Analysis, *Third Edition*

LePAGE and BILLARD · Exploring the Limits of Bootstrap

LEYLAND and GOLDSTEIN (editors) · Multilevel Modelling of Health Statistics

LIAO · Statistical Group Comparison

LINDVALL · Lectures on the Coupling Method

LINHART and ZUCCHINI · Model Selection

LITTLE and RUBIN · Statistical Analysis with Missing Data, *Second Edition*

LLOYD · The Statistical Analysis of Categorical Data

MAGNUS and NEUDECKER · Matrix Differential Calculus with Applications in Statistics and Econometrics, *Revised Edition*

*Now available in a lower priced paperback edition in the Wiley Classics Library.

†Now available in a lower priced paperback edition in the Wiley–Interscience Paperback Series.

MALLER and ZHOU · Survival Analysis with Long Term Survivors

MALLOWS · Design, Data, and Analysis by Some Friends of Cuthbert Daniel

MANN, SCHAFER, and SINGPURWALLA · Methods for Statistical Analysis of Reliability and Life Data

MANTON, WOODBURY, and TOLLEY · Statistical Applications Using Fuzzy Sets

MARCHETTE · Random Graphs for Statistical Pattern Recognition

MARDIA and JUPP · Directional Statistics

MASON, GUNST, and HESS · Statistical Design and Analysis of Experiments with Applications to Engineering and Science, *Second Edition*

McCULLOCH and SEARLE · Generalized, Linear, and Mixed Models

McFADDEN · Management of Data in Clinical Trials

* McLACHLAN · Discriminant Analysis and Statistical Pattern Recognition

McLACHLAN, DO, and AMBROISE · Analyzing Microarray Gene Expression Data

McLACHLAN and KRISHNAN · The EM Algorithm and Extensions

McLACHLAN and PEEL · Finite Mixture Models

McNEIL · Epidemiological Research Methods

MEEKER and ESCOBAR · Statistical Methods for Reliability Data

MEERSCHAERT and SCHEFFLER · Limit Distributions for Sums of Independent Random Vectors: Heavy Tails in Theory and Practice

MICKEY, DUNN, and CLARK · Applied Statistics: Analysis of Variance and Regression, *Third Edition*

* MILLER · Survival Analysis, *Second Edition*

MONTGOMERY, PECK, and VINING · Introduction to Linear Regression Analysis, *Third Edition*

MORGENTHALER and TUKEY · Configural Polysampling: A Route to Practical Robustness

MUIRHEAD · Aspects of Multivariate Statistical Theory

MULLER and STOYAN · Comparison Methods for Stochastic Models and Risks

MURRAY · X-STAT 2.0 Statistical Experimentation, Design Data Analysis, and Nonlinear Optimization

MURTHY, XIE, and JIANG · Weibull Models

MYERS and MONTGOMERY · Response Surface Methodology: Process and Product Optimization Using Designed Experiments, *Second Edition*

MYERS, MONTGOMERY, and VINING · Generalized Linear Models. With Applications in Engineering and the Sciences

† NELSON · Accelerated Testing, Statistical Models, Test Plans, and Data Analyses

† NELSON · Applied Life Data Analysis

NEWMAN · Biostatistical Methods in Epidemiology

OCHI · Applied Probability and Stochastic Processes in Engineering and Physical Sciences

OKABE, BOOTS, SUGIHARA, and CHIU · Spatial Tesselations: Concepts and Applications of Voronoi Diagrams, *Second Edition*

OLIVER and SMITH · Influence Diagrams, Belief Nets and Decision Analysis

PALTA · Quantitative Methods in Population Health: Extensions of Ordinary Regressions

PANKRATZ · Forecasting with Dynamic Regression Models

PANKRATZ · Forecasting with Univariate Box-Jenkins Models: Concepts and Cases

* PARZEN · Modern Probability Theory and Its Applications

PEÑA, TIAO, and TSAY · A Course in Time Series Analysis

PIANTADOSI · Clinical Trials: A Methodologic Perspective

PORT · Theoretical Probability for Applications

POURAHMADI · Foundations of Time Series Analysis and Prediction Theory

PRESS · Bayesian Statistics: Principles, Models, and Applications

PRESS · Subjective and Objective Bayesian Statistics, *Second Edition*

PRESS and TANUR · The Subjectivity of Scientists and the Bayesian Approach

*Now available in a lower priced paperback edition in the Wiley Classics Library.
†Now available in a lower priced paperback edition in the Wiley–Interscience Paperback Series.

PUKELSHEIM · Optimal Experimental Design

PURI, VILAPLANA, and WERTZ · New Perspectives in Theoretical and Applied Statistics

PUTERMAN · Markov Decision Processes: Discrete Stochastic Dynamic Programming

* RAO · Linear Statistical Inference and Its Applications, *Second Edition*

RAUSAND and HØYLAND · System Reliability Theory: Models, Statistical Methods, and Applications, *Second Edition*

RENCHER · Linear Models in Statistics

RENCHER · Methods of Multivariate Analysis, *Second Edition*

RENCHER · Multivariate Statistical Inference with Applications

* RIPLEY · Spatial Statistics

RIPLEY · Stochastic Simulation

ROBINSON · Practical Strategies for Experimenting

ROHATGI and SALEH · An Introduction to Probability and Statistics, *Second Edition*

ROLSKI, SCHMIDLI, SCHMIDT, and TEUGELS · Stochastic Processes for Insurance and Finance

ROSENBERGER and LACHIN · Randomization in Clinical Trials: Theory and Practice

ROSS · Introduction to Probability and Statistics for Engineers and Scientists

† ROUSSEEUW and LEROY · Robust Regression and Outlier Detection

* RUBIN · Multiple Imputation for Nonresponse in Surveys

RUBINSTEIN · Simulation and the Monte Carlo Method

RUBINSTEIN and MELAMED · Modern Simulation and Modeling

RYAN · Modern Regression Methods

RYAN · Statistical Methods for Quality Improvement, *Second Edition*

SALTELLI, CHAN, and SCOTT (editors) · Sensitivity Analysis

* SCHEFFE · The Analysis of Variance

SCHIMEK · Smoothing and Regression: Approaches, Computation, and Application

SCHOTT · Matrix Analysis for Statistics

SCHOUTENS · Levy Processes in Finance: Pricing Financial Derivatives

SCHUSS · Theory and Applications of Stochastic Differential Equations

SCOTT · Multivariate Density Estimation: Theory, Practice, and Visualization

* SEARLE · Linear Models

SEARLE · Linear Models for Unbalanced Data

SEARLE · Matrix Algebra Useful for Statistics

SEARLE, CASELLA, and McCULLOCH · Variance Components

SEARLE and WILLETT · Matrix Algebra for Applied Economics

SEBER and LEE · Linear Regression Analysis, *Second Edition*

† SEBER · Multivariate Observations

† SEBER and WILD · Nonlinear Regression

SENNOTT · Stochastic Dynamic Programming and the Control of Queueing Systems

* SERFLING · Approximation Theorems of Mathematical Statistics

SHAFER and VOVK · Probability and Finance: It's Only a Game!

SILVAPULLE and SEN · Constrained Statistical Inference: Inequality, Order, and Shape Restrictions

SMALL and McLEISH · Hilbert Space Methods in Probability and Statistical Inference

SRIVASTAVA · Methods of Multivariate Statistics

STAPLETON · Linear Statistical Models

STAUDTE and SHEATHER · Robust Estimation and Testing

STOYAN, KENDALL, and MECKE · Stochastic Geometry and Its Applications, *Second Edition*

STOYAN and STOYAN · Fractals, Random Shapes and Point Fields: Methods of Geometrical Statistics

STYAN · The Collected Papers of T. W. Anderson: 1943–1985

*Now available in a lower priced paperback edition in the Wiley Classics Library.

†Now available in a lower priced paperback edition in the Wiley–Interscience Paperback Series.

SUTTON, ABRAMS, JONES, SHELDON, and SONG · Methods for Meta-Analysis in Medical Research

TANAKA · Time Series Analysis: Nonstationary and Noninvertible Distribution Theory

THOMPSON · Empirical Model Building

THOMPSON · Sampling, *Second Edition*

THOMPSON · Simulation: A Modeler's Approach

THOMPSON and SEBER · Adaptive Sampling

THOMPSON, WILLIAMS, and FINDLAY · Models for Investors in Real World Markets

TIAO, BISGAARD, HILL, PEÑA, and STIGLER (editors) · Box on Quality and Discovery: with Design, Control, and Robustness

TIERNEY · LISP-STAT: An Object-Oriented Environment for Statistical Computing and Dynamic Graphics

TSAY · Analysis of Financial Time Series

UPTON and FINGLETON · Spatial Data Analysis by Example, Volume II: Categorical and Directional Data

VAN BELLE · Statistical Rules of Thumb

VAN BELLE, FISHER, HEAGERTY, and LUMLEY · Biostatistics: A Methodology for the Health Sciences, *Second Edition*

VESTRUP · The Theory of Measures and Integration

VIDAKOVIC · Statistical Modeling by Wavelets

VINOD and REAGLE · Preparing for the Worst: Incorporating Downside Risk in Stock Market Investments

WALLER and GOTWAY · Applied Spatial Statistics for Public Health Data

WEERAHANDI · Generalized Inference in Repeated Measures: Exact Methods in MANOVA and Mixed Models

WEISBERG · Applied Linear Regression, *Third Edition*

WELSH · Aspects of Statistical Inference

WESTFALL and YOUNG · Resampling-Based Multiple Testing: Examples and Methods for *p*-Value Adjustment

WHITTAKER · Graphical Models in Applied Multivariate Statistics

WINKER · Optimization Heuristics in Economics: Applications of Threshold Accepting

WONNACOTT and WONNACOTT · Econometrics, *Second Edition*

WOODING · Planning Pharmaceutical Clinical Trials: Basic Statistical Principles

WOODWORTH · Biostatistics: A Bayesian Introduction

WOOLSON and CLARKE · Statistical Methods for the Analysis of Biomedical Data, *Second Edition*

WU and HAMADA · Experiments: Planning, Analysis, and Parameter Design Optimization

YANG · The Construction Theory of Denumerable Markov Processes

* ZELLNER · An Introduction to Bayesian Inference in Econometrics

ZHOU, OBUCHOWSKI, and McCLISH · Statistical Methods in Diagnostic Medicine

*Now available in a lower priced paperback edition in the Wiley Classics Library.
†Now available in a lower priced paperback edition in the Wiley–Interscience Paperback Series.